Engineered Tribological Composites

Other SAE books of interest:

Brake Design and Safety, 3rd Ed.
By Rudolf Limpert
(Product Code: R-398)

Brake Technology Handbook
By Karlheinz Bill and Bert J. Breuer
(Product Code: R-375)

Dictionary of Materials and Testing, 2nd Ed.
By Joan Tomsic
(Product Code: R-257)

Engineered Tribological Composites:

The Art of Friction Material Development

By Roy L. Cox

SAE International

Warrendale, Pennsylvania, USA

SAE International®
400 Commonwealth Drive
Warrendale, PA 15096-0001 USA

E-mail: CustomerService@sae.org
Phone: 877-606-7323 *(inside USA and Canada)*
724-776-4970 *(outside USA)*
Fax: 724-776-0790

ISBN 978-0-7680-3485-1
SAE Order Number R-401
DOI 10.4271/R-401

Library of Congress Cataloging-in-Publication Data

Cox, Roy L.
Engineered tribological composites : the art of friction material development.
p. cm.
ISBN 978-0-7680-3485-1
1. Composite materials. 2. Friction materials. 3. Automobiles--Materials. I. Society of Automotive Engineers. II. Title.
TA418.9.C6C696 2011
620.1'18--dc23
2011035718

To purchase bulk quantities, please contact:
SAE Customer Service
E-mail: CustomerService@sae.org
Phone: 877-606-7323 *(inside USA and Canada)*
724-776-4970 *(outside USA)*
Fax: 724-776-0790

Visit the SAE International Bookstore at http://books.sae.org

Dedication

This book is dedicated
to Marie, for teaching me to dream the impossible;
to Deena, for the greatest gift a man could have;
and to Greyson and Mattie,
who teach me the truest meaning of love.

Table of Contents

Preface

Nature speedily extinguishes every motion upon earth, and seems to diffuse a principle of silence and repose...
—John Leslie

AS JOHN LESLIE HAS ELOQUENTLY STATED, friction is an omnipresent force that works to slow every motion on earth. In addition, it seems that friction suffers from a dissociative identity disorder—commonly being viewed as the nemesis of all moving components in modern equipment and yet also seen as a critical safety component in today's transportation industry.

Without friction, tires would not grip roads, clutches and synchronizer rings would not transmit torque, and brake linings could not stop a vehicle. Therefore, for our purposes, friction is a necessary force that must be predictable and fully controllable during use to provide the necessary functionality required by these special areas of automotive system design.

Why a Book on Friction Materials?

Despite more than one hundred years of progress in virtually all aspects of the transportation industry, the world of friction material development remains a dark, obscure world often regarded as a field of study akin to alchemy, witchcraft, or other suitably arcane black art. This perception is based on the lack of knowledge available to the general public, which is for the most part due to deliberately imposed secrecy necessary to protect commercial interests. Although this level of secrecy has been effective at protecting commercial interests, it has also been insulating and has led to the inevitable consequence of reinvention, which leads to limited growth and increased cost.

It must be noted that the primary desire of this work is not profound, nor is it deeply philosophical. The purpose of this text is to prevent reinvention by providing a single reference containing commonly recorded findings from various sources of previous research efforts and associated art. It is an attempt to review and gather more than one hundred years of public domain information, including published friction material compositions, manufacturing techniques, and a collection of esoteric oddities, with the purpose of directing researchers toward more fruitful areas of research. It is hoped that readers will arm themselves with this information and thus avoid some of the pitfalls experienced by our predecessors.

Acknowledgments

To write a book that condenses a century of information into a readable form is a task that cannot be achieved by a single individual. I am indebted to the entire editorial staff of SAE International, especially Martha Swiss for her kind attention and seemingly endless patience regarding my numerous questions about the most mundane topics. It is the experience and knowledge of the editorial staff that enabled me to improve this text in many ways. I remain convinced that without them, this little book would have become hardly more than an effective sleep aid.

There were many reviewers whose comments enabled me to significantly improve several areas of the book. Their assistance allowed me to stretch beyond the boundaries of my personal knowledge and grow to new levels. I am very thankful for their help.

I would specifically like to thank Dr. Alan Hase for his constructive comments that have given me insight into recent studies of abrasion and underlying wear mechanisms. His work provides a foundation that will undoubtedly lead to further advances in the understanding of third body layer formation and will remain an area of increased interest in the years to come.

Likewise, reviewers within our industry provided a perspective that encouraged me to see the richness of our industry's history from a global perspective. This direction enabled me to expand greatly on specific sections pertaining to the earliest days of friction material development and the advent of the motorcar.

In addition, I am indebted to the kind instruction given by Dr. Peter J. Blau and Dr. Jayashree Bijwe. These men accepted the challenge of reviewing specific sections of the text that were unquestionably laborious to read. It is their many comments and corrections on specific aspects of tribology that have helped me to improve these sections greatly. For this I am eternally grateful.

And at the risk of leaving someone out, I would also like to acknowledge a circle of friends who share a similar passion, albeit one that few "normal people" would understand. For countless reasons, these few were at times my mentors, guides and occasionally, my analysts!

Ken W., "Kazu," Ron, Steve, Greg, Rob, "Mo," Chris, Mark, and Raleigh… Thank you all very much…

Introduction

Iteration, like friction, is likely to generate heat instead of progress.
—George Elliot

THE HISTORY OF MANKIND'S ATTEMPTS TO harness the beneficial attributes of friction began when our ancestors first determined that rubbing two bits of wood together with sufficient force would result in combustion. Simplistically, a fundamental appreciation of friction can still be gained by quickly rubbing our hands together to generate warmth. However, the phenomena that we call friction is a maddeningly complex process that has only recently become an area illuminated by modern scientific experimentation through the use of sophisticated analytical techniques.

The level of complexity inherent in all observed frictional processes is further confounded by the use of heterogeneous composite materials, which contribute their own anisotropic properties to an already large number of system variables. This increases the number of factors that must be considered during the design phase of product development.

Furthermore, the number of factors can quickly become convoluted when tribological effects are amplified by variation in the structure of the composite, and during operation in a plethora of environments by countless operators of varying skill levels. Therefore, an effective study of tribological processes associated with sliding composites comprised of mixtures of diverse materials is a daunting mission rife with challenges at many levels. It must be noted that the level of difficulty does not arise from "bad science" but comes about due to the multitude of approaches and the diversity in compositional formulations.

Each of these historical approaches can introduce "noise" that makes effective analysis difficult. For instance, there are a number of sources in the available literature that illustrate seemingly contradictory responses related to the use of a given raw material as a frictional additive. These conflicting reports are often in opposition until it is realized that the discrepancies might be the result of other factors including purity of raw materials, raw material source, and perhaps more importantly, the method of manufacture of the final composite.

Each of these subtle factors can inflate error in measurement and necessitates that the designers of tribological composites have a thorough understanding of the influence of the various manufacturing methods as well as an understanding of the chemical interactions of each compositional ingredient. Both are fundamental necessities for success in today's friction material industry.

A Word on Perspective

This text is written from the perspective of a chemist whose intent is to examine many of the interactions produced by thermally induced changes in the molecular structure and chemical makeup of the compositional additives. It is readily admitted that the day-to-day operation of a tribological composite will involve a great deal of mechanically induced stress. The effects of the imposed mechanical stress are more than simple combinations of applied forces that physically deform the components. Stress imposed on the system also affects the tribological responses during operation, especially at elevated temperatures. As such, the influence of both mechanical and chemical factors will be examined whenever possible.

Today's modern friction material industry has become an elaborate mixture of perspectives derived from the efforts of groups from railway, automotive, industrial, and aerospace industries. Each of these groups has approached the manufacture and compositional tailoring of friction materials from various angles throughout the past century. Based on the sheer size of the available literature, an encyclopedic treatment is not feasible and would easily span multiple volumes covering many topics that are irrelevant in today's world. For this reason, an attempt has been made to limit the discussion to the more relevant topics with a focus on the compositional makeup and effects of ingredients as they relate to the wear and friction performance of the final product.

At the most fundamental level, this text was written for all who would like to know more about a truly fascinating and immeasurably complex application of composite materials. If the reader's occupation involves day-to-day product development or advanced research, or if interest in this subject is simply academic in nature, the information gathered in these pages is intended to promote growth and will hopefully allow for further improvements to be made in the field of engineered tribological composites.

Chapter 1
Engineered Tribological Composites

The world is wide, and I will not waste my life in friction when it could be turned into momentum.
—Frances Willard

FOR NEARLY ONE HUNDRED YEARS, AUTOMOBILE producers have relied on engineered tribological composites in the form of clutch facings and brake linings to provide high levels of friction during operation. The compositions, commonly known as friction materials, range from simple leather-covered blocks of wood to complex mixtures of functional additives bound together in an advanced polymeric matrix.

The matrix is an essential component of the composite. It provides structural rigidity, allowing the composite to withstand the mechanical forces present during operation. Additionally, the matrix material, as well as the fiber and additives, each play a role in the friction and wear behavior of the composite. Therefore, the composite designer is required to understand each raw material and its effect on the final performance of the product.

Likewise, manufacturing can also have an impact on product performance. Each of the steps in the manufacturing process presents the potential for variation in the structure of the composite. Considering the potentially large number of possible raw material combinations and numerous process variables, a functional definition is necessary to provide a starting point.

Functional Definition

The term "engineered tribological composite" includes any manufactured composite material designed to generate friction when operated against a mating member within a specified tribosystem. The composite is required to have a predictable life expectancy under a given set of operating conditions and can be used in a host of vehicles, equipment, or machinery. It can operate fully immersed in a lubricating fluid (e.g., wet friction system) or in the uncontrolled open atmosphere (e.g., dry friction system).

Because of variations in naming conventions, the terms linings, segments, pads, or facings may be used interchangeably depending on the end use application and are considered equivalent for our purposes. It is assumed that at a superficial level, clutch systems are typically designed to operate as energy transfer devices, and brakes are designed to be machines that convert kinetic energy into other forms, primarily heat. Each material, and the tribosystem within which it operates, is often unique. Therefore, each system should be examined as an isolated material pair operating within its own unique environmental constraints.

Operational Objectives

Regarding operational objectives, tribological composites must be able to provide a defined resistive force effectively and consistently under a given set of conditions. Many modern vehicle and equipment designers also impose other restrictions on performance that are typically focused on meeting pre-established comfort expectations. This typically manifests itself in the form of a requirement for the material to operate free from excessive vibration or other forms of unacceptable noise that could lead to negative customer perceptions.

The view that a brake lining must give a high friction output was at one time the single most critical component of the design phase; however, this view is not entirely accurate in today's automotive industry. To be successful today, the designer is required to satisfy multiple and often opposing criteria while concurrently meeting cost targets within predefined manufacturing constraints to supply a global marketplace with a specifically engineered product.

Herbert Frood and Ferodo

To establish a historical starting point, the development of engineered tribological composites is typically considered to have begun with Herbert Frood, a native of England, during the early part of the last century. Frood was an inventor for a number of years [1-1] and by 1900 his interest in friction materials emerged. In 1901he was awarded a patent [1-2] for the use of wooden brake blocks machined to fit within a metal holder. The blocks were intended for use on early automobile brake designs where the metal holder provided a durable rigid support structure that forced the wooden friction facing against the surface of the rubber tires to generate the resistance necessary to effectively stop the vehicle. The wooden blocks were formed to have a keyed shape as shown in Fig. 1.1.

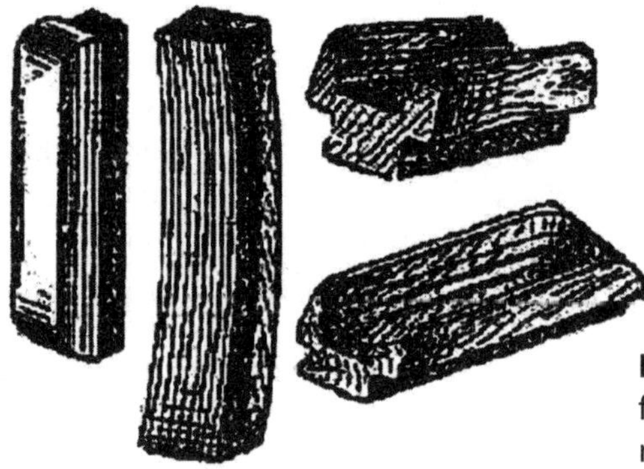

Fig. 1.1 Image taken from Great Britain patent no. 190018292.

Although these early brake linings were not true "engineering composites" like modern friction materials, the assembled product was composite in construction.

Frood's design, using wood as the active sacrificial tribosurface and metal as a rigid supporting structure, is a design feature analogous to modern brake and clutch materials that are rigidly bonded or mechanically attached to metallic carrier plates. Frood's composite construction design is also notable in that it is the first example of a brake system that allowed for effective replacement of the friction material without replacing the entire brake assembly.

Commercial Friction Materials

By 1906, Herbert Frood and Daniel Holmes [1-3], who had begun commercial sales as the Herbert Frood Company, developed a composite brake lining material formed from layers of a fibrous material bonded together by a "cementitious substance." The fibrous materials were shredded fabric, previously saturated with a solution of solvated resin and rubber. This composite also contained metal scraps as a friction enhancing additive– one of the first known examples of an additive being incorporated to induce a specific outcome during operation. Fig. 1.2 shows this early friction material concept.

Within a few years, Frood moved away from cotton fabric and began using asbestos fabric. Additionally, the manufacturing method evolved to incorporate basic compression molding techniques [1-4] for producing friction materials from woven asbestos. It had been determined that molding friction materials under heat and pressure was a necessary enhancement that effectively consolidated the composition and ensured a consistent level of compressibility.

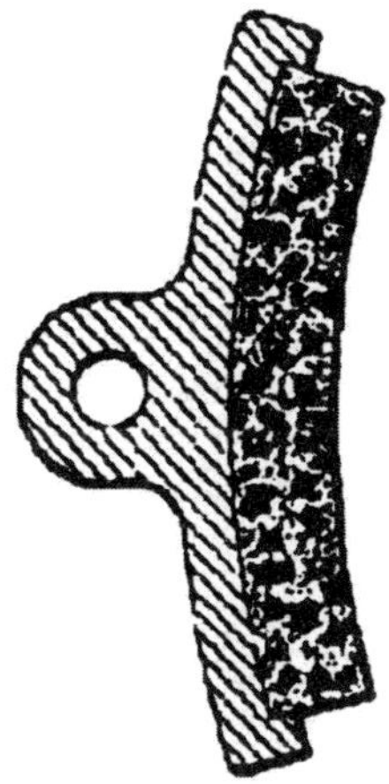

Fig. 1.2 Image taken from Great Britain patent no. 190606330.

According to Frood, the compressibility of linings made without the added step of compression molding was excessive and required the vehicle operator to frequently adjust the brake system to accommodate for compaction during use. The process of compression molding eliminated the need for frequent adjustment, making Frood's early linings very popular.

By 1920, the Herbert Frood Company had experienced substantial growth and changed its name to Ferodo [1-5], an anagram in honor of Herbert's wife Elizabeth (E. Frood). Within a few years, Ferodo's materials became widely accepted in this burgeoning friction material industry and were considered by Henri Perrot [1-6] to be "the European standard" for excellence. The name Ferodo would ultimately become synonymous with brakes in many parts of England and today Ferodo [1-7] remains a popular brand of automotive friction material in Europe and around the world.

The Birth of an Industry

Frood was not alone in the friction material industry. A number of entrepreneurs seized on the need for supplying friction materials to car manufacturers during the infancy of the automotive industry. For example, in 1929 Akebono began producing friction materials at a time when the Japanese automotive industry had produced only a total of 400 vehicles [1-8]. From these humble beginnings, Akebono [1-9] has grown to become a global supplier that enjoys a large portion of the automotive friction materials market, supplying both original equipment manufacturers and automotive aftermarkets with modern asbestos-free friction materials.

Likewise, Nisshinbo [1-10] began in 1907 as the Nisshin Cotton Spinning Company. Over approximately forty years, Nisshinbo increased in size and built several more operational sites. By 1947, Nisshinbo had expanded into non-textile areas, including automobile brakes. Today, Nisshinbo has operations specifically tailored to the manufacture and testing of friction materials for automotive brakes that serve the global automotive industry from locations in the United States and Asia.

During the time that Ferodo was growing in England, German lawmakers issued a regulation specifying that every vehicle must have redundant braking systems (i.e., two independent braking systems for every vehicle). This regulation quickly led to the development of new materials to replace the existing wooden blocks covered with leather commonly used at this time. Around 1912, a small company in Coswig named "Kirchbach'sche Werke" began to manufacture friction linings and by 1920 the company had begun selling molded brake linings under the name Jurid [1-11]. Today Jurid [1-12] is a leading brand of high-performance friction materials supplied to Germany and the rest of the world.

It should also be noted that during the early part of the last century, companies with a history of manufacturing belting began to expand into the friction materials market. Most notably, W. Willson Cobbett Limited incorporated in 1879 and began supplying Henry Ford with material in 1908. The company also supplied woven belting for use as a cone clutch material to Rolls Royce in 1909. W. Willson Cobbett Limited changed its name to the Scandinavia Belting Company in 1911. Because of the successful application of their belting in the transmissions of the very popular Model T, within three years the Scandinavia Belting Company was selling Henry Ford literally miles of their material for use as a friction material [1-13].

Like many early materials of this era, the material from the Scandinavia Belting Company was woven from yarn spun from carded asbestos and cotton that had been impregnated with bituminous resin [1-14].

These early linings did not form a significant part of the company's business for a number of years. However, by the late 1920s, increases in vehicle production drove demand to new heights. The company acquired the British Asbestos Company in 1925, resulting in a name change to BBA (British Belting and Asbestos) with commercial sales under the trademark Mintex (a contraction of MINeral TEXtiles) [1-15]. As the years progressed, BBA founded a distribution arm in Germany known as Textar. The company began autonomously producing brake linings in the mid-1930s and Textar remains a popular brand in Europe today.

A separate chapter of BBA's history began much earlier with Small and Parkes, another belting manufacturer located in Manchester. Small and Parkes eventually changed its name to Don International [1-16] and became a subsidiary of Cape Industries, a leading supplier of asbestos and asbestos products. Don International was a major producer of asbestos clutch and brake linings in the United Kingdom for many years prior to being sold to BBA in the mid-1980s. This sale resulted in the formation of Mintex-Don. Today Textar, Mintex, and Don survive in a portfolio of brands retained by TMD Friction Holdings GmbH.

Modern Engineered Tribological Composites

From these beginnings, modern engineered tribological composites have emerged as exceptional entities that have been developed to meet very specific tribological requirements. They incorporate a wide range of constituents, especially since the elimination of asbestos. In spite of the diversity of techniques and methods, many of today's products share a similar design philosophy to linings developed during the early days of the friction material industry.

For example, it has been known since the beginning of the automobile industry that no pure substance or single material [1-17] is effective as a tribological material across multiple platforms and modes of operation. A "one size fits all" material does not and cannot exist because of differences in vehicle design, operator influences, and geographic effects that expose the composite to varying local ambient conditions during use. Therefore, each tribological composite that has been successfully introduced and gained acceptance as a commercially viable product has been through a rigorous design cycle to achieve validation as the material of choice. This iterative process involves evaluation under a range of conditions to realize the desired balance of cost, friction output, and wear resistance. Furthermore, because these materials are also expected to have a high coefficient of friction and yet remain free from excessive noise or vibration during the entire life of the product, they are often subjected to a systematic regimen of tests to determine the noise propensity of each composition under extremes of temperature, speed, and pressure.

The Art of Compromise

It is this amalgamation of opposing design goals that presents a continual balancing act for the engineer who strives to refine each composite to achieve the desired outcome. In all cases, this process of design and redesign will result in a measure of compromise that yields a final product capable of achieving a specific design target (such as an elevated nominal coefficient of friction), but at the expense of another area of performance (such as wear resistance). Regardless of the design, some form of compromise cannot be avoided during the development of tribological composites.

Tribology

The term tribology is a moderately new word that began with a working group under the direction of Peter Jost. Jost's group had been assigned to investigate the state of lubrication education and research in England with the directive of preparing an opinion statement on the needs of industry [1-18]. This working group decided that no suitable term was available and, consequently, introduced the new term "tribology" to describe aspects of lubrication, bearing technologies, and associated fields of study.

By March 1966, the working group had produced a report, which has become known as the "Jost Report," with suggestions regarding the need for improving education in the area of tribology. The outcome of this investigation was a critical need for an effective scientific approach toward the education, identification, and resolution of numerous tribological issues that affected various industries.

It was determined that friction and the associated wear of machine components was responsible for a vast number of equipment failures, lost productivity because of downtime, and subsequent maintenance and repair. At the time, the cost implications of tribologically induced failures were estimated to be valued at more than 500 million British pounds [1-19].

Today friction and wear-related issues have been estimated to cost 1–2 % of the gross domestic product [1-20]. A reduction in the friction of automotive powertrain components is estimated to generate savings exceeding 100 billion U.S. dollars. Therefore, tribology is more than an academic curiosity. It is a field of study that has emerged as a global necessity for all industrialized nations.

A Cross-Functional Approach

Modern tribology is not a single science with rigid boundaries, but requires a cross-functional scientific approach that engages aspects of chemistry, physics, surface analysis, and material science. These approaches are combined and might use a combination of simple and advanced analytical techniques to study interactions between surfaces during sliding in order to clarify transitions that occur at the interface.

The inherent complexity of a tribosystem using composite materials containing multiple constituents of different size, shape, hardness, and reactivity makes an understanding of even the most basic laws of friction increasingly difficult, if not impossible, to model effectively from a single perspective.

An Absence of Mathematical Models

Because of the immense complexity of composite tribology, there is an acknowledged absence of mathematical models in this book. This absence is due to the fact that many of the models found in the available literature simply do not work across the full range of operational conditions in which automotive tribological composites are expected to perform. Therefore, they are not applicable for all applications under all operating conditions and, consequently, have not been included in this book.

References

1-1. Frood, H. 1898. An improved holder for bicycles. Great Britain Patent 189722802, filed October 5, 1897, and issued August 13, 1898.

1-2. Frood, H. 1901. Improvements in Brake Blocks for Use on Road Vehicles, and for Other Purposes. Great Britain Patent 190018292, filed October 15, 1900, and issued October 5, 1901.

1-3. Frood, H., and D. Holmes. 1907. Improvements in Brake Blocks and Manufacture Thereof. Great Britain Patent 190606330, filed March 16, 1906, and issued March 14, 1907.

1-4. Frood, H. 1913. Improvements in Brake Blocks or the Like. Great Britain Patent 191219952, filed September 2, 1912, and issued October 2, 1913.

1-5. Ferodo Limited. 1922. "Befestigung von Bremsbelägen, Futtern o. dgl. an ihrem Grundkörper." German Patent 361997, filed June 29, 1920, and issued October 20, 1922.

1-6. Perrot, H. 1924. "Four-Wheel Brakes." SAE Paper No. 240015. Warrendale, PA: Society of Automotive Engineers.

1-7. Ferodo. Registered trademark of Federal-Mogul Friction Products Limited Company, 20 St. Mary's Parsonage, Manchester, United Kingdom, Registration No. 0102837, February 23, 1915.

1-8. Akebono Brake Corporation. "History." Accessed January 31, 2011. http://www.akebonobrakes.com/company/history.asp.

1-9. Akebono. Registered trademark of Akebono Brake Industry Co., Ltd., Koami-cho Chu-ku, Tokyo, Japan, Registration No. 1256025, November 1, 1983.

1-10. Nisshinbo Company. "Nisshinbo History." Accessed January 31, 2011. http://www.nisshinbo.co.jp/english/press/pdf/2003/ar2003-no5.pdf.

1-11. Honeywell. "Honeywell Bremsbelag GMbh History." Accessed January 31, 2011. http://www.honeywell.com/sites/bremsbelag-en/history.htm.

1-12. Jurid. Registered trademark of Honeywell International, Morristown, New Jersey, Registration No. 2445492, April 24, 2001.

1-13. TMD Friction Company. "TMD History." Accessed January 31, 2011. http://www.tmdfriction.com/en/company/history.

1-14. "Asbestos and Certain Asbestos Products: A report on the Supply of Asbestos and Certain Asbestos Products." Presented to Parliament in pursuance of Section 9 of the Monopolies and Restrictive Practices Act 1948 (printed January 23, 1973).

1-15. Mintex. Registered trademark of TMD Friction Services GmbH Corp., Leverkusen, Germany, Registration No. 0668264, October 14, 1958.

1-16. Grace's Guide: The Best of British Engineering. "Small and Parkes." Accessed January 31, 2011. http://www.gracesguide.co.uk/wiki/Small_and_Parkes.

1-17. Whitcomb, L., and T. Fay. 1909. "The Modern Trend of Brake Design." SAE Paper No. 090001. Warrendale, PA: Society of Automotive Engineers.

1-18. Hansard Millbank Systems. "The Jost Report." Accessed July 24, 2011. http://hansard.millbanksystems.com/written_answers/1966/aug/11/jost-report#S5CV0733P0_19660811_CWA_176.

1-19. Jost, P. 2006. "Tribology: How a Word Was Coined 40 Years Ago." *Tribolology and Lubrication Technology* 3: 24-29.

1-20. Nieuwsbank. "New Friction Technology." Accessed April 1, 2010. http://www.nieuwsbank.nl/en/2009/03/25/R010.htm.

Chapter 2
Friction: One Aspect of Tribology

Scientists give us gobbledygook about friction and molecules. But they don't really know.

—*Ray Bradbury,* Fahrenheit 451

TRIBOLOGY MAY SOUND LIKE GOBBLEDYGOOK, but the term was derived from the Greek tribos (τριβοσ,) meaning "to rub." Therefore, at its most elementary level, tribology is a scientific study of rubbing objects together. This may seem simple and intuitive; however, friction processes are not simple and far from intuitive. When two surfaces come into contact, there are a number of interactions that occur. It is helpful to divide these interactions into two large classes: mechanical and chemical.

Mechanical interactions produce irreversible physical changes to both surfaces because of plastic deformation of the asperities. After sliding has been initiated, heat is generated and modifies the compositional makeup of the materials near the surface by promoting chemical reactions between the surfaces and the operating atmosphere. In oxygen atmospheres, these reactions are typically oxidation and combustion. In wet friction systems, liquid additives can induce other reactions leading to the formation of surfaces with unique chemical attributes. A common practice involves the use of sulfur-containing additives that form a metallic sulfide film on the metallic mating surface.

This combination of physical and chemical changes alters future tribological responses by producing a surface that is uniquely different from the initial surface. This further exacerbates the analysis of frictional processes by introducing a time component. Therefore, tribological composite designers must also be aware of the operating history and how it can influence future results.

It should be quickly evident that friction and wear processes are inherently linked together and that friction is one aspect of the field of tribology. To gain insight into the world of modern tribology, an understanding of friction must be established. To illustrate the extent of growth, and to highlight critical advances in the study of friction that have been made over the intervening years, it is necessary to have a short historical review upon which to build.

A Brief History of Friction

Exploration of frictional processes was first recorded by Leonardo da Vinci [2-1, 2-2]. His writings describe the use of ropes, pulleys, inclined planes, and assorted simple machines to determine two fundamental principles of friction behavior. From his studies, he concluded that the force of friction would increase as the mass increased. In other words, heavier loads gave higher levels of frictional resistance during sliding. His second principle was that the force of friction was independent of the apparent contact area. Leonardo da Vinci's work was recorded in his notebooks and was not widely known for many centuries.

Guillaume Amontons

The person historically viewed as the founding father of friction was Guillaume Amontons [2-3]. Amontons was a Parisian who lost his hearing while a teenager. He studied physics, which led to his rediscovery of Leonardo da Vinci's two principles of friction. Amontons' concept of frictional resistance to motion was based on the supposition that friction was the amount of work required to lift interlocking asperities over each other during sliding. He further suggested that the coefficient of friction was universal in nature and had a value of roughly 0.33. This concept of friction, which has become known as the roughness hypothesis, was widely accepted for a number of years as the fundamental mechanism for sliding friction between two bodies in contact.

Bernard Forest de Bélidor

Bernard Forest de Bélidor, a French professor of artillery at the school of Fére-en-Tardenois, worked in several areas of science including hydraulics and construction [2-4], but his studies on friction were also very important. Bélidor used inclined planes as well as a number of novel methods, including sandwich style friction test methods like the one shown in Fig. 2.1:

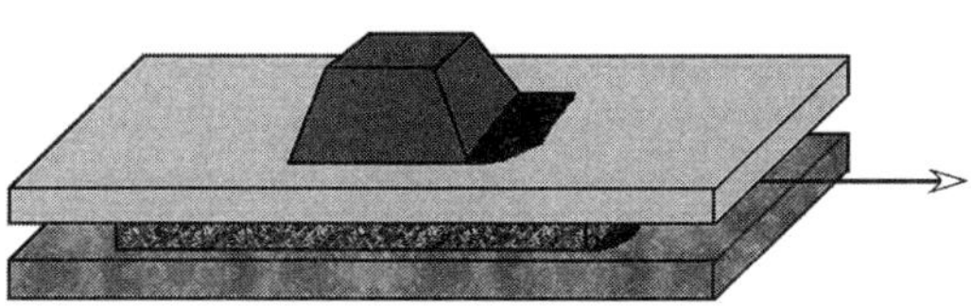

Fig. 2.1 A friction experiment as illustrated by Bélidor.

Bélidor envisioned interlocking spherical asperities on surfaces as the mechanism for resistance to sliding motion. From this concept, Bélidor was able to calculate a value of approximately 0.35 for the coefficient of friction between surfaces. Bélidor's work was very influential and resulted in the publication of a friction textbook that was later used in the instruction of another prominent tribologist, Charles Augustin de Coulomb [2-5], whose work is discussed later in this book.

Leonhard Euler

Leonhard Euler was a contemporary of Bélidor. Euler's work on friction [2-6] was examined by Blau [2-7] and Sahoo [2-8]. Unlike Bélidor, Euler's model for friction was based on surfaces composed of triangular asperities. The asperities were assumed to have a slope θ. The coefficient of friction would be equal to the tangent of the angle of these asperities as shown in Eq. 2.1.

$$\mu = \tan\theta \tag{2.1}$$

Euler's observations of the motion of a block on an inclined plane also led him to distinguish between static and kinetic friction (i.e., the force required to sustain motion during sliding is less than the force to initiate motion). A general conclusion was that static friction is always higher than sliding (kinetic) friction; however, exceptions to this rule have also been observed.

Charles Augustin de Coulomb

Charles Augustin de Coulomb made significant contributions in many areas of science while actively serving in the military. It was during his work on a construction project to build a fort entirely of wood at Rochefort in 1779 that Coulomb conducted research into the field of friction. It has been reported that he considered the role of adhesion between surfaces as a potential mechanism behind friction [2-9]. However, he was unable to reconcile that frictional resistance had been well-established to be independent of the geometric area of contact and thus adhesion, which was based on the area of contact between two bodies, could not be a major contributing factor to frictional resistance.

Therefore, Coulomb was forced to adopt the only tenable position. He explored the behavior of friction under the roughness hypothesis with a number of elegant experiments including ropes and pulleys as well as loaded sleds and plane surfaces. His findings, recorded in the *"Théorie des Machines Simple"* [2-10] won him the Grand Prix from the

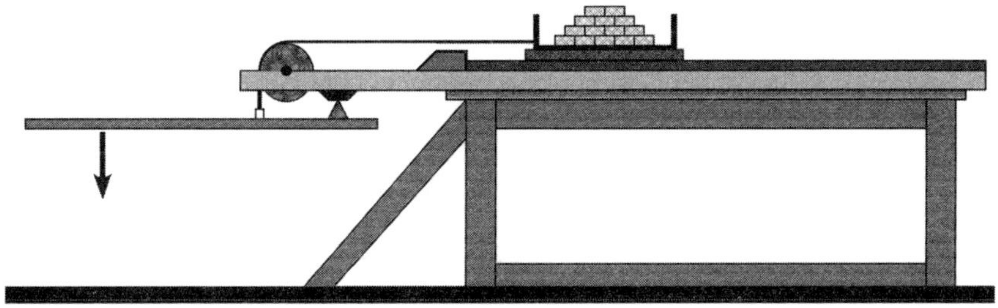

Fig. 2.2 An early friction test stand from Coulomb's *Théorie des Machines Simples*.

Academy of Sciences in 1781. In this text, he demonstrated that a wide range of materials behave in a manner consistent with the findings of Amontons. One illustration by Coulomb shows his keen experimental insight by using a loaded sled, a simple lever, and a pulley as a means for exploring friction. This method compares the load in the sled against the applied force of a lever needed to induce sliding motion. An example of this early friction test stand is shown in Fig. 2.2.

The Third Law of Friction

Coulomb also developed what is commonly known as the "third law of friction," which states that friction during sliding is independent of velocity. This "law" is often broken, and a number of exceptions have been found. Typically, exceptions are found at very low speeds and also at very high sliding speeds. For instance, at high sliding speeds, the friction level observed can show a gradual decrease as velocity increases. This is often seen in automotive braking applications on high-performance vehicles.

The Three Laws (and a Caveat!)

From Coulomb onward, the basic tenants of friction were summarized in three fundamental laws that were repeatedly tested and verified. These laws were considered sufficient for use by the engineering community of the time and are as follows:

1. Friction is a function of the applied normal load and changes proportionally to changes in the normal load.
2. Friction between two bodies is independent of the geometric (apparent) area of contact.
3. Friction is independent of sliding velocity.

It must be noted that in the last few decades these "laws" have been shown to have numerous exceptions and, therefore, should be considered limited in their scope of application.

Sir John Leslie

In the beginning of the 19th century, Sir John Leslie began to question the "laws of friction." Leslie, a Scottish physicist, pointed out that the roughness hypothesis for friction required work to lift a surface up along an asperity and that, after further motion, the body in motion would reclaim potential energy by dropping down the other side of the asperities [2-11]. If this mechanism was correct, the energy required to climb asperities would be equal to that reclaimed and thus be balanced, which implied sliding friction would not generate heat. Because sliding easily produces heat, the roughness hypothesis did not explain friction adequately, forcing Leslie to conclude that further insight was needed to fully understand frictionally generated heat.

A Teacup and Modern Tribology

Insight can come about in a rather odd fashion and at the most peculiar of times. For example, it was not until the early part of the twentieth century when Lord Rayleigh (John William Strutt, the third Baron of Rayleigh) observed that a few drops of water applied to the saucer would prevent the teacup from sliding. William Hardy took this idea and began a series of experiments to study the sliding behavior of loaded convex glass dishes on lubricated glass surfaces.

Approximately one year after Lord Rayleigh commented on his teacup observations, Hardy and Hardy published research showing that the glass surfaces would adhere together and be damaged when the tangential load increased to a critical value. It was noted by Hardy and Hardy that the range of the force of cohesion between the surfaces was "so small that all surfaces, no matter how carefully trued, may be considered to touch only at the summits of the elevations" [2-12]. Over a period of ten years, Hardy continued to explore the principles of boundary layer lubrication with various fluids. He published a number of interesting findings with his colleagues Doubleday [2-13] and Nottage [2-14, 2-15].

Bowden and Tabor

Decades later, Bowden and Tabor, who regarded the work of Hardy as the major stimulus for the modern approach to the subject of friction, took the concept of friction in an entirely new direction.

The efforts of Bowden and Tabor, and their many students, spanned several years and generated numerous publications on various aspects of the subject of friction. One of the best known publications, *The Friction and Lubrication of Solids* [2-16], combines many aspects of their work that led to the development of a new hypothesis for the underlying mechanism of friction—a hypothesis now known as the "adhesion model of friction."

The Adhesion Model of Friction

Bowden and Tabor observed sliding behavior between numbers of different metal pairs ranging from very soft metals such as indium to hardened steel. Regardless of the material combinations used in their experiments, the same conclusion regarding friction between metals was reached: surfaces tended to adhere and pluck out material from the region of contact. Based on these findings, it was determined that Amontons' laws, which were based on a fundamental assumption of elastic contact between surfaces, were not adequate for describing friction generated during the sliding of metals.

The Real Area of Contact

Bowden and Tabor further concluded that surfaces in contact, regardless of how smooth, were only touching at a few discrete locations, which reflects the previously mentioned thoughts of Hardy. Their reports presented the idea that contact between bodies is limited to a very small "real area of contact" [2-17]. This counterintuitive point of view opposed the previously held concept of a geometric or apparent area of contact. But experiments using electrical conductivity measurements revealed the "real area" of contact between metal pairs was several orders of magnitude smaller than the apparent area of contact [2-18]. In some cases, the real area of contact was found to be as little as 1/10,000 of the geometric area of contact.

Junction Formation

The physical manifestation of a vastly smaller area of contact under a finite load was an understanding that, due to the reduced area of contact, the applied load was distributed over a few small points and, therefore, it introduced high stress at these points. Bowden [2-19] stated that "even with the lightest loads" the stress at the tips of the asperities is capable of causing the hardest metals to flow plastically. Therefore, the concept of friction based on elastic or reversible deformation processes was not supported by their experiments and mandated a re-evaluation of the roughness hypothesis.

Bowden and Tabor further concluded that metallic asperities undergo plastic flow resulting in the formation of physical junctions between the two bodies. The junction formation mechanism was likened to the process of "spot welding" two discrete points together.

Based on the irregularities of a surface, the distribution of asperity sizes across the geometric landscape, and the combination of the applied load, surfaces in contact would form a number of connective bridges between the surfaces. The connective bridges were tiny in comparison to the geometric area of contact, but large relative to the molecular scale of dimensions. These newly formed junctions were considered the underlying cause of the observed resistance to motion known as friction.

It is logical that higher loads will force more asperities into contact and thus produce more junctions than lower loads. Further loading will continue to increase the real area of contact until it approaches the size of the geometric area of contact. Likewise, it should be understood that to separate two bodies in contact under a given load would require the application of a tangential force sufficient to shear all of the junctions formed during contact. If the applied (or normal) load was large enough, the force required to shear a real area of contact equal to the apparent area of contact would theoretically approach the bulk shear properties of the weaker of the two metals, which provides a quantifiable physical limit for frictional processes.

An overall conclusion is that friction is a dissipative process requiring constant energy input to continue relative motion. The degree of energy loss in the Bowden and Tabor model is assumed to be the result of plastic deformation of junctions and explains the generation of heat that occurs during sliding motion. This model has been tested extensively during the last fifty years and has been found to be highly repeatable under a host of conditions for a vast number of material pairs.

The Abrasive Model of Friction

Bowden and Tabor also examined abrasion. It was determined that the level of friction developed during sliding is not a mutually exclusive event limited to adhesive processes, but combines adhesional and abrasive components as shown in Eq. 2.2 where P_f is the plowing, or abrasive, component of friction and F_a is the adhesional component of friction.

$$\mu = P_f + F_a \tag{2.2}$$

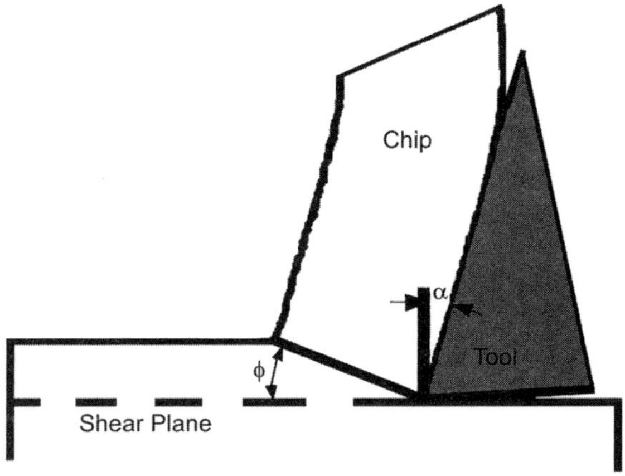

Fig. 2.3 Diagram illustrating chip formation process taken from Ernst and Merchant.

Both components are dependent on the load applied to the system. Higher loads would not only increase the interaction between asperities by forming a higher concentration of junctions but also force hard asperities deeper into softer materials increasing the abrasive component of friction. Bowden and Tabor's work on abrasion did not occur in isolation and mirrors studies by other research groups, most notably that of Ernst and Merchant [2-20, 2-21].

Ernst and Merchant

Ernst and Merchant explored the mechanisms of chip formation and associated friction phenomena during machining processes with interest in improving the surface finish of machined surfaces. Fundamentally, a machining process generates wear debris, or chips, by an exaggerated form of mechanical abrasion. It was observed that chip formation originates at the nose of the cutting tool and extends along the work surface at an angle of φ between the plane and surface, as seen in Fig. 2.3.

From their efforts, Ernst and Merchant developed a force diagram relating the friction level observed during machining by the expression shown in Eq. 2.3.

$$\mu = cot(2\varphi - \alpha) \tag{2.3}$$

Here μ is the coefficient of friction between the tool and the chip, which is a function of the shear angle (φ). The shear angle defines the plane of shear and can be generally assumed to be consistent throughout the bulk metal piece, but in reality the plane of shear might undergo a number of material dependent changes due to fracture irregularities, voids, and inclusions of other materials.

The plane of shear will vary as the rake angle (α) changes. It has been determined that smaller rake angles will produce lower coefficients of friction, which was desirable to improve machining efficiency. However, the observed results demonstrated higher frictional resistance than predicted. This led Ernst and Merchant to conclude that mechanisms other than pure abrasion might be operating simultaneously during the sliding process.

Based on the available data, the authors reported that surface roughness effects were negligible and adhesion between the tool and workpiece must be present. This combination of adhesional forces and plowing forces generated by the edge of the tool moving through the surface of the metal were therefore sufficient to explain the differences noted during their studies.

Application of the Abrasive Model to Friction Materials

The properties of the materials involved and the geometric attributes of hard asperities are important considerations that must be examined early in the development process, especially during abrasive selection. The selection criteria of any abrasive constituent should include an analysis of the size, shape, and physical character of the particles. Shape and size factors affect the rake angle of the particle during sliding and influence abrasion of the mating member. Physical characteristics affect the durability of the particles and impact the life expectancy of any tribological composite. Each of these must be examined to select an appropriate abrasive for friction materials.

The Two Dominant Forms of Friction

As Bowden and Tabor have stated, friction between metal pairs is a combination of both adhesion and abrasion processes. An increasing number of excellent papers are being written on the various aspects of friction. These introduce a number of subtly distinct friction mechanisms. However, for our purposes these two fundamental forms are sufficient.

This should not be viewed as a limited perspective, but as an attempt to focus future research toward a few factors that can effectively dominate experimental results. Therefore, abrasive friction mechanisms and adhesional mechanisms are considered the major modes of operation in dry tribosystems that occur in today's engineered tribosystems. Fully immersed systems are typically dominated by the properties of the fluid and operate under a different set of models, but might also be affected by these factors, especially under high application forces.

Friction Transitions

For the individual responsible for developing an engineered tribological composite, a key point to remember is that raw materials in the composition are not static during use. Many of the material combinations used in the manufacture of a tribological composite will undergo a series of transformations during use that change the performance for the remainder of the effective life of the composite. A thorough examination of friction transitions has been conducted by Blau [2-22] and is recommended as a useful resource.

As an illustration of the potential effects produced by friction transitions, a tribological composite can be viewed as a mixture of several ingredients that vary in their respective physical properties. During operation the composite will experience compressive and tangential loads that result in extensive plastic deformation at the asperity level of contact. This produces irreversible mechanical changes to the asperities and forms particles of wear debris by fracturing or dislodging a portion of the materials from the bulk of the composite. In addition, thermal energy developed during sliding not only affects mechanical characteristics (e.g., softening), but can also introduce chemical changes such as oxidation.

In other cases, debris particles generated during sliding can become load-bearing entities, provide additional points on which the normal load can be supported, and increase the real area of contact. Other particles can be ejected from the system and have no further effect. In other cases, particles of wear debris can be very small and become trapped at the surface, leading to the accumulation of wear debris forming plateaus of compacted wear debris.

Plateau Formation Mechanism

The process of plateau formation was described by Eriksson and colleagues [2-23]. It is a process whereby surface transformations during use will produce a series of larger load-bearing regions, known as plateaus, at the friction interface. These regions of contact affect both friction and wear during subsequent operation because of changes in the real area of contact. Eriksson reported the mechanism as an accumulation of wear particles behind a fiber at the interface, notably steel fiber. The fiber prevents particles from being eliminated from the interface and therefore prevents wear.

Additional studies by Österle and colleagues [2-24] reported that accumulation of wear debris does not necessarily require fibers to be present and can also occur in the vicinity of locally hard particles, such as silica. Whether it is a hard particle or a fiber, the formation process is an accumulation of finer particles that are prevented from flowing out of the interface region during sliding.

Rigney [2-25] took the concept of trapped wear debris and described the physical action of the particles during sliding as a form of mechanical mixing. During mechanical mixing, sliding surfaces and trapped wear particles experience attrition leading to further modification of the surface topography that might also aid in the formation of plateaus.

Friction Transitions in Metallic Elements

Transitions occur in all friction systems including those with metallic friction elements. Bowden and Rowe [2-26] stated that the combination of normal and tangential forces during sliding of clean metallic surfaces increases the size of asperity junctions. An increase in the size of a junction increases the real area of contact and thus requires more energy to rupture the larger junction, which manifests itself as an increase in the coefficient of sliding friction. This phenomenon might be beneficial in explaining why certain metallic friction materials sometimes show a pronounced rise in their friction output during use, especially at high temperatures.

A composite material, whether bound in a metallic or polymeric matrix, serves only to amplify the potential number of transitions that might occur during sliding. These transitions and the general observation of friction behavior will also depend on operational parameters, the compositional nature of the materials during sliding, and in some limited cases, the manufacturing method employed to physically produce the composite. There is a great deal of work that remains to be done, and a thorough understanding of induced raw material transitions is necessary to enhance the friction and wear behavior of modern tribological materials.

Suggested Reading

It is readily admitted that this section is not intended to provide a comprehensive treatise on friction and the various modes or mechanisms by which it can be seen, but rather to provide a simple perspective that introduces many of the salient points. For those who are interested, there are a number of excellent resources on this subject of friction such as Blau [2-7], Sahoo [2-8], Williams [2-9], Persson [2-27], Bhushan [2-28], Rabinowicz [2-29], and Suh and Saka [2-30].

References

2-1. Leonardo da Vinci. 1478-1581. *Codex Alanticus*. Milan, Italy: Biblioteca Ambrosiana.

2-2. Leonardo da Vinci. 1478-1581. *Codex Arundel*, folios 40v, 41r. London: British Museum.

2-3. Amontons, G. 1699. *Histoire de l'Académie Royale des Sciences avec les Mémoires de Mathématique et de Physique*. Paris, France.

2-4. Bélidor, B. 1737. *L'architecture Hydraulique, ou L'art de Conduire, d'élever et de ménager les eau*. Paris, France: Jombert.

2-5. Schofield, A. 2005. *Disturbed Soil Properties and Geotechnical Design*. London: Thomas Telford Publications.

2-6. Euler, L. 1750. "Sur la Friction des Corps Solides." *Memoires de l'academie des Sciences de Berlin* 4:122-132.

2-7. Blau, P. 1995. *Friction Science and Technology*. New York: Marcell Dekker.

2-8. Sahoo, P. 2005. *Engineering Tribology*. New Delhi, India: PHI Learning Private Limited.

2-9. Williams, J. 2005. *Engineering Tribology*. New York: Cambridge University Press.

2-10. Coulomb, C. 1821. *Théorie des machines simples, en ayant égard au frottement de leurs parties et a la roideur des cordages*. Nouvelle Ed. Paris: Bachelier, Libraire, Quai des Augustins.

2-11. Leslie, J. 1804. "An Experimental Inquiry into the Nature and Propagation of Heat." Printed for J. Mawman, No. 22. London: Poultry.

2-12. Hardy, W., and J. Hardy. 1919. "Note on Static Friction and on the Lubricating Properties of Certain Chemical Substances." *Philosophical Magazine*, 6 (38):.32-39.

2-13. Hardy, W., and I. Doubleday. 1922. "Boundary Lubrication: The Paraffin Series," Proceedings of the Royal Society of London, A 100, pp. 550-574.

2-14. Hardy, W., and M. Nottage. 1926. "Studies in Adhesion - Part I," Proceedings of the Royal Society of London, A 112, pp. 62-75.

2-15. Hardy, W., and M. Nottage. 1928. "Studies in Adhesion - Part II," Proceedings of the Royal Society of London, A 118, pp. 209-229.

2-16. Bowden, F., and D. Tabor. 1950. *The Friction and Lubrication of Solids*. London: Oxford University Press.

2-17. Bowden, F., et al. 1943. "The Ploughing and Adhesion of Sliding Metals." *Journal of Applied Physics*, 14:.80-91.

2-18. Bowden, F., and D. Tabor. 1939. "The Area of Contact Between Stationary and Between Moving Surfaces," Proceedings of the Royal Society of London, A 169, pp.391-413.

2-19. Bowden, F. 1955. "Recent Studies on Metallic Friction." *The Chartered Mechanical Engineer* 2:.86-101.

2-20. Ernst, H., and M. Merchant. 1941. "Chip Formation, Friction and High Quality Machined Surfaces." *Surface Treatment of Metals* 29: 299-378. New York: American Society of Metals.

2-21. Ernst, H., and M. Merchant. 1940. "Surface Friction of Clean Metals - A Basic Factor in the Metal Cutting Process," Proceedings of Special Summer Conference on Friction and Surface Finish, MIT Report #15, pp.76-101.

2-22. Blau, P. 1989. *Friction and Wear Transitions of Materials*. New Jersey: Noyes Publications.

2-23. Eriksson, M., et al. 1997. "Surface Characterization of Brake Pads After Running Under Silent and Squealing Conditions." *Wear* 232:163-167.

2-24. Österle, W., et al. 2001. "Chemical and Microstructural Changes Induced by Friction and Wear of Brakes." *Wear* 251:1469-1476.

2-25. Rigney, D. 1992. "Some Thoughts on Sliding Wear." *Wear* 152: 187-192.

2-26. Bowden, F., and G. Rowe. 1956. "The Adhesion of Clean Metals," Proceedings of the Royal Society of London, A 233, pp.429-442.

2-27. Persson, B. 2000. *Sliding Friction: Physical Principles and Applications*, 2nd ed. Berlin: Springer-Verlag.

2-28. Bhushan, B. 1999. *Principles and Applications of Tribology*. New York: John Wiley and Sons.

2-29. Rabinowicz, E. 1995. *Friction and Wear of Materials*, 2nd ed. New York: John Wiley and Sons.

2-30. Suh, N., and N. Saka, eds. 1981. *Fundamentals of Tribology. Cambridge:* The MIT Press.

Chapter 3
Sliding Wear

A gem cannot be polished without friction...

—Chinese Proverb

TO BRING OUT THE FIRE AND BRILLIANCE of gemstones, they must be shaped and polished. During polishing, even the hardest gemstones wear to yield the highly prized lustrous finish desired by consumers. Without frictionally induced wear, gemstones would be little more than pretty rocks.

Likewise, an engineered tribological composite will inevitably experience wear processes during operation. These wear processes tend to arise from physical and chemical transitions at the sliding interface during use and will alter the topology of the surface by serving to generate, accumulate, compact, and/or liberate particles from both materials. In many cases these events occur simultaneously and typically result in the formation of a thin layer of severely transformed material at the friction interface.

The thin layer, commonly known as the "third body layer" or "transfer film," is of singular importance to the friction material industry because it can stabilize the coefficient of sliding friction and enhance the wear resistance of friction materials, thus prolonging their effective lifespan. Because the third body layer is formed by wear, a basic understanding of wear processes is necessary for success in the challenging field of tribological composite development.

Four Common Modes of Wear

Common automotive brake linings operate in an uncontrolled environment and are subjected to four basic modes of wear: adhesive, abrasive, chemical, and delamination wear. Wet friction systems also experience these modes but have an additional mode of erosion because of particulates in the cooling fluid. Because wet friction systems typically incorporate filtration methods to extract particulates from the fluid, the role of erosive wear is minimized and the four mechanisms of wear are generally anticipated to dominate the other mechanisms at work. During sliding wear the different modes of wear can occur simultaneously and should not be considered mutually exclusive, nor should they be limited to a given tribosystem.

The Complexity of Sliding Wear Processes

If we consider an automotive brake system, the surface of the metallic mating member has been machined and possesses a measurable roughness. Likewise, the surfaces of friction material composites are often ground during manufacture, which also results in a surface that is measurably rough. As sliding contact begins, the initial mechanism for wear is mainly abrasive [3-1]. Upon further sliding, a transition from abrasive to adhesive wear occurs [3-2] and the rate of wear begins to shift toward a mild wear regime [3-3] because of the formation of a transfer film.

The transfer film is a layer of compacted wear debris that effectively transforms the surface topography by filling interstices between the asperities [3-4]. It generally forms during the "running-in" or "burnishing" phase of the operating lifespan of the composite. During this timeframe, the friction levels are typically stabilized and materials generally operate for an extended period of time without undue concern.

Hooten [3-5] examined the protective film that forms on the surfaces of metal ceramic composites during operation in high-energy friction applications. It was concluded that an optimum film thickness will provide a stable coefficient of friction with adequate transfer of forces across the interface. On the other hand, if the film thickness becomes excessive, hydrodynamic lubrication occurs and a rapid drop in friction is observed.

Likewise, Suh [3-6] reported that an excessively thick transfer film might begin to delaminate and flake away from the underlying substrate. This is a result of recurring compressive stress imposed on the system that exceeds the cohesive strength of the particles composing the transfer film. Flakes of wear debris from the failed third body layer might introduce heavily oxidized materials into the third body system and could result in a transition to a more abrasive wear regime.

Alternatively, the third body layer can flake away revealing clean metallic surfaces that are now subject to chemical attack from the atmosphere (e.g., oxidation) or from lubricants and other reactive additives present in the tribosystem. In the case of automotive brakes (i.e., dry friction systems), the oxidation of metals can have a significant impact on performance and can introduce abrasion. On the other hand, small amounts of surface oxide formation can also aid in the regeneration of a stable third body layer by enhancing the topography of the surfaces and promoting adhesion. Each system is dependent on the composition of the composite as well as the operating temperatures [3-7].

Transitions such as these are often difficult to predict and usually manifest themselves in the form of unwanted friction variation, excessive wear, or unacceptable noise and vibration characteristics. These generally occur at the onset of a transition from one regime to another. Therefore, the four modes of wear and transitions between them are of critical importance to the designer of any tribological composite.

Abrasive Wear

According to Suh [3-6], abrasive wear is considered the most significant form of the four modes of wear in our discussion. He assigned this level of importance based on comparative analysis of wear coefficients from several wear processes. Suh found abrasive wear to have a larger coefficient of wear, sometimes by an order of magnitude, in comparison to the wear coefficients for other forms of wear. This approach may seem of interest only to the academic; however, abrasive wear is more than an academic concern.

Abrasive wear has significant financial implications because it can account for as much as 50% of the total wear encountered in industrial machinery [3-8, 3-9]. The required extensive maintenance expenses associated with abrasive wear necessitate effective lubrication and provide sufficient justification for the vast amount of research that has been devoted to the development of advanced lubricants.

In automotive wet friction systems, many of today's coolant fluids are the result of extensive studies investigating a vast array of chemicals including extreme pressure additives, pour point depressants, antioxidants, and wear reducing agents. Knowledge of these additives and their impact on a given composite is important for those who work in the field of immersed friction materials. However, a full coverage of the chemistry of lubricating coolants is beyond the scope of this text and will be discussed only as needed.

Defining Abrasive Wear

Definitions for all wear terms will be taken from ASTM's Standard Terminology Relating to Wear and Erosion [3-10]. Consequently, abrasive wear is defined as "wear because of hard particles or hard protuberances forced against and moving along a solid surface."

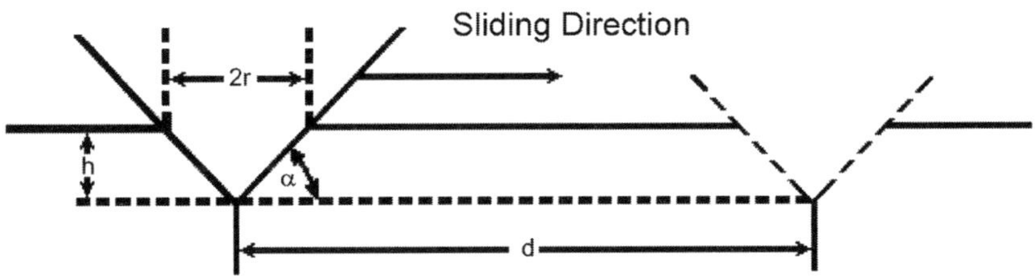

Fig. 3.1 Simple model of abrasion.

A hard particle becomes a small indenter when forced into contact with another surface. During sliding, the passage of the embedded indenter through the material produces abrasion. A commonly discussed model for abrasion employs a cone penetrator under load as shown in the Fig. 3.1.

In this model, the conical penetrator makes contact with the surface at an angle α. The contact angle, similar to the rake angle described by Ernst and Merchant in the previous chapter [2-19, 2-20], is a function of the geometric properties of the asperities. If the penetrator has large internal angles, the penetration behavior would differ significantly from a penetrator with small internal angles under the same normal force. Likewise, spherical, cubical, or cylindrical asperities produce different contact angles, penetrate to differing depths, and require varying levels of tangential force to move relative to the surface.

Based on the hardness of the materials and the normal load applied, the depth of indention (h) will vary. With the conical penetrator, sliding will produce a groove with a triangular cross section through the material a given distance (d). From this information, the area of material plowed out of the surface (A_p) can be related as a function of the contact angle (α) and the diameter of the cone (2r) as shown in Eq. 3.1.

$$A_p = r^2 \tan \alpha \qquad (3.1)$$

Plowing

Abrasive processes can be subdivided into two forms: plowing and cutting. The two forms are subtly distinct but must be understood to fully appreciate the generation of wear debris during abrasive wear processes. Plowing is thus defined as "the formation of grooves by plastic deformation of the softer of the two surfaces in relative motion."

Brostow and coworkers [3-11] used stationary polymeric materials to study groove formation during scratching by a moving, hard penetrator. During sliding, the penetrator displaced material and produced ridges along the sides of the groove. It was concluded that plowing simply relocates material around the groove, and that liberation of wear particles from the bulk material does not occur in all cases.

Cutting

Abrasion by cutting is analogous to common machining operations [3-12]. The cutting tool is a large macro-indenter and the workpiece is the softer of the two materials. Cutting has been studied extensively [3-13, 3-14] and typically results in the formation of a chip or ribbon of material ejected from the working surface.

Chip formation is the outcome of fracture of the workpiece [3-15] because of high shear forces along the plane of sliding. Chips physically exist as discrete particles, whereas ribbons are continuous lengths of material. The fracture behavior is dependent on the nature of the materials involved (e.g., the ductility of the materials) and depends on the operating parameters of the system (such as the relative velocity of the two surfaces).

Archard's Wear Equation

A discussion of abrasive wear would be incomplete without mentioning the work of Archard [3-16]. Archard examined wear processes of metal pairs using a hemispherical slider sized to give the same radius as the contact area. He related the rate of wear to applied load and contact geometry of the indenter as shown in Eq. 3.2.

$$W = \frac{KP}{3a} \tag{3.2}$$

In this formula, W is the volume of material removed, P is the load, and a is the radius of contact that determines the real area of contact. The constant K was incorporated as a probability factor to express the potential that an asperity interaction would result in the formation of a wear particle. The constant 1/3 is a geometric constant based on the hemispherical slider, which means that other geometries would have different values and assumes that wear particles are hemispherical with the same radius as the contact area.

Czichos [3-17] took the Archard wear equation a bit further by considering effects of elastic versus plastic contact as defined by the plasticity index of the materials. He demonstrated that wear volume is proportional to applied load and sliding distance. Therefore, larger loads and longer sliding distances give higher wear rates, and the only material property important to the discussion of abrasive wear processes was the yield pressure of the softer material. It was assumed the indenter was the harder of the two materials in contact.

From Czichos' consideration of the Archard wear equation, the yield pressure may be considered equal to the indention hardness of the softer material, which in turn simplifies the equation for wear to reveal an inverse hardness relationship (i.e., harder materials typically resist abrasion better than softer materials).

Abrasive Wear in Friction Materials

Abrasive wear processes in automotive friction materials involve not only cutting and plowing processes, but also tearing processes [3-18]. The addition of this subtly distinct mode of wear is based on the heterogeneity of commercial friction materials, which often contain isolated particulates of rubber. Other compositions might have a dual-phase matrix system comprised of mixtures of soft, rubbery compounds in combination with a hard glassy polymeric component such as phenolic resin.

This range of variation in material properties makes classification challenging. Oehl and Paul [3-18] recommended that designers of friction materials make every attempt to find a balance where excessive abrasion is avoided, but allow for the generation of sufficient abrasion to efficiently remove deposits formed on the surfaces during higher temperature operation. This balancing act is commonly achieved by blending materials of various sizes, shapes, and hardness, which also imparts much of the heterogeneity common in modern commercial engineered tribological composites. To be successful, the designer must understand the abrasive selection process.

Abrasive Selection Process

The control of abrasion in a friction material sliding against a mating metal is accomplished by the addition of hard particles to the mixture. Hard particle additions were studied by Rabinowicz and Mutis [3-19]. They determined that the most important aspect in controlling abrasion is the use of proper selection methods focused on correctly sizing the particles, with the added consideration of hardness of the material as well as its inherent fracture toughness. These three attributes affect the ability of the particles to remove material during sliding, especially when bound in a rigid polymeric matrix.

Particle Size Concerns

The selection of the correct particle size can have a significant effect on the wear behavior of a composite. Increases in volumetric wear rates of materials are a function of the grit size of the abrasive additive and have been shown to increase as the particle size increases. The increase in wear was also found to have an upper limit based on specific factors involving the penetration behavior of the particles and the rate of change in the volumetric wear rates. The rate of change in wear rates was found to diminish as particle size increased. In essence, an asymptote is approached as the result of change in contact angle. It specifies an upper limit on the size of abrasive particles where further increases in size give no significant increases in material removal rates.

Particle Shape Concerns

Likewise, the shape of the particle is also of fundamental importance [3-20]. Sharp angular particles will have vastly different penetration characteristics than spherical particles. Highly angular particles often have very small angles of contact when placed in contact with a surface and, for a given load; will tend to cut through surface films more readily, increasing the probability of direct contact between surfaces. This generates higher rates of wear when compared to other particle morphologies.

Fracture Behavior Concerns

In addition to size and angularity, the fracture behavior of a particle should also be considered. Particles that are prone to splinter easily under compressive stress will generate more cutting surfaces at the interface. This increases particle angularity with a consequent decrease in the size of the particles at the interface. Both factors affect abrasion rates significantly and can lead to transformations from effective two-body abrasion to three-body abrasion mechanisms.

Abrasive Wear Contaminants

Abrasive wear can also result from unwanted and deleterious contaminants in the system. These can be present as contaminants in raw materials. They can develop during use because of chemical transitions, or be acquired during use (e.g., road salt during winter driving). In all cases, abrasive wear is a highly aggressive form of wear that must be carefully avoided to achieve the desired lifespan of operation.

Adhesive Wear

Adhesive wear processes were considered by Burwell [3-21] to be a fundamental form of wear that existed when two surfaces were in contact. Adhesive wear occurs regardless of the presence of lubricants and is generally considered a mild form of wear in comparison to other forms. Likewise, Rabinowicz [3-22] deemed adhesive wear to be of utmost importance. He further added that adhesive wear cannot be eliminated, but only reduced. Therefore, adhesive wear processes are present in all tribological systems.

The Junction Formation Process

Adhesive wear begins with the formation of a physical junction between two surfaces. Recalling the work of Bowden and Tabor [3-23], a junction is formed when two surfaces interact to form a physical linkage between the surfaces. Interaction between the surfaces will occur during the application of a sufficiently large normal force that brings them into intimate contact. Junctions are easily formed between soft materials, but they also form during contact between very hard surfaces.

Fuller and Tabor [3-24] explored interactions present when contact between crossed cylinders of hard materials occurred. They reported that adhesive forces between very hard surfaces could be correlated to the roughness of the two solid surfaces. It was determined that adhesive forces increased as the roughness of the surfaces decreased (i.e., smoother surfaces gave high adhesion values than rougher surfaces). Adhesive force was quantified by measurement of the relative "pull off" force between the cylinders and provided strong evidence in support of the formation of an actual "bond" between the two surfaces.

Junctions formed during contact will increase in size under the application of a tangential force. This has been illustrated by a simple two-dimensional model proposed by McFarlane and Tabor [3-25] and can be illustrated as shown in Fig. 3.2.

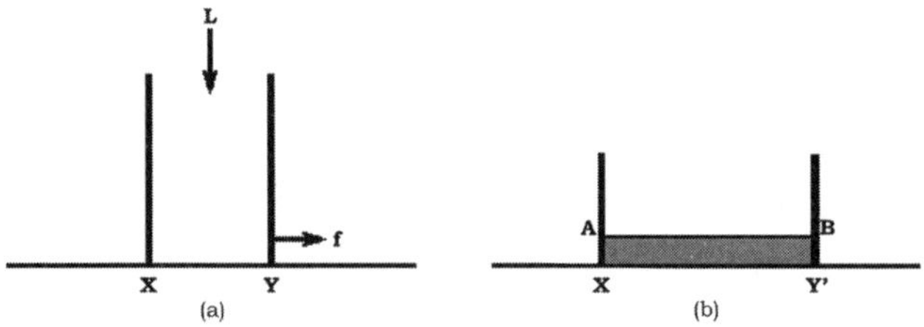

Fig. 3.2 Two-dimensional model illustrating junction formation (a) and growth during sliding (b).

A junction (XY) forms under a normal load (L) as shown in Fig. 3.2(a). The application of a tangential force (f) increases the size of the junction (XY′) as shown in Fig. 3.2(b). It was also noted that the highest probability for junction rupture was expected to occur in the shaded region defined by XY′AB.

Tabor [3-26] states that junction growth under a combined tangential and normal stress is a process of plastic deformation that continues until a critical size limit has been reached. After the junction grows to this size, strain causes failure to occur. The strain can be the result of further increases in tangential stress or of elastic recovery after the normal force has been reduced.

Three Types of Junction Failure

Regardless of the failure mode, rupture particles of wear debris are formed during the junction process. There are three potential types of junction failure. Failure can occur at the boundary between the original points of contact. It can happen in the weaker of the two materials or in the stronger of the two materials.

Failure at the Boundary

Failure at the boundary of the junction is the most ideal situation. In this case, junction failure at the boundary would result in a scenario with little or no transfer of material. Therefore, the wear values will be low [3-27] and can actually fall within the range of "zero wear," which has been defined as a change in surface roughness "less than one-half of the original peak to valley roughness values" [2-21].

Failure in the Weaker Material

On the other hand, if the junction fails in the weaker (usually softer) of the two materials, particles of the weaker material will be plucked out of the bulk surface and transferred to the harder material. If this process continues unabated, the surface will become coated with particles of the weaker material, and friction levels usually increase. The increase in friction is a result of the transition of the tribosystem where the sliding couple, which was originally dissimilar materials, becomes one with a single material interface of the softer material rubbing against itself. Wear rates can be quite high in this scenario and should be avoided if possible.

Failure in the Stronger Material

The third mode of junction failure occurs when the stronger (usually harder) material fails and transfers to the weaker material. As counterintuitive as this may seem, evidence found by Rabinowicz and Shooter [3-27] demonstrated that hard metals can transfer to softer plastic materials such as PTFE. This forms a hard asperity (fragment of strong material) embedded in a weaker (softer) matrix and can act as a cutting tool resulting in significant abrasion to both materials.

Adhesion by Mutual Interpenetration

For materials that are very soft, highly ductile, or capable of creeping under load, the concept of mutual interpenetration [3-28] suggests that adhesion can also be the result of other mechanisms.

For instance, during contact a soft material under sufficient pressure might flow into interstitial voids at the surface of the mating material. This process forms a mechanically interlocked surface that effectively resists shearing and thus provides an observable increase in the resistance to motion (i.e., friction) between the two bodies. This mechanism was illustrated by Bikerman [3-29] and Elin [3-30] who visualized the process as shown in Fig. 3.3.

Formation of Adhesive Wear Debris

Whether the surfaces interpenetrate each other or form localized junctions between them, the act of separation is likely to result in the formation of wear particles. It then becomes a question of whether the particle will be ejected, transferred, or reattached to the original

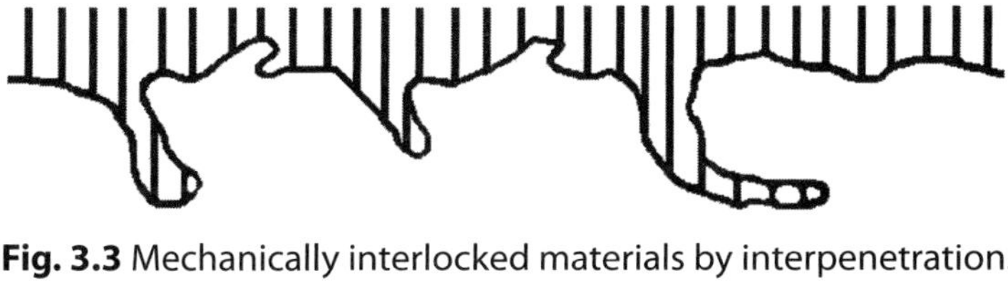

Fig. 3.3 Mechanically interlocked materials by interpenetration.

surface. The final destination of wear particles was systematically addressed by Rabinowicz [3-31, 3-32]. Additionally, the team of McFarlane and Tabor [3-33] generally agreed that adhesive wear debris particles are usually very small.

Small particles are low in mass but have high surface energy. The ratio of surface energy to mass allows the particle to cling to one of the surfaces or to other particles surrounding it. In addition, particles at the interface can fuse during sliding to form larger particles with increased mass. Larger particles or agglomerations of smaller particles have a reduced ratio of surface energy to mass. Therefore, they do not cling to the surface as tightly, which increases the probability that larger particles will be ejected.

Ejection by Release of Stored Energy

Alternatively, the application of normal loads to the surface results in two possible scenarios for asperity deformation: plastic or elastic. If the deformation mode is plastic, more particles of debris are created that can undergo the process of adhesion, cohesion, accumulation, and liberation just described. However, if the dominant deformation scenario is elastic, the particles under compression will store elastic energy, which can lead to the generation or liberation of wear particles after the normal load is reduced. The release of stored elastic energy can be visualized by a spherical indenter with radius (r) applied against a surface with normal force (F). After the normal force has been removed, the release of the stored elastic energy in the surface will push the indenter away. Relaxation of the surface produces an indentation with a radius larger than the indenter. This is illustrated in Fig. 3.4

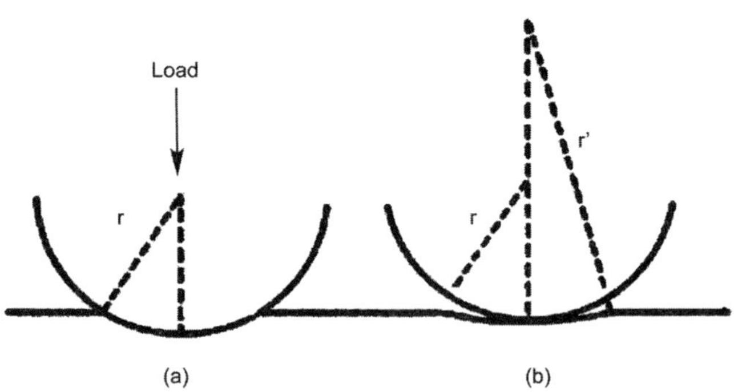

Fig. 3.4 (a) Spherical indenter under load, (b) Spherical indenter after load has been removed.

If the released energy is greater than adhesion of the particles, then particles can be released during the release of stored elastic energy, especially when the cohesive strength of bodies is less than the adhesive strength to the underlying substrate.

In support of the surface energy effects proposed by Rabinowicz, the team of Archard and Hirst [3-34] found that wear particles at the surface tended to agglomerate and form larger particles by the successive layering of these agglomerations. The lifespan of the larger agglomerations of particles was measured, and particles were found to be present at the interface even after several hundred revolutions of sliding. It was concluded that ejection of the particles would occur only when the particle had grown to be massive enough to overcome the attraction of the energy at the surface. However, the process is not as rapid as previously thought.

Chemical Wear

Chemical wear, which might also be called "corrosive wear," is defined as "wear in which chemical or electrochemical reaction with the environment is significant." This type of wear can be observed in the oxidation of a clean metal surface. As an example, iron will react with oxygen to form a layer of iron oxides at its surface. If water is present, the oxidized layer formed can be a mixture of iron oxides and hydroxides. This illustration might seem simplistic, but chemical wear produces a new surface different from the bulk material and is dependent on the environmental conditions in which the tribological composites operate. Virtually any substance that reacts with one or both of the two tribomaterials present in the system can introduce chemical wear and can be an area of concern.

Metal Sulfides as Chemical Wear Agents

Friction couples in modern automotive brake systems generally employ a mating member of cast iron. The friction material composite typically contains one or more substances capable of reacting with iron, especially in the presence of moisture or during operation at elevated temperatures. These materials are exemplified by metallic sulfide compounds. Designers have been incorporating them into friction compositions for over a century [3-35, 3-36]. Metal sulfides can be harnessed to achieve a blend of high-friction output with reduced wear rates across a wide range of operating conditions. They provide an excellent illustration of the benefits of chemical wear agents in modern tribological composites.

Sulfide Films

Bowden and Tabor [3-21] pointed out that sulfide films formed by the reaction of sulfur with iron prevent contact from occurring between asperities of the two surfaces. They noted that, in spite of the fact that the film prevents contact; sulfide films do not necessarily give low friction values. This was confirmed by Johnson [3-37] who noted that ferrous sulfide films give higher coefficients of sliding friction and do not impart significant improvements in wear resistance. Findings such as these are not exclusive, nor are they always positive.

Negative Aspects of Chemical Wear

Bely [3-28] described the negative aspects of chemical wear resulting from sulfur liberation at the friction interface. Metal sulfide lubricants operating in atmospheres with high humidity were found to form a solution of metal ions in sulfuric acid capable of attacking tribomaterials and causing severe damage.

The reaction of metallic sulfides with water is an equilibrium reaction. Increases in the acidic content promote further dissolution of the metal sulfide, which liberates more acidic species into solution. In essence, the reaction can lead to excessive corrosion of the surfaces where sliding action between surfaces can remove the superficial corrosion layer [3-38] allowing the underlying material to be attacked by any remaining reactive species present in the system. This promotes further degradation of the surfaces and leads to higher wear rates.

Positive Aspects of Chemical Wear

In light of these statements, the reader might be convinced that chemical interactions between additives and metals should be avoided at all cost. However, chemical wear is an essential part of modern lubrication technology. Rabinowicz [3-21] reported that reactive lubricants, such as metal sulfides in a tribological composite, typically produce the smallest wear particles and therefore should dramatically reduce wear rates.

It is widely known that chemicals containing chlorine, sulfur, and phosphorous encourage corrosive wear at the interface of certain metal pairs. Their use as extreme pressure lubricants enables many modern lubricating fluid systems to provide longer operation. However, the use of these agents is not without concerns. To effectively use active lubricant compounds, the reader must fully appreciate that "the difference between a medicine and a poison is often merely a matter of concentration" [3-39].

Concentration Control is Critical to Success

Successful implementation of a reactive lubricant lies in the ability to precisely control the concentration of the additive at the interface. It was reported by Buckley [3-40] that as little as 0.45% sulfur can reduce wear rates of iron on iron systems. He suggested a mechanism where iron sulfide layer smoothes the topography by coating the surface with a tenacious film and prevents direct metal-to-metal contact, which in turn reduces wear. In this case, the correct use of a reactive sulfide lubricant is based on the amount of available sulfur and must be sufficient to form a layer that is not excessively thick, yet is continuous along the entire surface.

Delamination Wear

Even in highly lubricated systems, wear will occur after prolonged sliding and is generally observed as pitting of the surfaces. The formation of pits, and the associated flat "platelet" wear particles, was explained by Suh [3-6] as the result of repetitive stress that induces fatigue of the materials and is called delamination or fatigue wear.

The mechanism underlying delamination wear begins by the transfer of the normal force from one surface through the lubricant layer, if present, into the other material. The transfer of load produces subsurface damage in the form of very small and highly localized zones of plastic deformation that cause tiny cracks to form. The cracks generally form around a naturally occurring flaw in the material, or other suitable stress riser. The "micro-cracks" can then propagate to form a much larger crack, which tends to extend, usually along an axis parallel to the sliding direction. During extension and propagation of the crack, the material is weakened to the point that a large wear particle will be ejected from the surface and a pit is formed on the surface.

Subsurface Deformation

Davies [3-41] examined the effects of sphere size under varying intensities of load to improve armor plating. He demonstrated that the maximum pressure was localized at the center of the contact area; however, the principal stress reached a maximum below the center of the area of contact and was actually located in the "bulk"

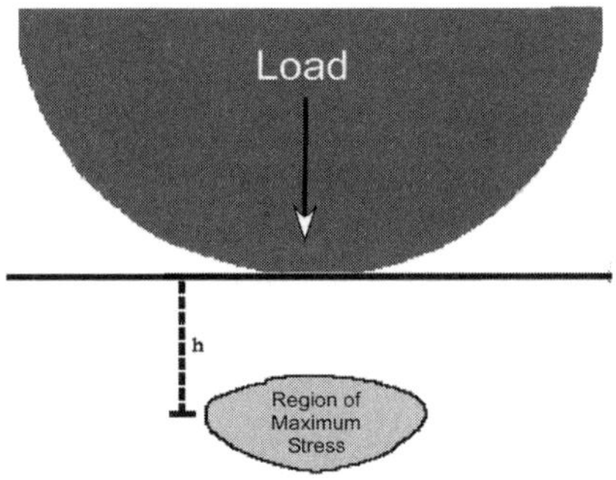

Fig. 3.5 Diagram of principal stress during static loading (sphere on plane) based on Davies.

of the material. This can be seen graphically in Fig. 3.5, which shows the region of maximum stress at a given depth (h) beneath the penetrator.

Davies' results were confirmed [3-28] by researchers who examined the contact mechanics of a sphere against a semi-infinite plane. It was concluded that the contour of contact has a maximum stress below the surface. A maximum tensile stress for this system can be defined mathematically as shown in Eq. 3.3.

$$\sigma_{yp} = \frac{3N}{2\pi a^2}\left[\frac{1-2u}{3} + f\frac{\pi}{8(4+u)}\right] \tag{3.3}$$

Here u is Poisson's ratio, f is the coefficient of friction, N is the normal load, and *a* is the area of contact. From this equation, the highest tangential stress concentration is a function of the coefficient of friction of the system.

Delamination Wear Is Not of Lower Importance

Burwell and Strang [3-42] listed delamination wear as lower in importance than the other forms of wear. They stated that surface fatigue is expected to occur less frequently than other types of wear but, in spite of this, the authors conceded that delamination processes can contribute to the overall wear behavior of certain materials.

In the case of a system operating with an amorphous third body layer subject to various degrees of compressive and tensile stress, the importance of delamination wear is essential. In dry friction regimes, the third body layer cannot withstand extremes of localized pressure and will tend to fracture, fragment, and delaminate from the interface. This has a significant impact on the wear rate of the composite and should be avoided when possible by ensuring that a stable and cohesive third body layer is formed at the friction interface.

Third Body Layers

The development of a stable, wear-resistant third body layer is the "main objective" that must be accomplished during the development process of an automotive friction material [3-18]. This level of importance is based on the view that a third body layer provides a means to transform mechanical energy into other forms and protects the surfaces from extensive damage.

During the transformation of mechanical energy into other forms, the third body layer modifies wear debris particles, provides friction, reduces wear, and stabilizes friction levels to a more consistent value, which might actually be lower than desired. In other cases, stabilization of the coefficient of friction can provide a lowered propensity for noise because of a reduction in friction- induced vibrations. There are many modern sources that elaborate on the concept of a third body layer [3-43, 3-44], but one of the earliest examples was Beilby [3-45].

Beilby's Layer

Near the turn of the last century, Beilby determined that wear debris at the surface can alter surface finish. He proposed that surfaces are altered by the accumulation of wear debris in the grooves, pits, and vacancies between asperities on the surface. Altered surfaces are therefore considered to be a layer of amorphous material distinct from the bulk material. In many cases, the surface layer appears to have experienced flow, much like a viscous liquid, suggesting that a rheological phenomenon might have been active during sliding. It was also noted that surfaces could be altered by a single pass sliding event.

More recently, Buckley [3-40] used an extensive array of analytical techniques to examine tribological surfaces altered by sliding. He reported that tribological surfaces possess a fine grain structure almost lacking in grain boundaries. The unique structure was suggested to have been formed by the repeated crystallization of the solid surface that occurred when deformation of the surfaces was large enough to lower the transformation temperature of the materials. As a result of repetitive thermal cycling because of sliding, an amorphous layer forms at the interface.

In brake systems, the effects of sliding have resulted in changes to the materials. Rhee and DuCharme [3-46] analyzed brake drums and found the cast iron microstructure had been significantly changed during braking. Most notable of their findings was the discovery that graphite flakes were absent in the interface region and had been replaced by a layer of heavily deformed metal.

The "Friction Heat-Affected Layer"

Worn organic brake linings were examined using a combination of different analytical methods and found to have undergone chemical and physical changes in the subsurface region because of thermal energy (according to Jacko [3-47]). This layer, termed the "friction heat-affected layer" or FHAL, exhibited significant changes in both inorganic and organic constituents of an asbestos-based friction material. The FHAL was later shown by Kim and Jang [3-48] to be critically important for the stabilization of the friction coefficient at elevated temperatures. They pointed out that the heat-affected layer is continuously regenerated during sliding.

The proposed mechanism for regeneration of the FHAL begins with heat generated at the interface being transmitted into the bulk of the material where thermal degradation of the constituents occurs near the surface. This assists in replenishing the surface destroyed by wear processes and maintains the thickness of the heat-affected layer by converting the underlying bulk material into a new FHAL, which implies FHAL thickness can be maintained during the entire operational life of the brake lining.

The interface region is not distinct and the transition between the friction heat-affected layer and non-transformed bulk material is gradual. However, the FHAL is subsequently transformed into the thin third body layer, which is often only a few microns in thickness [3-49]. Third body layers at the surface are dynamic regions in a state of constant generation and depletion because of cyclic processes of wear and FHAL formation.

Velocity Accommodation in the Third Body Layer

Berthier [3-43] and Godet [3-44] used the term "velocity accommodation" in an effort to describe the ability of the third body layer to accommodate differential speeds present during sliding. In other words, the relative difference in speeds of the moving surfaces results in a velocity gradient between the bodies that must be absorbed. It was concluded that the third body layer is able to effectively dissipate the total kinetic energy of the system into other forms (such as heat) by this mechanism.

Anyone who has been involved in the study of liquid lubricants is familiar with ccncept of velocity accommodation by a third body during hydrodynamic lubrication regimes. In fact, liquid lubricants are only one example of an artificial third body that controls friction behavior by way of rheological responses to mechanical forces. Unlike liquid lubricant systems with only one mode of velocity accommodation (e. g., hydrodynamic lubrication) there are a total of twenty velocity accommodation mechanisms in dry friction systems.

The Twenty Accommodation Mechanisms of the Third Body Layer

The twenty accommodation mechanisms occur as four modes of operation at five different accommodation sites. Specifically, the velocity can be accommodated at either of the two rubbing solids (first bodies), in the third body layer (between the two first bodies), or at the junctions between first and third bodies (screens). Screens are essentially an interlayer between the surface and the third body layer.

Each of the five sites accommodates the velocity differential by a combination of one of four modes: elastic, fracture, shear, and rolling. Therefore, the possible number of velocity accommodation mechanisms is dependent on the site and mode that is operational at any given moment. It is also feasible to think that all twenty modes might be operating at the same time.

The Third Body Layer in Action

Brake systems have friction materials that are rigid non-moving surfaces operating against a rotating surface of metal (e.g., a brake disc or drum). The operation of the brake requires the non-moving friction members to be forced into contact with the rotating metal member. The third body dissipates the systems kinetic energy across a bed of wear debris at the sliding interface through processes that transfer and transform the energy. This is done without allowing extensive direct contact between the surfaces and thus gives the third body layer the ability to prevent excessive wear.

Another Perspective

A different perspective of third body layers is that of a fluidized bed of material trapped at the interface. This view by Iordanoff and Berthier [3-49] visualizes the third body concept as a tribological circuit with material flowing between surfaces. There are a number

Fig. 3.6 Transient bridge formation between surfaces proposed by Iordanoff and Berthier.

of potential outcomes with this concept. For instance, during flow of the layer through the circuit, particles can be ejected, resulting in wear. Alternatively, particles can be detached from the surfaces during motion and added to the flow, thus allowing for continual regeneration of the layer to occur.

It was also during this study that Iordanoff and Berthier gave an explanation of a phenomenon where the particles in the third body layer form a transient bridge between the surfaces. This bridge serves as a physical connection through a continuous network of particles. It creates an instantaneous increase in the contact strength-resisting motion and provides a mechanism for friction generation. This effect is illustrated in Fig. 3.6.

Additional Complexity with Third Body Layers

Chateauminois and Briscoe [3-50] concluded that third body layers provide protection against wear as a result of the ability to accommodate much of the imposed relative displacement by shear within the layer. The intensity of shear stress transmitted through the third body layer increases as the normal load increases. This was described as a change in the state of the layer because of application of higher normal forces transforming the third body into a highly viscous "pseudo-liquid" that effectively prevents contact, but induces parasitic drag, which increases resistance to sliding.

Third body layers resemble fluids in many respects and can also be affected by temperature [3-51, 3-52], applied pressure [3-53], and humidity [3-54]. A number of rheological models have been developed along these lines of thought. Hou [3-55] reported that if the third body layer is easily sheared, it will behave as a solid lubricant and exhibit a low friction value. On the other hand, if the layer is very viscous, or highly resistant to shear, it can be used to augment frictional properties by increasing resistance to motion.

Concerns with Third Body Layers

Third body layers are not without their unique set of problems. One of the more prevalent concerns involves incomplete film formation. This concern is justified by findings showing a patchy third body layer to be one of the dominant causes of judder, an offensive form of vibration transmitted from the brake components into the steering wheel.

Nagayoshi [3-56] reported that third body layers can have a pronounced effect on the noise propensity of a braking system. It was found that groan, a form of low-frequency noise, was the result of preferential deposition of material around the periphery of a brake disc. As a result of preferential deposition at high temperature, the third body layer grew thicker in hotter regions and did not form as readily in the cooler regions resulting in a patchy corrugated surface film, which induced intermittent friction responses producing brake noise.

Third Body Layer Summary Comments

One point must be made concerning third body layers. The formation of a third body layer happens within a very small time frame. Godet used the term "quasi-immediately" to describe how quickly a third body might be formed. Singer [3-57] and Myshkin [3-58] both agreed with Godet and stated that frictional resistance begins at an early stage during sliding. Singer added that the third body can form at a detectable level before noticeable wear occurs. In essence, a third body can be formed as early as the first sliding contact between surfaces, which is reminiscent of the comments by Beilby.

Third body layers are not hypothetical constructs but are in reality a vitally important factor in the proper design and function of a tribological composite. These layers rely on a combination of wear processes to generate a cohesive third body that completely covers the interfacial area between system components. Control of these layers can be achieved by blending lubricants (both active and non-active forms) with mild to moderately hard abrasive materials of the proper size and shape. If used in the correct combination, the final composite will provide a reservoir able to generate, regenerate, or replenish the third body layer as needed. When this is accomplished, wear and friction performance can be greatly improved.

Transgranular Fracture and the Ultimate Origins of Wear

Another aspect of third body formation that has only recently begun to emerge is the concept of an "original element of wear." Mishina [3-59] and Hase [3-60] examined sliding pairs of metals using modified pin-on-flat devices integrated with metallographic microscopes and atomic force microscopy techniques. Observations of the interface revealed the development of slip lines in single crystals near the sliding surface. The slip line density was higher at the surface, but was shown to propagate deeper into the crystals. The interaction of the slip lines at the junction of the two materials generated finely divided wear elements called the original element of wear.

Hase provided a four stage model in which wear begins with surface asperities making contact (i.e., an adhesive contact). Slip lines occur in the asperity as a result of stress at the region of contact and a junction is formed as slip lines from compressive stress interact with frictionally induced shearing stress. The slip lines cross, resulting in transgranular fracture along the lattice of the crystals, and the contact interface of the asperities is pulled away, producing a wear particle by mutual transfer and growth. This model might be the underlying mechanism for third body formation. More work is definitely needed in this area of composites tribology.

Concluding Remarks about Wear and Friction

Wear processes and friction phenomena are intimately mixed together in a convoluted mess of variables that do not always obey the designers' best intentions. Some factors will affect wear rates but might not affect friction output. Contrariwise, efforts to reduce friction do not always improve wear resistance, and dramatic increases in wear have also been observed. These sentiments have been repeated by a host of researchers, but one of the more adamant statements made by Bely [3-28] recorded that a "universal theory of friction as well as simplified, ill-founded interpretations of frictional phenomena are no effect."

This may seem harsh to many readers, but many of the models used to predict friction or wear are often limited to a very narrow operating window between two very precise pairs of materials and tend to lose applicability in the world of automotive friction generating composites.

Furthermore, Kim and colleagues [3-61], during a study of the curing behavior of phenolic resins, determined that no systematic predictive relationship exists between the physical properties and the friction characteristics of a brake lining. This lack of evidence suggests that all friction properties must be verified by testing and logically leads to the conclusion that friction and wear properties can be understood only by a systematic approach of raw material selection, process optimization, and a thorough understanding of the operational environment.

References

3-1. Georges J. 1985. "Nature of the Surface and its Effect on Solid-State Interactions." In *New Directions in Lubrication, Materials, Wear, and Surface Interactions: Tribology in the 80s*, edited by W. Loomis. New Jersey: Noyes Publications.

3-2. Sinha S., and S. Biswa, 1992. "Friction and Wear Behaviour of Continuous Fibre as Cast Kevlar - Phenolic Resin Composite." *Journal of Material Science* 27: 3085-3091.

3-3. Fusaro, R. 1990. "Self-Lubricating Polymer Composites and Polymer Transfer Film Lubrication for Space Applications." TM102492. Washington, DC: National Aeronautics and Space Administration.

3-4. Lancaster, J. 1965. "Lubrication by Transferred Films of Solid Lubricants." *Tribology Transactions* 8:146-155.

3-5. Hooton. N. 1969. "Metal Ceramic Composites in High Energy Friction Applications." *Bendix Technical Journal* 2: 55-61.

3-6. Suh, N. 1986. *Tribophysics*." New Jersey: Prentice-Hall.

3-7. Swartzfager, D. et al. 1998. "Relationship of the Transfer Film Formed under Low Pressure Drag Conditions and Torque Variation." SAE Paper No. 982255. Warrendale, PA: Society of Automotive Engineers.

3-8. Miyoshi, K. 2001. "Solid Lubrication Fundamentals and Applications: Abrasion-Plowing and Cutting." TM-D2001-210749. Washington, DC: National Aeronautics and Space Administration.

3-9. Eyre, T. 1976. "Wear Characteristics of Metals." *Tribology International* 10: 203-212.

3-10. ASTM International. 2005. *Standard Terminology Relating to Wear and Erosion*. ASTM G40-05. West Conshohocken, PA: ASTM International.

3-11. Brostow, W. et al. 2007. "Grooves in Scratch Testing." *Journal of Materials Research* 22: 2483-2487.

3-12. Rozeanu, L., and O. Preotescu. 1957. "A Qualitative Separation of Wear Factors." *Revue de Mecanique Appliquee* 2:115-127.

3-13. Ernst, H., and M. Merchant. 1941. "Chip Formation, Friction and High Quality Machined Surfaces." In *Surface Treatment of Metals 29: 299-378. New York:* American Society of Metals.

3-14. Ernst, H., and M. Merchant. 1940. "Surface Friction of Clean Metals - A Basic Factor in the Metal Cutting Process," Proceedings Special Summer Conference on Friction and Surface Finish, MIT Report #15, pp.76-101.

3-15. Astakhov, V., and S. Shvets.2004. "The Assessment of Plastic Deformation in Metal Cutting." *Journal of Materials Processing Technology* 146:193-202.

3-16. Archard, J. 1953. "Contact and Rubbing of Flat Surfaces." *Journal of Applied Physics* 24: 981-988.

3-17. Czichos, H. 1978. *Tribology: A Systems Approach to the Science and Technology of Friction, Lubrication, and Wear*. Amsterdam: Elsevier.

3-18. Oehl, K. H., and H. G. Paul. 1990. "Bremsbelage fur Strassenfahrzeuge." Verlag Moderne Industrie AG. Germany: Landsberg Lech.

3-19. Rabinowicz, E., and A. Mutis. 1965. "Effect of Abrasive Particle Size on Wear." Wear 8: 381-390.

3-20. Burwell, J.1958. "Survey of Possible Wear Mechanisms." Wear 1: 119-141.

3-21. Rabinowicz, E. 1995. *Friction and Wear of Materials*, 2nd ed. New York: John Wiley and Sons.

3-22. Bowden, F., and D. Tabor. 1950. *The Friction and Lubrication of Solids*. Oxford University Press, London, 1950.

3-23. Fuller, K., and D. Tabor. 1975. "The Effect of Surface Roughness on the Adhesion of Elastic Solids," Proceedings of the Royal Society of London, A345, pp. 327-342.

3-24. McFarlane, J., and D. Tabor. 1950. "Relation Between Friction and Adhesion," Proceedings of the Royal Society of London, A 202, pp. 244-253.

3-25. Tabor, D. 1959. "Junction Growth in Metallic Friction: The Role of Combined Stresses and Surface Contamination," Proceedings of the Royal Society of London, A 251, pp.378-393.

3-26. Bowden, F., and D. Tabor. 1973. *Friction: An Introduction to Tribology*. New York: Anchor Press.

3-27. Rabinowicz, E., and K. Shooter. 1952. "The Transfer of Metal to Plastics during Sliding," Proceedings of the Physical Society, B 65, pp.671-673.

3-28. Bely, V. et al. 1976. *Friction and Wear in Polymer-Based Materials."* New York: Pergamon Press.

3-29. Bikerman, J. 1976. "Adhesion in Friction." Wear 39: 1-13.

3-30. Elin, L. 1959. *Mutual Interpenetration of the Surface Layers of Metals as a Cause of Wear during Incomplete Lubrication* 100:879. USSR: Izd. Akademia. Nauk.

3-31. Rabinowicz, E. 1958. "The Effect of Size on the Looseness of Wear Fragments." *Wear* 1: 4-8.

3-32. Rabinowicz, E. 1953. "On the Looseness of Wear Fragments." *Journal of Applied Physics* 24: 367.

3-33. McFarlane, J., and D. Tabor. 1950. "Adhesion of Solids and the Effect of Surface Films," Proceedings of the Royal Society of London, A 202, pp.224-243.

3-34. Archard, J., and W. Hirst. 1957. "An Examination of a Mild Wear Process," Proceedings of the Royal Society of London, A 238, pp.515-528.

3-35. Mayall, T., and L. Mayall. 1889. Anti-friction Compound. United States Patent No. 384920; filed June 21, 1887, and issued June 19, 1889.

3-36. Miyoshi, K. 1996. "Solid Lubrication Fundamentals and Applications: Introduction and Background." TM-107249-Ch1. Washington, DC: National Aeronautics and Space Administration.

3-37. Johnson, R. et al. 1948. "Friction of Solid Films on Steel at High Sliding Velocities." Technical Note No. 1578, Washington, DC: National Advisory Committee for Aeronautics.

3-38. Bisson, E. 1968. "Various Modes of Wear and Their Controlling Factors." TM-X-52426. Washington, DC: National Aeronautics and Space Administration.

3-39. Thorp, C. 2008. *Pharmacology for the Health Care Professions*. New York: John Wiley and Sons.

3-40. Buckley, D. 1981. *Surface Affects in Adhesion, Friction, Wear and Lubrication*. Amsterdam: Elsevier Scientific Publishing.

3-41. Davies, R. 1949. "The Determination of Static and Dynamic Yield Stresses Using a Steel Ball," Proceedings of the Royal Society of London, A 197, pp.416-432.

3-42. Burwell, J., and C. Strang. 1952. "On the Empirical Law of Adhesive Wear." *Journal of Applied Physics* 23:18-28.

3-43. Berthier, Y. et al. 1989. "Velocity Accommodation in Friction." *Tribology Transactions* 32: 490-496.

3-44. Godet, M. 1990. "Third-Bodies in Tribology." *Wear* 136: 29-45.

3-45. Beilby, G. 1903. "Surface Flow in Crystalline Solids under Mechanical Disturbance," Proceedings of the Royal Society of London, 17, pp.218-225.

3-46. Rhee, S., and R. DuCharme. 1973. "The Friction Surface of Gray Cast Iron Brake Rotors." *Wear* 23: 271-273.

3-47. Jacko, M. 1978. "Physical and Chemical Changes of Organic Disc Pads in Service." *Wear* 46: 163-175.

3-48. Kim, S., and H. Jang. 2000. "Friction And Wear of Friction Materials Containing Two Different Phenolic Resins Reinforced with Aramid Pulp." *Tribology International* 33: 477-484.

3-49. Iordanoff, I., and Y. Berthier. 1999. "First Steps for a Rheological Model for the Solid Third Body." In *Lubrication at the Frontier*, edited by D. Dowson et al. Amsterdam: Elsevier.

3-50. Chateauminois, A., and B. Briscoe. 2003. "Nano-rheological Properties of Polymeric Third Bodies Generated within Fretting Contacts." *Surface Coating Technology* 163-164: 435-443.

3-51. Wirth, A., and R. Whitaker. 1992. "An Energy Dispersive X-Ray And Imaging X-Ray Photoelectron Spectroscopical Study of Transfer Film Chemistry and Its Influence on Friction Coefficient." *Journal of Physics D: Applied Physics* 25: A38-A43.

3-52. Wirth, A. et al. 1994. "A Fundamental Tribochemical Study of The Third Body Layer Formed during Automotive Friction Braking." *Wear* 179: 75-81.

3-53. Aleksendric, D., and C. Duboka. 2006. "Prediction of Automotive Friction Material Characteristics Using Artificial Neural Networks - Cold Performance." *Wear* 261: 269-282.

3-54. Yoon, E. S. et al. 1998. "Tribological Behavior of Sliding Diamond-Like Carbon Films Under Various Environments." *Wear* 217: 262-270.

3-55. Hou, K. et al. 1997. "Rheological Model of Solid Layer in Rolling Contact." *Wear* 211: 134-140.

3-56. Nagayoshi, T. et al. 2004. "Groan Caused by Generation of Striped Pattern of Transfer Film on a Ventilated Rotor," Technical Paper No. 2004-01-2788. Warrendale, PA: SAE International.

3-57. Singer, I. 2007. "Role of Third Bodies in Friction and Wear." Report A711264. Washington, DC: U. S. Naval Research Laboratory.

3-58. Myshkin, N. 2000. "Friction Transfer Film Formation in Boundary Lubrication." *Wear* 245: 116-124.

3-59. Mishina, H. 1998. "Surface Deformation and Formation of Original Element of Wear Particles in Sliding Friction." *Wear* 215: 10-17.

3-60. Hase, A., and H. Mishina. 2009. "Wear Elements Generated in The Elementary Process of Wear. *Tribology International* 42: 1684-1690.

3-61. Kim, M. et al. 1991. "Study on the Curing of Phenol-Formaldehyde Resol Resins by Dynamic Mechanical Analysis." *Industrial & Engineering Chemistry Research* 30: 798

Chapter 4
Tribological Systems

Foolproof systems don't take into account the ingenuity of fools.

—Gene Brown

A TRIBOLOGICAL COMPOSITE OPERATES WITHIN a defined system and will be limited by the boundaries imposed by the operational constraints of that system. Modern clutch systems operate in a wet, or liquid lubricated, friction regime. Annular brake systems can also operate in similar wet friction systems but, like the majority of automotive braking systems, can also be dry systems operating in an uncontrolled atmosphere. Each of these systems is subject to the whims of its unique operational environment and will invariably operate under distinctively different circumstances.

In an effort to understand the use of tribological composite, an examination of common braking systems is necessary. In many cases the transposition from brakes to clutches is not drastically different and many of the same concerns face both systems. For those who are interested in a deeper understanding of the various systems, the author recommends this short list of works for further study [4-1, 4-2, 4-3, 4-4].

The Snubbing Chain

The "snubbing chain brake" was reported by White [4-5] to have been used in the Roman Era by charioteers who would wrap a chain around the hub of an axle as a means to slow the vehicle. This primitive design is shown in Fig. 4.1.

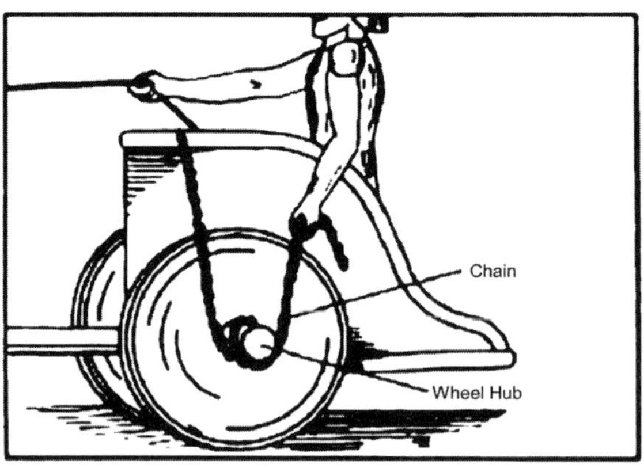

Fig. 4.1 Image of snubbing chain brake taken from White.

Operation of the snubbing chain required the driver to pull upward to generate a constrictive force between the chain and the hub of the axle. The illustration from White shows that the chain was wound around the hub in such a way as to be self-energizing and acted to draw the chain tightly around the knave of the wheel as a result of the rotation of the wheels. Stenberg [4-6], whose work predates that of White by nearly 30 years, also described the concept of a snubbing chain and its use during antiquity. However, recent communication with Olson [4-7] suggests that Stenberg and White may have misinterpreted some of the terms used by the ancients. The term "trochopedes," which literally means "wheel shackle," might have been the original concept described by Stenberg and White.

According to Olson, the use of trochopedes has also been recorded in a description of the travels of Herodes Atticus, an orator, and leads to the belief that trochopedes would have been a block of wood forced between the spokes of the wheels and the body of the chariot. Olson points out that the term "epocheus" was also recorded by the historian Simaristes in relation to brakes for early wheeled vehicles, but little information can be found to support the use of these terms.

In spite of the semantics involved in a study of ancient languages, there is a significant lack of surviving physical evidence that would provide weight to the fact that the chariots of antiquity were equipped with brakes of any type; however, the idea is conceptually sound.

The Snubbing Cable

According to Schultz [4-8], the snubbing cable brake concept was first used by Gottlieb Daimler. It consisted of a wire cable wrapped around a brake drum rigidly mounted on the axle of the vehicle. In Daimler's design, the cable could be drawn tightly against the drum surface by applying pressure to a lever. Like the snubbing chain, the cable could be wrapped around the drum in such a way as to be self-energizing to increase the constrictive force and magnify the operator's effort.

A similar description by Crane [4-9] showed the use of a snubbing chain during the era of the horse-drawn omnibus, a common mode of transportation in Paris during the last part of the 19th century. The buses were equipped with shoe brakes that pressed against the rear tires and, in addition to the use of shoe brakes, the greater portion of the braking force was provided by several turns of rope wound around bronze capstans on the hubs of the axles. The effect

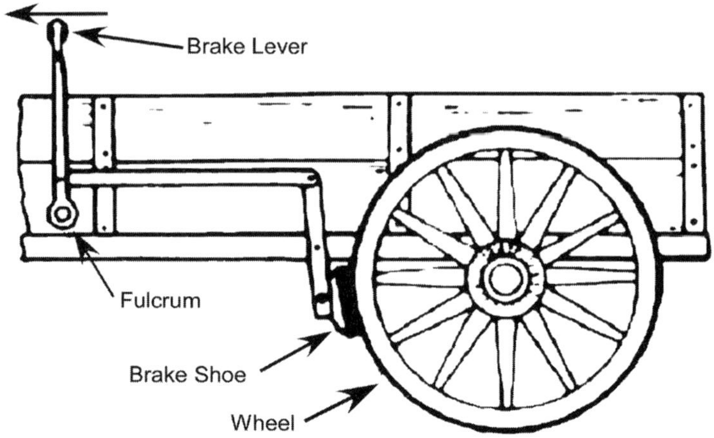

Fig. 4.2. Wagon equipped with a spoon brake from White.

of the wheel rotation would draw the rope tighter and provide the necessary amplification of output.

The Spoon Brake

Spoon brakes are simple mechanical brake systems that date to approximately 1838 [4-10] and are said to be the invention of Kirkpatrick MacMillan, a Scottish blacksmith. This type of brake was commonly found on horse-drawn wagons and used a series of pivoted levers connected by a beam to apply a block of material forcibly against the exterior surface of the wheel. Additionally, if the levers were locked into position forcing the block against the wheel, then the wagon could be parked at rest. An illustration of the "spoon brake" as described by White is shown in Fig. 4.2.

Early spoon brake designs were also common on many early bicycle designs. Cochran and Cochran [4-11] patented a spoon brake for bicycles in 1893, which used a brake rod to force the shoe into contact with the rear tire of the bicycle. The brake lining, made of vulcanized rubber, was curved to fit around the periphery of the tire. Despite the widespread application of spoon brakes on bicycles and animal-drawn vehicles, they were not an ideal design for many applications.

Issues with the Spoon Brake

There were concerns with the use of spoon brakes as early as 1880. Consider, for example, the story of Nauticus [4-12], a sailor who reportedly rode an early tricycle a distance of more than 1400 miles (approximately 2200 kilometers). During his travels on this feat of endurance, he found that spoon brakes were effective except when mud on the tires became a lubricating agent. As amusing as this may seem, his basic concern is still valid. Any foreign substance such as mud or water, at the interface between spoon and tire, prevents effective contact and alters the performance characteristics. This can also occur in today's modern disc brake systems, producing an effect known as "wet fade" or "water fade" [4-13] with a precipitous drop in the friction level.

Richardson [4-14] also pointed out that spoon brakes not only suffered from a loss of effectiveness when the roads were "slimy and slippery," but tended to trap dirt between the tire and the brake. Trapped dirt at the interface abraded the surface of the tires, generating excessive wear and forcing frequent replacement of the tires.

Successful Application of the Spoon Brake

Spoon brakes were also successfully applied to a number of applications including trains [4-15] and aircraft [4-16]. One highly successful form of the spoon brake was a pneumatically actuated design for use as a shoe brake on railcars. This design was illustrated by Todd [4-17] and is shown in Fig. 4.3.

The success of the spoon brake can be attributed to its application where the brake linings operate against a hard, wear resistant, metal member. This method was later adapted successfully for use on bicycles by Valentine and Valentine [4-18]. The ability to successfully apply the spoon brake concept has allowed it to survive until modern times.

The spoon brake concept remains in operation today, especially in rail applications. It relies on friction materials that are either cast

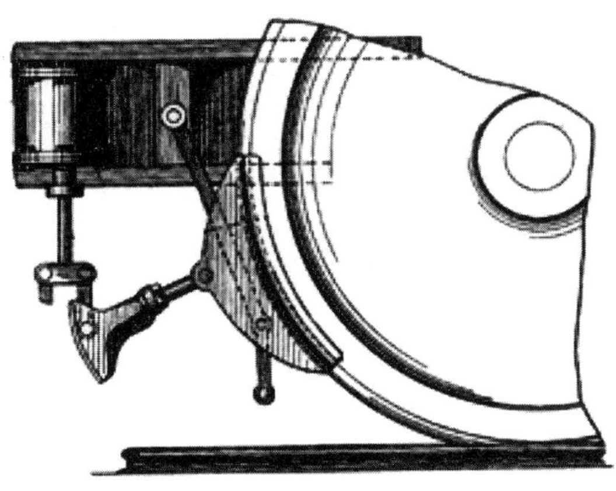

Fig. 4.3 Early spoon brake for use on railway applications by Todd.

iron or composition brake shoes made from various mixtures of ingredients. Interestingly, modern spoon brakes can be actuated by pneumatic systems much like the one described by Todd, but can also be hydraulically actuated systems.

The Band Brake

The band brake is a variant of the snubbing chain. The exact date of invention is lost in the fog of history; however, one of the first recorded uses in North America dates back to 1838. The invention by Kimball [4-19] used a simple band brake design to clamp against the beam of the loom to provide the necessary amount of friction to ensure "evenness in the cloth." By 1870, band brakes were being applied to carriages and other forms of "land conveyance."

Kosinski [4-20] patented the use of a band brake on both axles of a carriage to improve braking. The actuation method involved the use of a bell crank to draw the bands tightly around drums mounted on the axles. An interesting point of Kosinski's design is that it employs band brakes on both axles that operate in opposite directions—a simple solution for effective braking in both forward and reverse direction.

Band Brakes on Automobiles

If we examine the band brake concept from a purely automotive perspective, we find that one of the first recorded uses was by Ransom Olds [4-8], the founder of the REO Motor Car Company. Olds equipped his early automobile designs with a very effective band brake design, and by 1904 many automobile manufacturers had adopted his band brake for use in their own vehicles.

High Torque Output of Band Brakes

The effectiveness of the band brake design is based on its unique self-energizing nature where the torque output is an exponential function of the coefficient of friction and the included angle of the band (i.e., wrap angle). According to Orthwein [4-21], the torque output of a band brake anchored at one end and extending over a wrap angle of φ is shown in Eq. 4.1

$$\mathbf{T = Fr(1 - e^{-\mu\varphi})} \tag{4.1}$$

where F is the force applied to the end of the band, r is the radius of the drum, φ is the included angle, and μ is the coefficient of friction. From this relatively simple relation, the non-linear output of band brakes can be seen to provide significantly higher torque when compared to the other forms of early braking systems.

Band Brakes as Parking Brakes

Since early brake systems were mechanically actuated and thus were prone to substantial mechanical losses, the use of band brakes was common in parking brake designs. Designers used a simple solution of locking devices to hold the levers rigidly in place, which kept the brake bands in contact against the drums and prevented the brakes from unwinding.

An alternative method, which was used as early as 1903, was to install a small diameter band brake on the propeller shaft (i.e., drive shaft) to park the vehicle. The primary benefit was the location. Rather than locating the band brake at the larger diameter wheels, the addition of a band brake between the gearbox and the differential, as illustrated by Hasluck [4-22] and La Brie [4-23], provided a simple and robust solution to holding the car stationary by placing a high torque output brake on a small diameter shaft.

Issues with Band Brakes

Despite the effectiveness of this design, the band brake was not without flaws. One major flaw with the design is based on the fear that if the driver were to stall while going up a hill, the brakes would unwrap, and the driver would roll backward down the grade. This was also a problem when the car was parked on an incline, but was alleviated by parking brake designs such as those described above.

Another issue that plagued early band brakes was the constant exposure to the environment that rendered them subject to the whims of nature. Even with the most fastidious care and attention, band brakes were susceptible to contamination from the elements (i.e., water and mud) or from early automotive lubricants (i.e., axle grease). Therefore, the drivers of early vehicles equipped with band brakes were required to expend a great deal of time adjusting and cleaning the brakes to ensure effective operation.

Modern Use of the Band Brake

In spite of the weaknesses inherent in the design, band brakes survived. Today, more than 100 years later, we find that the band brake concept once used on the drive shaft as a parking brake continues to find a home in automotive power train applications.

Fujii and colleagues [4-24] teach the use of a double-wrap band brake design to improve the performance and operation of an automatic transmission. Fujita and Yanagihara [4-25] focused on the miniaturization of a very conventional looking band brake design and included it in the power train components. One specific intention of their design was to provide static holding in automatic transmission systems.

Today, band brakes can be found in service as the primary braking devices for small recreational vehicles and in some industrial application. However, their use in automotive applications has become limited to the realm of an automatic transmission component, and growth is not expected outside of this area of design.

The Transition to Internal Expanding Brakes

The internal expanding drum brake is comprised of a friction lining mounted to a carrier, which was oddly called the "dog" [4-26] in some early patent literature. The carrier fits within the confines of a drum and is operated by forcing the friction material against the interior surface of the drum, thus creating frictional resistance and the necessary braking force. The technological leap from band brakes to drum brakes was not a difficult one, but is also not well defined because of many overlapping design philosophies.

Page [4-27] examined a number of brake designs that were being employed by automobile manufacturers during the early part of the twentieth century. As part of the discussion surrounding these designs, he points out that some brake designs incorporated an external band brake for dynamic braking and an internal expanding band brake to ensure that the vehicle remained stationary when parked. One notable example of this can be seen in a cam-actuated internal band brake design by Pomeroy and Oliver [4-28].

Other examples of the overlap in design philosophy that led to the transition from band brakes to drum brakes can be seen in the patent work of Adams [4-29] and Mortimer [4-30]. In the design by Adams, an external band brake lined with a friction material was used in combination with an internal two-shoe drum brake. The two shoes were fixed at a single pivot point and actuated by a rotating cam that spread the shoes outward against the drum. Springs were shown in this design and were intended to aid in returning the internal shoes to a fully retracted position. The design outlined by Adams is shown in Fig. 4.4.

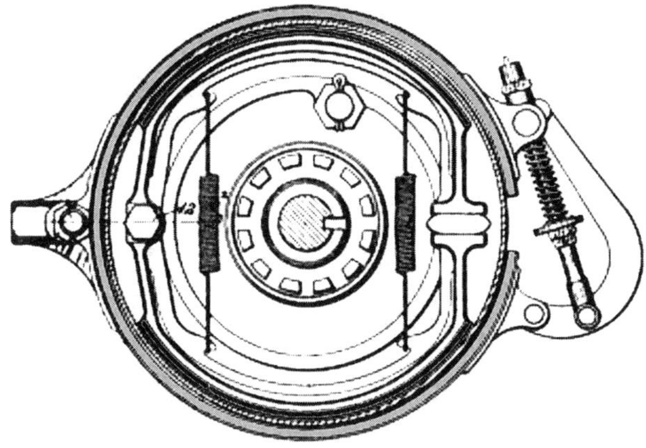

Fig. 4.4 Dual external and internal brake system by Adams.

Fig. 4.5 Cam-actuated two-shoe drum brake design by Persson.

As the transition period continued, the design of brake systems was eventually simplified and the use of both internal and external band brakes was eliminated from consideration. The eventual successor was the internal expanding drum brake concept.

The Internal Expanding Drum Brake

Early internal expanding drum brakes followed a number of design paths, but one of the earliest designs was the cam-actuated two-shoe brake design by Persson [4-31], which is shown in Fig. 4.5.

The shoes were formed, presumably by casting, to have a recess that holds a friction producing material (with wood listed as the preferred material). It should also be noted that the brake shoe design shows a reinforcing rib to ensure stiffness of the shoes—a design practice that continues to this day.

Throughout the next several decades, the literature is widely populated with a host of drum brake designs. Many of the designs relied on the simplicity of a two-shoe design; however, multiple shoe designs were also common.

Multiple Shoe Designs

There are a number of three-shoe design concepts such as those by Schmitz [4-32] or Price [4-33]. These designs used mechanical actuation to engage three brake shoes and depended on the links between the shoes to transfer the actuation force from one shoe to the other. Another multiple shoe design by Schmitz [4-34] took the concept one step further and used a combination of four shoes.

In general, multiple shoe brake designs were intended to provide enhanced braking in both directions and a universal attempt at amplification of the driver's ability to exert sufficient force to ensure adequate braking. The benefits of multiple shoe designs were usually not realized and were typically offset by their complexity and higher cost.

Two-Shoe Brake Designs

Ultimately, the simpler system began to emerge as the clear winner and two-shoe internal expanding brake designs remain in use today. Mechanically activated cam designs were very common in the early years of the automotive industry, but were replaced by hydraulic systems. However, cam-operated shoes are still in service.

Modern cam-actuated designs are usually limited to air-actuated systems that employ a pneumatic cylinder connected to an extension rod that forces rotation of an S-shaped cam. These are generally referred to as S-cam brakes and are usually found on commercial vehicles in the United States.

Six Types of Internally Expanding Drum Brake Designs

As mentioned previously, there were a number of brake designs throughout the past century. For the sake of space, many of these can be combined and typically fall into one of six major types of internal expanding drum brakes. These designs are:

- Leading-trailing
- Two-leading
- Two-trailing
- Dual cylinder
- Uni-servo
- Duo-servo

The terms leading and trailing are directional indicators that describe the shoe relative to the direction of rotation. These terms have been in place since the early days of the industry and have been nearly impossible to eradicate. It should also be noted that the terms leading and trailing can refer to the same shoe if the direction of rotation changes. To aid in understanding the six types of drum brake designs, they are shown in Fig. 4.6. The direction of rotation for each is indicated by an arrow.

Leading-Trailing Drum Brake

The most common two-shoe brake design is the leading-trailing brake design, which is also known as a simplex brake. This symmetrical design uses a combination of identical shoes that were originally mounted at a fixed anchor point but that, in modern

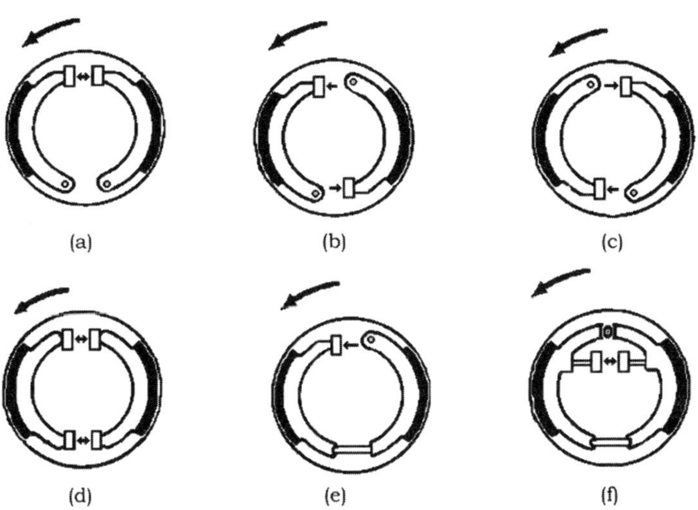

Fig. 4.6 Six types of drum brake designs.

designs, are allowed to rest or float against a rigid pivot point. The actuation of this design forces both shoes into contact with the drum simultaneously, and the symmetrical design provides a system that has no difference in output with respect to rotation.

One benefit to this design is interchangeability of the shoes. Over the course of time, the leading shoe might wear more rapidly than the trailing shoe, and these can be switched to increase the replacement interval. This also helps in reducing manufacturing costs by using identical design philosophies to decrease the complexity involved with multiple shoe types.

Leading-trailing internally expanding drum brake designs do not generate the highest torque output of the six designs, but the design is also not as sensitive to friction variation, and the output is very predictable. Additionally, leading-trailing designs are very effective as parking brakes and are currently used in many automotive applications.

Two-Leading (Two-Trailing) Drum Brake

The two-leading and two-trailing shoe designs are essentially mirror images of each other; however, their torque output is vastly different. In the forward direction the two-leading shoe is self-energizing with both shoes, and delivers a high torque output. However, this design is very sensitive to variation of the coefficient of friction. Slight changes in the material or the operating conditions can have a significant impact on braking efficiency.

When operated in the reverse direction, the two-leading shoe design is effectively transformed into a two-trailing design and becomes wholly de-energizing. This gives very poor torque output. Based on the torque sensitivity in the energized direction and the poor torque performance in the de-energized direction, the two-leading (and two-trailing) designs did not see widespread commercial success.

Dual Cylinder Drum Brake

The dual cylinder brake design was also named the Twinplex brake (originally a registered trademark of the Bendix Aviation Corporation [4-35] that dates back to 1940). This design uses dual hydraulic cylinders to force both shoes into contact with the drum. It was intended to eliminate the loss of output when operated in the reverse direction. The Bendix design was not the first attempt to actuate both shoes simultaneously, as it was preceded by a design by

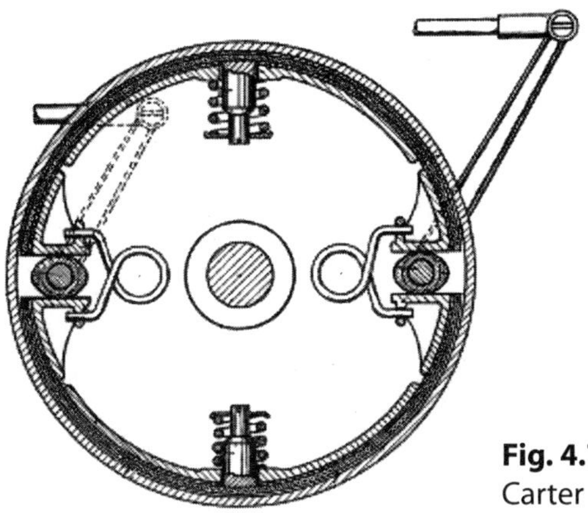

Fig. 4.7 Dual cam design by Carter and Rosenburg.

Carter and Rosenburg [4-36]. The Carter and Rosenburg design used mechanical actuation by employing dual levers to operate twin cams and dual return springs to provide the same braking effect regardless of direction. This unique design is shown in Fig. 4.7.

Whether dual cam or dual cylinder actuation is employed, this design offers a very effective brake design when used as a dynamic brake or as a parking brake. Additionally, the non-directional behavior and high torque output made this one of the most effective internal expanding drum brake designs. However, the primary issue with this design was its complexity and higher cost. Over time, the benefits were simply not justifiable for automotive use and the design was replaced by lower cost options.

Uni-Servo Drum Brake

The uni-servo brake design used a primary shoe, or leading shoe, that was actuated by the piston and a secondary shoe, or trailing shoe, which was rigidly mounted at the trailing end. The shoes were connected between the trailing end of the primary and the leading end of the secondary shoe. During braking the applied force to the primary and the energizing nature of the drum brake would drive the secondary shoe into contact with the drum at an increased level of force because of the leverage effect of the linkage between the shoes.

This type of design was typified by very powerful amplification and gave very high torque output when operated in the forward direction. Because of the nature of the design (the rigidly mounted secondary and the linked primary), operation in the reverse direction was characterized by abysmally low levels of frictional performance.

The uni-servo design was also prone to a number of issues including poor modulation and rapid wear of the secondary brake lining. Even though this brake is not considered a wildly successful design, the amplification of the forces because of the servo action made the uni-servo a very effective design for parking brakes.

Duo-Servo Drum Brake

The duo-servo brake design overcame the directional effects of the uni-servo design by actuating both shoes simultaneously, thus providing high torque output in both forward and reverse directions. This seems ideal; however, these brakes were generally condemned as being overly aggressive and difficult to modulate because of sensitivity. Even the slightest changes in the coefficient of friction would result in significant changes in the torque output. Therefore, the commercial use of duo-servo drum brake designs suffered much the same fate as the uni-servo design.

Brake Effectiveness Factors

To illustrate the power of the internal expanding brakes, many designers would calculate the brake effectiveness factor (BEF) for each design. There are several methods of analysis including the work of Acres [4-37], the staff of Ferodo [4-38], and Fazekas [4-39]. Each of these groups used geometric analysis methods to determine the "shoe factor" for a given brake design. Even though the term shoe factor is widely used in these sources, it must be remembered that it has no influence on the duty that a brake can perform.

The BEF is strictly a comparison of the pressure at the tip of the shoe and the tangential brake effort provided by this level of pressure. A shoe with a high BEF factor requires less effort (applied pressure) than a design with a lower shoe factor. This technique was very critical during the design phase of mechanically actuated braking systems, but is not as critical in today's automotive market where power amplification is available through vacuum assisted systems. To illustrate the different shoe factors available from the various drum brake designs, consider Fig. 4.8.

Today, internally expanding drum brakes are typically floating shoe leading-trailing designs. This design offers acceptable performance levels and moderate wear performance, and is relatively low in cost. Another application of this type of drum lining design has been adapted to fit within the hat portion of a disc brake producing the drum-in-hat design that uses the benefits of disc brakes for dynamic

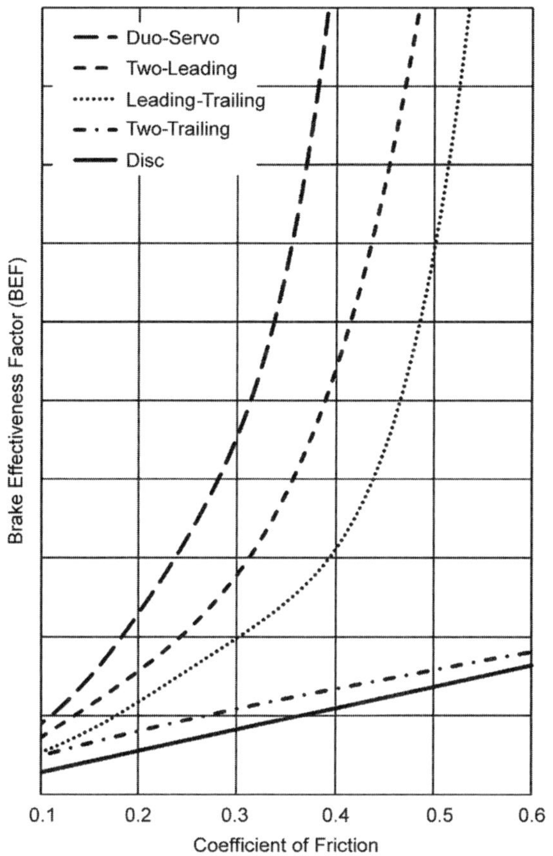

Fig. 4.8 Brake effectiveness factor for several brake designs.

braking and the effectiveness of a drum brake for parking. In the future, the demand for drum brakes is expected to diminish because of increasingly stringent stopping requirements.

The Disc Brake

There are two forms of the disc brake. The first form is an annular design that shares many of the same design features as a modern clutch system. This type of brake uses a thin ring of friction material that is forced against a metal mating member to generate friction and slow the vehicle. One of the earliest uses of this design in automotive applications appears as early as 1927 [4-40]. Over the course of time the annular disc brake was also applied to airplanes [4-41], and railcars [4-42].

Annular Disc Brakes

Annular disc brakes, which are also known as full circle brakes, did not completely fade away. Modern use is generally limited to applications with an immersed friction system, such as those employed on off-highway vehicles. The immersed friction system, or wet brake, can be very effective and astonishingly durable when sized properly. On the negative side, they are often more complicated because of fluid handling concerns. Therefore, general acceptance in the passenger car segment of the automotive industry has not been forthcoming.

Another common feature in the world of modern wet disc brake systems is the use of multiple friction plates comprised of a system of rotors and stators [4-43] as opposed to a single disc and pressure plate. These designs emerge occasionally as a concept brake for automotive applications and have recently been presented as a high-performance braking system [4-44]. However, the complexity, lack of extensive historical success, and overall cost tend to limit their acceptance.

The Spot-Type Disc Brake

The second type of disc brake is a "spot-type brake" that uses small pads or pucks of friction material that operate against a portion of the surface of a rotating disc. This type of brake design is actually the first recorded form of a disc brake. It was patented by Lanchester [4-45, 4-46] as a concept for use within an automotive transmission, and he might have imagined it being useful as a parking brake. Lanchester's designs, as well as many of the other early spot-type brakes, did not have the same torque generation capacity as band or drum brakes. They were prone to wear quickly because of high local pressures and were therefore quickly replaced by more robust drum brake designs.

In approximately 1933, the American Cable Company [4-47] revisited the spot-type disc brake design and introduced a mechanically actuated version, which is shown in Fig. 4.9.

This early disc brake used friction pads to cover an arc of approximately 90° of the available friction surface. It is not known how successful this design became; however, the literature suggests that it was in service for a few years and lasted to approximately 1945 [4-48]. Ultimately, this type of spot brake also gave way to the drum brake as the dominant form of automotive brake.

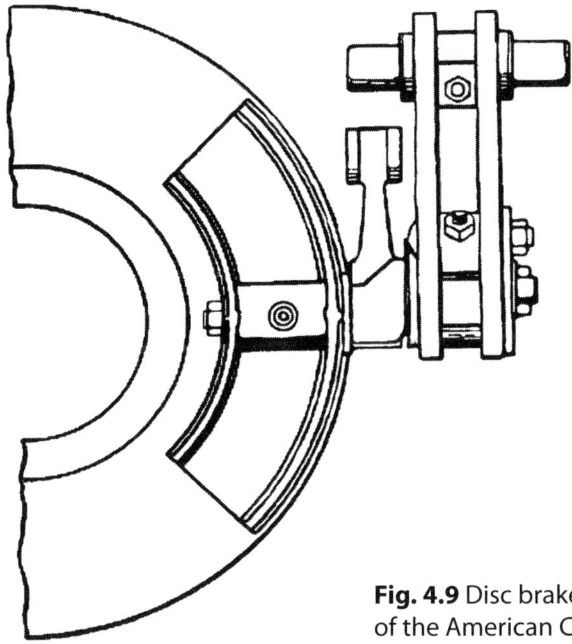

Fig. 4.9 Disc brake design concept of the American Cable Company.

Modern Disc Brake Designs

Over the years, the disc brake has seen a number of design iterations that eventually resulted in the modern hydraulic disc brake commonly use around the world. This design uses a rigidly mounted caliper that fits over the disc. It contains a single piston that forces replaceable brake pads, lined with friction material, into contact with the surface of the disc. The size, orientation, and number of pistons can vary widely, but the fundamental design remains the same and relies on the clamping forces generated by two opposing friction materials. Today many modern disc brake systems use a floating caliper design such as the one shown in Fig. 4.10.

The operation of a floating caliper disc brake begins by the application of pressure to the backside of the piston, forcing the inboard shoe against the surface of the disc. The reaction force causes the caliper to slide in the opposite direction and draws the outboard pad into contact. The clamping action of the pads provides the stopping power of this type of brake. There are a number of other designs in use, but this one is typically lower in cost and is therefore widely used.

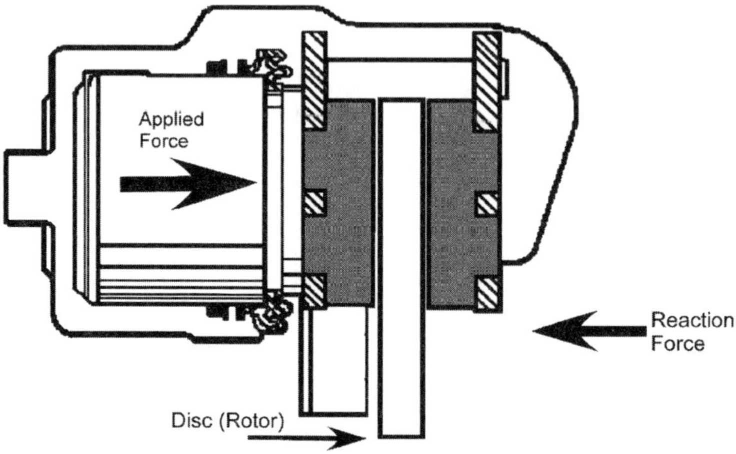

Fig. 4.10 Floating caliper disc brake.

The success of the disc brake can be attributed to various factors including lower weight and a more linear torque output that is less sensitive to friction variation. Fade performance of disc brakes is also profoundly better than with drum brakes, and maintenance is usually considered easier to perform.

Without going into excessive detail or listing numerous examples, it is clear that the disc brake has been in a state of constant evolution since the earliest part of the twentieth century. These designs have also expanded beyond traditional automotive use and are now being installed as control devices in a number of industrial applications.

In addition, many of the defects associated with the early disc brake designs have been eliminated and certain troublesome issues (such as noise) have general solutions that are reliable and more readily available at lower costs than previously experienced. As such, the disc brake design has emerged as the current leader in the world of brake systems and will continue to play a dominant role in the automotive and industrial markets for years to come.

The Multi-Disc System

The multi-disc system is a variant of the annular disc brake system that can be used as either a brake or as a clutch. It is a concept that predates most of the conventional brake designs. Perhaps the earliest example of a multi-disc system [4-49] was as an improvement in the manufacture of windlasses. The friction-controlled windlass used a multiple disc system where the friction surfaces were made from discs of teak and operated against galvanized iron stators. Over the years, the multi-disc system was widely used in a number of applications of various sizes including a small design for use in motorcycles (see Ellett [4-50]). One of the more successful and commonly used small diameter designs of a multiple disc system is that of a clutch system comprised of very thin friction discs that are used in many automatic automotive transmission systems [4-51]. However, perhaps the most common use of the multi-disc brake system is in aircraft braking systems.

Multi-Disc Aircraft Brakes

As a case in point, modern commercial aircraft require a braking system that is lightweight and capable of withstanding extreme excursions of temperature for a brief period of time. It is not uncommon for an aircraft brake lining to experience temperatures up to 1000°C (1832°F) [4-52].

Awasthi and Wood [4-53] reported that during the braking of a commercial passenger airplane, the kinetic energy levels can reach 670 MJ and surface temperatures during braking can reach as high as 3000°C (5432°F). Based on the higher energy flux and an absolutely critical need for safety, the materials in a multi-disc aircraft brake must be able to rapidly absorb this energy within a matter of seconds. One combination of materials that has shown success is the carbon-carbon multi-disc configuration described by Thebault [4-54] and shown in Fig. 4.11.

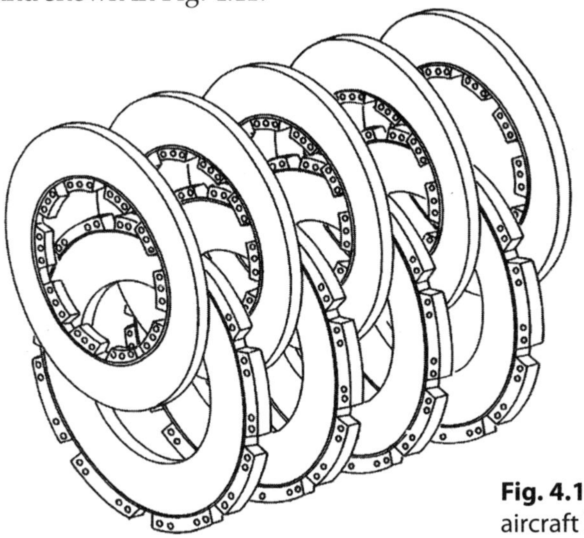

Fig. 4.11 Carbon-carbon aircraft brake from Thebault.

The unique properties of carbon-carbon systems will be touched on in later chapters, but the use of carbon-carbon materials in a multi-disc brake system offers an effective means of safely stopping the aircraft with a low weight, highly heat-resistant friction material. In addition, the carbon-carbon materials can provide the ability to absorb the heat that occurs during landings and slowly release the heat over time by acting as an efficient heat sink.

Multi-disc brake and clutch systems offer a number of improvements over conventional single disc and spot-type disc brakes, but are more complex. Generally, the costs are also prohibitive and the use of multiple disc tribosystems is limited to a few applications that cannot employ conventional disc brake systems.

References

4-1. Baker, A. 1986. *Vehicle Braking*. Warrendale, PA: Society of Automotive Engineers.

4-2. The Technical Staff of Ferodo, Ltd. 1961. *Friction Materials for Engineers*, 2nd ed. England: Ferodo Ltd.

4-3. Limpert, R. 1999. *Brake Design and Safety*. Warrendale, PA: Society of Automotive Engineers.

4-4. Breuer, B., and K. Bill. 2008. *Brake Technology Handbook*. R-375. Warrendale, PA: SAE International.

4-5. White, A. 1963. Brake Dynamics: *An Introduction to Vehicle Brake Testing at the Inspection Station Level*. Lee, New Hampshire: Research Center of Motor Vehicle Research of New Hampshire.

4-6. Stenberg, T. 1935. *The Brake Library - Volume 1: Brake Linings*. Akron, OH: Firestone Park Station.

4-7. D. Olson, personal communication regarding Athenaeus 3.99c, trochopedes, and Simaristes' use of epochleus/epocheus, 2009.

4-8. Schultz, M. 1985. "Brakes: A Century of Progress." *Popular Mechanics Magazine* 6:133-135.

4-9. Crane, H. 1923. "Some Notes on Brake Design and Construction." Technical Paper No. 230040, Warrendale, PA: Society of Automotive Engineers.

4-10. Metcalfe, P., and R. Metcalfe. 2006. *Engineering Studies*. Australia: Pascal Press.

4-11. Cochran, I., and W. Cochran. 1983. Brake for bicycles and similar vehicles. United States Patent No. 510407, filed September 1, 1892, and issued December 12, 1893.

4-12. Ridgway, W. 1880. *Nauticus on His Hobby Horse*. Piccadilly, London.

4-13. Berg, R., 1936. "Performance Characteristics of Friction Materials and Their Evaluation." Technical Paper No. 360073. Warrendale, PA: Society of Automotive Engineers.

4-14. Richardson, B. 1883. "A Model Tricycle." In *Good Words for 1883*, edited by D. MacLeod. London: Isbister and Company.

4-15. Ward, W. 1878. Improvement in car brakes. United States Patent No. 201313, filed September 1, 1877, and issued March 12, 1878.

4-16. MacBeth, C. 1916. A device for use in connection with aeroplane wheels. Great Britain Patent No. 125123, issued May 31, 1916.

4-17. Todd, A. 1883. Brake shoe. United State Patent No. 283438, filed June 6, 1886, and issued August 21, 1883.

4-18. Valentine, W., and W. Valentine.1897. Bicycle brake. United States Patent No. 581444, filed August 17, 1895, and issued April 27, 1897.

4-19. Kimball, S. 1838. Take up and let off. United States Patent No. 758, issued May 30, 1838.

4-20. Kosinski, W. 1870. Brake for land conveyance. United States Patent No. 108366, issued October 18, 1870.

4-21. Orthwein, W. 2004. *Clutches and Brakes: Design and Selection*, 2nd ed. New York: Marcel Dekker.

4-22. Hasluck, P. 1903. *The Automobile*. London: Cassell and Company.

4-23. La Brie, L. 1950. Braking system for motor vehicles. United States Patent No. 2532670, filed November 20, 1946, and issued December 5, 1950.

4-24. Fujii, T. et al. 2007. Double-wrap brake band assembly. United States Patent No. 7219774, filed March 6, 2006, and issued May 22, 2007.

4-25. Fujita, K., and H. Yanagihara. 2000. Band brake. United States Patent No. 6068093, filed December 10, 1997, and issued May 30, 2000.

4-26. Altmeyer, M. 1900. Brake. United States Patent No. 659997, filed April 16, 1900, and issued October 16, 1900.

4-27. Page, V. 1918. *The Modern Gasoline Automobile*. New York: Norman Henley Publishing Company.

4-28. Pomeroy, L., and A. Oliver. 1936. Band brake. United States Patent No. 2066308, filed February 11, 1935, and issued December 29, 1936.

4-29. Adams, F. 1925. Band brake for automotive vehicles. United States Patent No. 1554396, filed April 18, 1923, and issued September 22, 1925.

4-30. Mortimer, W. 1882. Carriage-brake. United States Patent No. 258793, filed March 24, 1882, and issued May 30, 1882.

4-31. Persson, O. 1904. Brake for vehicles. United States Patent No. 771929, filed May 21, 1903, and issued October 11, 1904.

4-32. Schmitz, R. 1923. Brake. United States Patent No. 1462722, filed June 8, 1920, and issued July 24, 1923.

4-33. Price, J. 1928. Brake shoe. United States Patent No. 1660135, filed May 7, 1925, and issued February 21, 1928.

4-34. Schmitz, R. 1920. Brake. United States Patent No. 1344598, filed February 1, 1919, and issued June 22, 1920.

4-35. Twinplex. Registered trademark of the Bendix Aviation Corporation, 401 Bendix Drive, South Bend, Indiana, Registration No. 0382778, November 12, 1940. (Trademark cancelled December 2, 2001).

4-36. Carter, G. and R. Rosenberg. Internal expanding brake. United States Patent No. 1385482, filed May 16, 1919, and issued July 26, 1921.

4-37. Acres, F. 1946. "Some Problems in the Design of Braking Systems." *Journal of Institute of Automotive Engineers* 15: 19-49.

4-38. The Technical Staff of Ferodo, Ltd. 1965. "Friction Lining Calculations -The Sizing of Brake and Clutch Facings." *Automotive Design Engineer* 4: 93-94.

4-39. Fazekas, G. 1958. "Some Basic Properties of Shoe Brakes." *Journal of Applied Mechanics* ASME 25: 7-10.

4-40. Price, J. 1927. Disk brake. United States Patent No. 1649185, filed April 13, 1925, and issued November 15, 1927.

4-41. Hofer, A. 1932. Brake for landing gear wheels of airplanes. United States Patent No. 1869479, filed September 8, 1928, and issued August 2, 1932.

4-42. Oelkers, A. 1945. Brake arrangement. United States Patent No. 2371108, filed October 19, 1940, and issued March 6, 1945.

4-43. Megginson, G. et al. 1990. Wet disc friction brake with casing rods extending through stator bores. United States Patent No. 4890699, filed June 5, 1989, and issued January 2, 1990.

4-44. Sollenskog, R. 2008. Performance disc brake system. United States Patent Application No. 2009/0211856, filed February 22, 2008.

4-45. Lanchester, F. 1901. Improvements in gear for the transmission of power. Great Britain Patent No. 190004806, filed March 13, 1900, and issued February 23, 1901.

4-46. Lanchester, F. 1903. Improvements in the brake mechanism of power propelled road vehicles. Great Britain Patent No. 190226407, filed December 1, 1902, and issued October 15, 1903.

4-47. Conner, E., and C. Johnson. 1933. Brake. United States Patent No. 1940845, filed August 30, 1930, and issued December 26, 1933.

4-48. Heldt, P. 1945. *The Automotive Chassis without Powerplant*. Nyack, New York: P. M. Heldt.

4-49. Harfield, M. 1869. Improvement in windlasses. United States Patent No. 93878, issued August 17, 1869.

4-50. Ellett, F. 1916. Brake for motor-cycles. United States Patent No. 1197113, filed May 6, 1916, and issued September 5, 1916.

4-51. Koivunen, E. 2000. Compact multi-ratio automatic transmission. United States Patent No. 6071208, filed May 18, 1999, and issued June 6, 2000.

4-52. Georgievskii, G. 1962. "The Influence of Various Ingredients on Friction Properties of Plastics." Friction and Wear in Machinery 16: 104-130. USSR: Izd. Akademia Nauk.

4-53. Awasthi, S., and J. Wood. 1988. "C-C Composite Materials for Aircraft Brakes." Advanced Ceramic Materials 3: 449-451.

4-54. Thebault, J. et al.1997. Method of applying an anti-oxidative coating on brake discs of a carbon-containing composite material. United States Patent No. 5686144, filed November 30, 1995, and issued November 11, 1997.

Chapter 5
Friction Material Manufacturing Methods

Almost all quality improvement comes via simplification of design and manufacturing.
—Tom Peters

KNOWLEDGE OF THE MANUFACTURING PROCESSES used to produce engineered tribological composites is often as closely guarded as the compositional makeup of the materials themselves. This view is surprising, as nearly every process has been disclosed in the patent literature a number of times by various groups in their respective countries. The reality behind many of the secret processes is that most modern friction materials are produced with techniques that are not new and, in fact, many were well-documented during the first half of the last century.

There are a large number of friction material types with diverse compositional philosophies and an equally large number of manufacturing methods. The earliest woven friction material compositions were produced by weaving methods similar to those employed by the textile industry. Other methods relied on non-woven techniques that are still in use in the manufacture of paper-based friction products. And still other methods, such as extrusion, calendaring, and molding, mirror processes that have been in existence for a number of decades. Each of these techniques gives a distinctive set of characteristics to the final product and generally serves to define the product at a fundamental level.

Woven Materials

Woven friction materials are the forerunner of all friction material composites. The process began by weaving fibers into fabric and treating the fabric (or layers of fabric) with a suitable binder. Woven friction materials were perfectly designed for use as a tribomaterial in early band and drum brake designs because of their inherent flexibility and because they could be produced with readily available materials. Many of the earliest examples describe friction materials as simple blends of fiber and binder, where the binder became the matrix and the fibers provided the strength.

Fiber Selection Criteria

For a friction material, a fiber must be thermally resistant and readily available at low cost, and possess high strength. In addition, the fiber must provide the finished material with consistent frictional properties and increased wear resistance. Therefore, the most important step in the process of manufacturing a woven friction material begins with the selection of a suitable fiber.

According to Nicholson [5-1], the first friction materials were produced by Herbert Frood by saturating woven belting from hair or cotton with a solution of bitumen. Because the majority of readily available materials were from natural sources, the selection of fibers suitable for weaving was focused on organic fibers derived from plants or animals. The only readily available fiber with a high degree of heat resistance was limited to varieties of asbestos, such as chrysotile and crocidolite. Therefore, the usual answer for a woven friction material involved a blend of organic fibers (such as flax) with inorganic fibers (namely asbestos).

Asbestos Yarn

Fibrous forms of asbestos are naturally occurring around the globe and specific grades remain commercially available in certain parts of the world. Due to the natural variation in the geologic processes that form asbestos, the fibers are sold in various grades and sizes. Asbestos grades vary with length and were historically subdivided into weaving and non-weaving grades. According to Stenberg, most commercial grades [4-6] ranged from approximately 0.12 inches (3 mm) to lengths of 6 inches (approximately 150 mm), and certain grades were even longer [5-2].

Long-fiber asbestos was ideal for weaving but has a smooth surface and a silky texture that allows the fibers to slide over each other during the weaving process. To overcome this issue, organic fibers (such as flax) were incorporated as a rough fibrous core that inhibited slip and assisted in manufacture. Cotton soon replaced flax and the combination of cotton and asbestos allowed for the production of a heat-resistant fabric perfectly suited for friction material applications.

The composition of an asbestos yarn varied widely depending on the manufacturer. Stenberg noted that commercially available grades of asbestos were classified based on their asbestos content and provided a scale for grading yarn. Grade A yarn was composed of approximately 80% by weight asbestos and 20% cotton, whereas grade AAAA was composed strictly of asbestos fibers (i.e., 100% by weight). Yarns with greater than 20% cotton were not suitable because of the degradation of cotton at elevated temperature.

In addition to cotton, producers of asbestos yarn would also enhance the process by using a core of non-ferrous metal wire to provide tensile strength during weaving. Typically, soft metals such as lead and zinc [5-3] were used and also provided specific tribological functionality to the final friction material.

World Bestos

There are a number of patented methods for the commercial production of asbestos-based yarn and the majority of them appear to have been assigned to the World Bestos Corporation [5-4, 5-5, 5-6]. In fact, World Bestos was awarded sixteen patents in 1926 that involved the production of asbestos-based yarn, weaving of this yarn into fabric, and using the woven yarn in a friction material fabric.

According to these patents, the preferred method for yarn production began by producing a thin sheet of paper from short-fiber asbestos. The paper was cut into thin strips and moistened with a starch solution, or other suitable adhesive processing aid, and then twisted around a core of wire or cotton. The twisted paper filaments could then be braided or wound with wire to produce a heavy yarn reinforced with metallic wire that was strong enough to weave into a friction fabric.

Stitched Fabrics

A single layer of fabric was typically woven with simple weave patterns. However, to produce a fabric of the desired thickness for a brake lining, several layers of fabric were often sewn together as shown in Fig. 5.1.

Fig. 5.1 Stitched layers of fabric for use as a friction material.

Modern Woven Friction Materials

The basic weaving process has changed little, at least in principle, during the last one hundred years. The majority of advancement in this process typically includes better equipment, such as computer controlled looms that allow for faster production, and a greatly increased rate of output.

In addition to advances in equipment, advancements in the areas of polymer and fiber science have also pushed the limits of fabric selection by adding fiber choices such as aramid, polybenzimidazole, polyphenylene sulfide, and carbon fiber to the list of candidates. Many of these synthetic fibers offer an exceptional blend of strength and thermal resistance that can be integrated into woven fabric, thus providing the desired balance of properties.

Woven Carbon Fiber

Carbon fiber in the form of a simple carbon fabric has been found to be useful in automotive clutch applications [5-7, 5-8]. This material was specifically targeted toward those applications where the friction material is repeatedly subjected to elevated temperatures because of continuous slip and, therefore, a high-temperature material capable of surviving under these rigorous conditions is required.

Carbon fabric is also an important reinforcement in the production of carbon-carbon friction elements and in fiber reinforced ceramic composite structures [5-9]. These highly sophisticated composite materials are examples of how advances in woven friction materials have survived to become the best choice for high-performance tribological applications.

The Future of Woven Friction Materials

Woven materials have stood the test of time, but suffer from one negative aspect—high cost. The cost of production is unacceptable for conventional automotive market segments. However, for those markets that need high strength and high thermal capacity, the selection of a woven carbon fiber reinforced composite is often the only choice. Other types of woven friction materials have virtually disappeared from use in all but a few, very limited industrial applications. Future growth in the area of woven friction materials is expected to remain slow until the advent of a low-cost manufacturing method for carbon fiber.

Folded and Compressed Materials

A specific form of woven material that was common in the early literature was known as a folded and compressed brake lining. These linings, typically for use in drum and band brake systems, were produced by coating asbestos fabric with rubber compounds. The basic process has been outlined [5-33, 4-5, 4-6] as coating asbestos fabric by feeding it through a set of calendar rollers that have previously been loaded with a rubber mixture. The compression of the rollers forces the rubber compound into the gaps between the weave, forming a rubberized fabric.

This rubberized fabric is then folded over to produce a laminated structure of the desired thickness. The laminated or folded lining is then compressed under heat and pressure to vulcanize the rubber compound, which simultaneously binds the layers of the fabric together. Due to the relatively thin nature of a rubberized fabric, there are a number of possible folding methods that can be used prior to compression. Stenberg illustrated several of these as shown in Fig. 5.2.

Folded and compressed linings were reported to be highly wear-resistant under general use. In addition, this class of friction materials developed a high coefficient of friction during operation. Despite these benefits, however, they were not recommended for severe duty applications where thermal abuse could result in delamination because of degradation of the bond between the layers.

One proposed solution for the delamination concern was to stitch the layers together prior to compression, which would eliminate the failure mode. As it might be imagined, this solution added complexity to an already labor-intensive and expensive method of

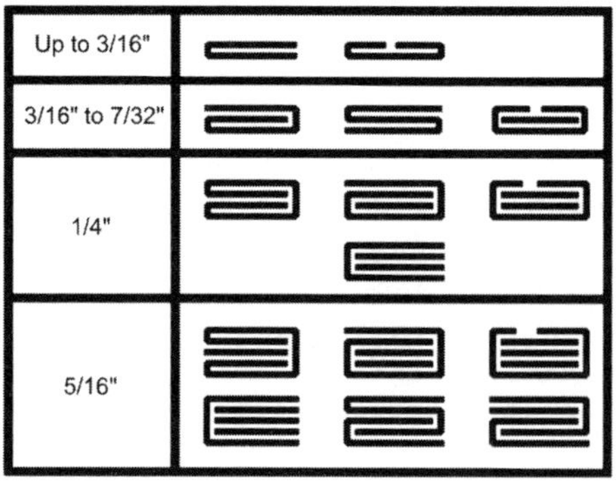

Fig. 5.2 Folding options for a folded and compressed friction lining.

manufacture. In spite of the cost and the potential for delamination, this process survives today and is used in a very small niche of select industrial friction applications.

Sheeter Materials

The sheeter process begins by compounding a rubber-based friction material and feeding the mixture into a sheeter mill (a specialized form of a rubber mill that uses a combination of a large-diameter and small-diameter roller). In addition to the size difference, the larger roller was usually heated [5-10].

The process of using a sheeter mill to produce sheets of fiber-filled rubber friction material was patented by Johns-Manville [5-11]. An image from one of their patents, which shows many of the salient features of the process, is shown in Fig. 5.3.

The Johns-Manville patent also shows the laminated structure of the material and the orientation of the fibers as a result of process. The laminated structure is a result of preferential adhesion to the heated roller and the continuous process of increasing the gap between the rollers to ensure that a fresh layer is deposited on the previously adhered layer. The thickness of the layers was only a few thousandths of an inch (tenths of a millimeter) and required the accumulation of many layers to produce a lining of the necessary thickness for use in braking systems.

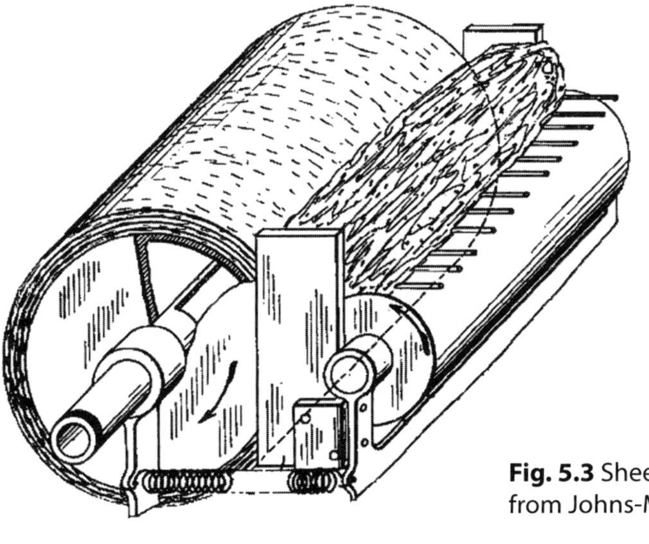

Fig. 5.3 Sheeter process from Johns-Manville.

Upon reaching the desired thickness, which could be as much as 0.5 in. (about 1.3 cm), the laminated stock was removed from the larger roller and cut to shape. The segments of friction material were then bonded to the shoes with an adhesive, or mechanically attached with rivets in some cases, and then finished by grinding the surface to the correct final dimension.

Virtually all available literature states that sheeter linings were one of the best types of friction material available. These materials provided exceptional frictional stability and were considered to be very wear-resistant in comparison to the other types of commercially available friction materials of this era.

The negative aspects of sheeter linings were based on the high cost of manufacture and the potential for delamination of the layers. Sheeter materials are not in widespread use today, but can be found in limited production by a few global friction suppliers for use in niche industrial applications.

Rolled Materials

The rolling process begins by producing a granular mixture with the texture of loosely aggregated pebbles. The granular mixture is fed into the material hopper of the rolling machine and forced into the nip of two rotating rollers, which are operating at differential surfaces speeds because of the difference in their diameter. The specific attributes of the equipment have been thoroughly described by Brockway [5-12] and Smith [5-13]. A simplified diagram is shown in Fig. 5.4 to show the essential nature of the process.

The primary difference between calendaring and rolling is the physical dimension of the final product. Rolling typically produces brake lining materials in the form of a continuous ribbon whose width is small in comparison to the large sheets of material produced during calendaring operations.

Unlike sheeter materials, the friction material produced from rolling operations is a single layer and therefore does not allow for delamination to occur. In addition, the final product does not have significant orientation of the fibers and the internal structure is irregular without a definite observable pattern.

Today the rolling process is still used for the manufacture of drum brake linings and continues to be reliant on the use of mixtures with a

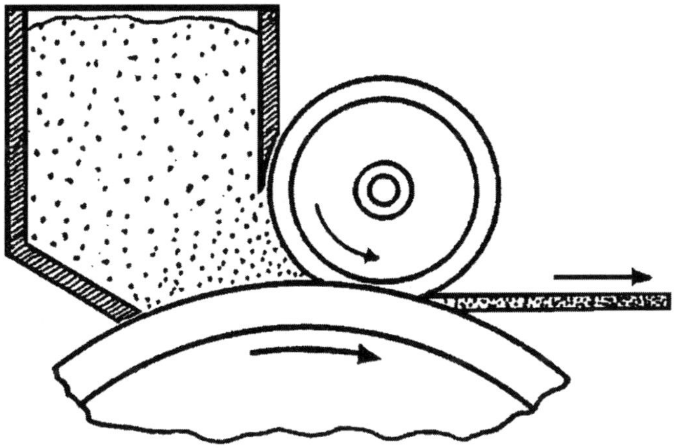

Fig. 5.4 Rolling machine for continuous production of strips of friction material.

high concentration of fibrous materials, which is necessary to impart a high degree of strength to the final product. One of the reasons for the survival of this process is the flexibility inherent in the operation of the rolling mill. Specifically, the speeds can be moderately high, allowing for a large amount of material to be formed into linings quickly, and the nip rollers can be various widths to provide a wide range of sizes that will fit virtually any size drum or band brake application.

Another aspect of the long-term success of the rolling process is based on the ability of the producer to form their materials into coils. Coils of friction material can be pre-cured to a semi-rigid form and supplied to facilities that re-line drum brakes by replacing worn friction material with new. Coils reduced the need for stocking different lengths of the same material and allowed facilities to simply cut off the necessary length prior to attachment and finishing. The coil process can still be used in limited cases, but the number of repair shops with the capabilities and expertise necessary to re-line drum brake shoes has greatly diminished in recent years.

One negative aspect of this process is the weakness of the rolled linings prior to cure. In many cases, the linings require a gentle touch to prevent damage and distortion during handling and transport. One remedy for this concern was the development of wire backed linings.

Wire-Backed Linings

A special form of rolled friction material was known simply as wire-backed linings. As the name suggests, these were produced with a reinforcing backing of wire gauze embedded in the surface of the underside of the material. An exceptionally good demonstration of this method was recorded by Smith [5-14]. In this example, wire gauze of the correct width is drawn from a spool and fed into the nip along with the mixture as shown in Fig. 5.5.

Wire-backed linings had increased strength that reduced the propensity for damage during handling. After the coils were fully cured, the embedded gauze backing provided the additional benefit of imparting significant structural reinforcement that enhanced the rivet-holding capacity of the brake linings.

Aside from the advantage of increased strength provided by the wire backing, the additional cost and the need for stocking several

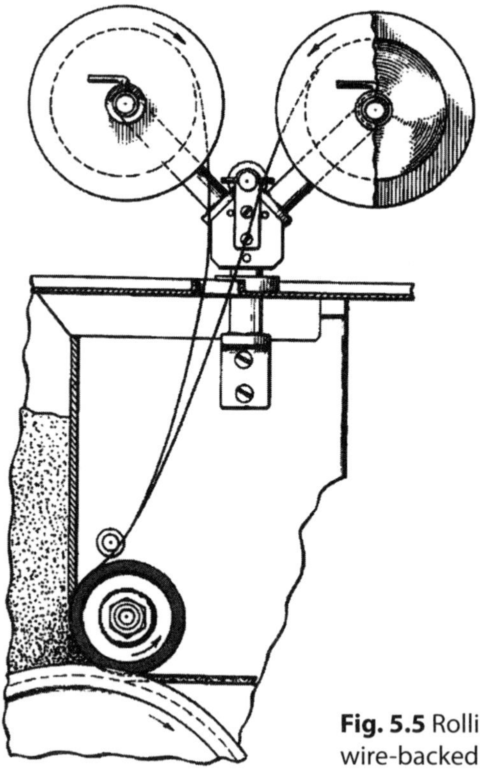

Fig. 5.5 Rolling machine for making wire-backed friction materials.

sizes of wire gauze were negative aspects of wire-backed linings. Today wire-backed friction material can still be available from global suppliers, but this type of friction material is not expected to see growth in the future.

Extruded Linings

In a three part series reporting on the manufacture of asbestos-based brake linings, Halstead [5-15] reports that the extrusion process was reliant on mixes made with solvents or solvated binder systems. The extrusion process begins by forcing a soft rubber mixture through an opening of the desired cross section to form a continuous strip of material of the desired thickness and width. The length was controlled by cutting the extruding strips into segments. These cut segments of soft material were dried and eventually vulcanized (or cured in curved molds) to match the diameter of the brake band or shoe.

One early example of an extruded friction lining was comprised of a mixture of rubber, litharge, sulfur, carbon black, and low-cost short-fiber asbestos in an appropriate solvent such as benzene [5-16]. The benefits of the lining were based on the single layer structure of highly flexible material that was not prone to delamination. One salient point to mention was the addition of a post-extrusion rolling step that was necessary to ensure the extruded linings were compressed to a target density.

Later examples, such as Cofek [5-17], used a polymeric binder composition incorporating phenolic resin compounds. The addition of phenolic resins to the rubber extruded brake linings was made possible by the use of selected drying oils that were compatible with both elastomer and phenolic resin. The benefit of the phenolic resin was an increase in the heat resistance of the final product as compared to a pure rubber matrix.

In general, extrusion processes tend to orient the fibers present in the composition, and the overall strength of the product is anisotropic. Most often, the strips of friction material are stronger in the alignment direction of the fiber and weaker in the direction transverse to extrusion. Today, extruded friction materials are still in limited production and are generally found on small low-energy industrial applications.

Extrusion and Molding

One variation on the concept of extrusion was first described by Price [4-33] in 1928. Price used soft, pliable friction materials that were still plastic to mold drum linings. During the process of molding, the soft pliable structure would penetrate the rivet holes in the brake shoe and form a mechanical attachment upon curing.

This concept has recently been revised and the combination process of extrusion during molding has been applied to the formation of a noise attenuating layer on brake linings. This process described by Kulis and colleagues [5-18] used a laminated combination of materials and novel processing techniques to form a dual-layer friction material composite. During molding, a molten material is forced through the holes in the steel carrier plate and into cavities in the base plate of the mold to form a coating on the back of the brake pad assembly. Upon curing, the molded composition forms an anti-vibration layer that provides a reduction in the noise output of the friction material during operation.

Paper Materials

Paper materials are produced by the same methods used in the manufacture of paper and thus derive their name from the process. To be more precise, these products can be classified as a non-woven fibrous mat, web, or felt. Products made by the paper making process are of varying grades and thicknesses, but all of them are by necessity very high in fiber content. Due to the excessively large number of potential citations regarding the paper-making process, our discussion will be limited to a few prominent examples illustrating some of the more important concepts that have been applied to the development of tribological composites.

Like conventional writing paper, the process begins by pulping fiber in a suitable solvent (typically water). The type of mixing equipment is referred to by a host of names such as a pulper or beater. These two types of pulping equipment are actually distinctly different, but a full discussion of these nuances is outside of the scope of this work. Therefore, all types will be broadly classified as pulpers.

The pulping operation involves dispersing fibers in an excess amount of solvent to form fibrous slurry with low solids content. The fibers can include a wealth of possible choices such as cellulose, asbestos, or aramid. The solvent can be limited to water, but can also include a number of other additives to enhance processing and fiber dispersion.

After the mixture has achieved the desired level of fiber dispersion, the slurry is gradually fed into a container, commonly known as the headbox, which is at the head of the paper machine. The mixture flows over a weir onto a continuous screen that traps the fibers at its surface. The screen, which can be flat or inclined relative to the headbox, allows water to be pulled out of the mixture. The removal of water can be accomplished by gravity, but vacuum is usually present to accelerate the process. The damp mixture is then dried and formed into sheets or rolled onto a mandrel for further processing.

For reference, the use of an inclined paper machine was fully described and illustrated by Chappel [5-19]. His design is shown in Fig. 5.6.

Types of Paper

The range of products made on a paper machine varied extensively and allowed early producers to make asbestos paper from short-fiber asbestos, which was not suitable for weaving operations. As we have mentioned previously, Nanfeldt [5-4] produced paper that was subsequently converted into yarn for the manufacture of woven friction materials. Another type of paper material, and possibly one of the oldest forms of paper-based friction materials, uses this technique to produce a thick sheet of paper that was ideally suited for clutch materials [5-20].

The compositions of paper friction materials are generally simple mixtures with a large preponderance of fibrous constituents.

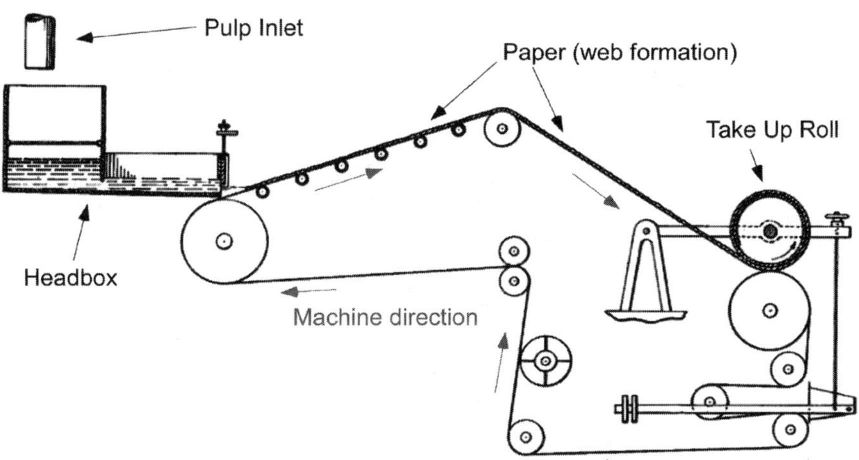

Fig. 5.6 Inclined paper machine by Chappel.

For instance, Steck and colleagues at Raybestos-Manhattan [5-21] developed a wet friction material containing 60% by weight asbestos fibers. In addition, the composition also included a small amount of barite powder and wood particulates. It is interesting to note that the wood particles were added to the composition to impart specific friction characteristics desirable to the end product and its performance.

To aid in processing, starch (or another suitable aqueous binder) could be added to prevent segregation of the materials during the pulping phase. The strength of the wet paper can be improved by adding small amounts of cellulose pulp, which also enhanced processing by increasing the degree of fiber entanglement. The final composition was formed into paper, dried, and then saturated with a heat resistant matrix material in solution. After curing the matrix, the final product could be finished to form a paper-based friction material of the correct size and shape for use in the preferred application.

Modern Paper Friction Materials

The principal characteristics of the process have not changed greatly and the true advancements have come about with the introduction of advanced polymeric fibers and friction modifiers. Today's modern wet friction material will often contain a blend of fibrous components that incorporates longer fibers for strength, shorter pulp fibers for web formation, and porosity enhancing additives that increase the permeability of the final product. These advances are then augmented by a wide range of synthetic resin choices that can be modified to provide superior thermal resistance and exceptional tribological performance.

Millboard

A very special form of paper friction material, which was stiffer and thicker than conventional paper, was given the name millboard. Since the products were composed mostly of asbestos fibers, the name millboard asbestos is common in many of the early literature sources.

Halstead [5-10] lists millboard friction materials as one of the oldest manufacturing methods for producing flat facings especially suited to the formation of annular friction rings.

Alternatively, Seigle [5-22] and Russell [5-23] used millboard products in a different way to form some very interesting friction concepts. In both of these cases, the millboard sheets were saturated and shaped while soft or wet to facilitate bending the thick paper into curved segments. Once formed, the curved segments could be dried and cured to produce a friction material for use in drum or band brake applications.

Today millboard asbestos and millboard non-asbestos products are available from global friction material suppliers. Many of these grades can be produced in sheets up to 2 inches (about 50 mm) in thickness. Overall, however, the use of millboard linings has greatly diminished. Furthermore, it is expected to be an area limited to wet friction material development, such as those in automatic transmissions or immersed brakes, and will not see a great deal of growth in years to come.

Saturation

Regardless of the form (thin paper, felt, blanket, woven fabric, or heavy millboard), early friction materials required the addition of a binder to give the product thermal and dimensional stability. The saturant was important to the generation of friction and enhanced the wear characteristics of the lining. There are a number of materials including asphalt, drying oils, rubber, and resins (both natural and synthetic), along with complicated mixtures thereof, that have been used to saturate friction composites. Each of these matrix materials will be touched on separately, but the process of saturation is considered to be an extension of the paper-making and woven linings manufacturing processes.

Saturation begins by feeding the friction material into a solution of the saturant and allowing the friction composite to absorb the desired amount of binder. Specific techniques could be as simple as dipping the materials into a solution, which was described by Werner [5-24] in 1918. Other methods could be vastly more complicated combinations of vacuum and oxidation by superheated air such as was reported by Howard [5-25] in 1926.

The methods and techniques often required each organization to empirically determine the correct dilution factors, feed rates, soak times, and temperature to give the essential depth of penetration of the saturant for their specific friction material composition. In other words, each material might have necessitated changes to the saturation process to achieve ideal performance, and the only method for determining this was by experimentation.

Modern Saturation Techniques

As recently as 1997, Pinkhasovich [5-26] employed a rubber modified phenolic resin solution to saturate fabric for use as a friction element. The unique fabric was a mixture of copper, glass, and basalt fibers where the volumetric concentration of basalt and copper was related to the coefficient of friction by a simple quadratic polynomial expression.

Another example of modern saturation techniques was described by workers at Toyota [5-27]. This group utilized a mixture of natural pulp fibers along with high-strength synthetic fibers to produce a friction material containing mineral fillers, such as diatomaceous earth. The paper from this mixture was then saturated with a resole phenolic resin blended with a selected silicone polymeric property modifier.

It was further reported that the blend of silicone with resole phenolic resin improved the thermal properties of the friction material and provided an elastomeric compliance to the structure of the composite. This added degree of flexibility, or compressibility, was deemed important to the final product for reducing the formation of transient hot spots at the friction interface during sliding.

Saturation processes are a well-documented phase of the manufacturing process that continues to be critical in the successful development of thin woven and non-woven tribological composites. Advances in fiber technology and polymer science are extending this type of product into regions of higher energy. This will be an area of exciting growth opportunity in the years to come.

Molded Materials

Molded (or rigid) materials encompass the largest majority of the modern friction materials market. Their use is prevalent in a number of industrial applications and has become the method of choice for production of automotive brake linings. One common approach to the manufacture of molded friction materials begins by combining all the ingredients with a powdered thermosetting novolak phenolic resin in what has become known as the dry mix process.

A typical novolak phenolic resin suitable for use is a powdered form [5-10] with moderately small particles (i.e., less than 200 mesh [approximately 75 microns]). The dry mix process combines the fibers, lubricants, abrasives, and filler together with the resin and can use solvents to aid in dispersing the resin. The solvent is then removed under vacuum, which leaves a wet sand type mixture with low volatility. However, in many cases, the use of solvent had been abandoned because of environmental concerns.

After mixing, the material is divided into small portions of a precise amount of material sometimes called a charge. The charge of mix can be compacted under ambient conditions to form a briquette that is subsequently place in a heated mold. According to White [4-5] this process known as preforming is sometimes skipped and the charge of mix can be added directly to the mold in a technique that has become known simply as "loose filling."

Typical molding operations use a combination of elevated pressure and temperature to melt the polymeric binder and further consolidate the composition into the final product. The molten binder coats the particles and fuses together, which effectively traps and cements the constituents to form the rigid friction material.

There are a number of types of molding operations, and each presents their own unique benefits. Regardless of the process employed, molding operations require the expense of obtaining tooling in the form of molds for both curing and preforming, where applicable. The molds are also generally designed for a single end-use application, requiring producers of molded friction materials to purchase a great deal of tooling in various shapes and sizes to produce the widest possible array of commercial friction products.

Despite the expense incurred, the major benefit of molded composites is that they are considered the most heat stable of all standard automotive friction materials and are especially suited for use in heavy duty, higher energy service conditions. This has made them the material of choice for many high energy vehicle and industrial applications.

Heat and Curve Method

A unique method for producing molded drum linings is the heat and curve method, which was originally developed by workers at Raybestos [5-28]. Their process began by preforming a sheet of mix under cold pressure until the thickness of the compacted material was approximately five times that of the final desired product. This preform did not have sufficient strength to handle easily and was formed on thin flexible sheets of metal to aid in transport and handling. The preform was first molded under heat and low pressure to form a flat sheet, which was then immediately transferred to a curved mold as shown in Fig. 5.7.

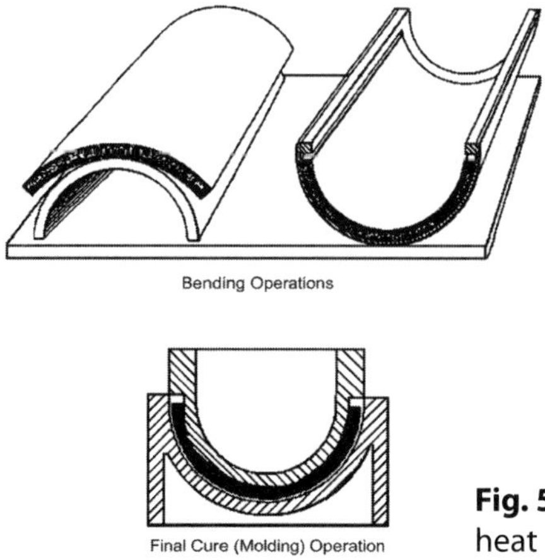

Fig. 5.7 Molded drum linings by heat and curve method.

The warm preformed sheet of mix was molded under elevated heat and increased pressure to ensure that the composite was effectively consolidated to the requisite density and to promote the polymerization of the resinous binder. The final product was a curved sheet of rigid heat-resistant molded friction material that matched the diameter of a specified drum brake, which could then be finished by cutting, machining, drilling, riveting or bonding to produce the final commercial friction lining assembly.

Another method by Rosner [5-29] used a sheet of millboard asbestos that had been saturated with a mixture of phenolic resin and a drying oil. The saturated sheet was then dried to eliminate the solvent, carefully bent to the required curvature and cured with heat and pressure to form a rigid molded millboard asbestos lining.

It is interesting to note that Rosner used a drying oil as a "frictioning agent," which implied the oil was not only an agent for improved flexibility, but also contributed to the friction output during use. This perspective can be contrasted with the work at Raybestos, which employed a sulfurized linseed oil that served to enhance flexibility but did not mention any effects on the friction output.

Compression Molding

There are three basic forms of compression molding used in the production of molded friction materials: flash molding, semi-positive molding, and positive molding. Carswell [5-30] discusses these forms of molding and uses simple illustrations to show the critical differences between them. This concept is illustrated in Fig. 5.8.

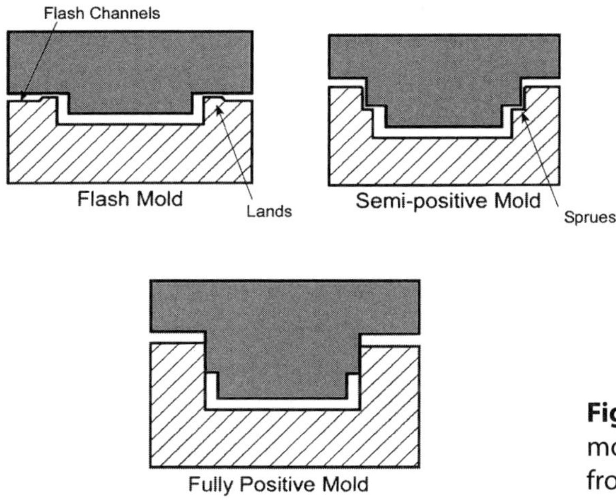

Fig. 5.8 Compression molding techniques from Carswell.

In general, these methods are very similar and can be used to produce molded friction composites that are excellent for use in automotive brake systems, especially for the production of disc brake pads. Typically, the overriding factor in deciding the type of molding operation to use in the manufacturing process is based on cost and typically increases from the lowest cost flash molding to the most expensive positive molding.

The benefits of a full positive mold include the ability to effectively control the thickness of the molded article. This can be combined with precise control of the charge weight to provide an effective means for controlling the final density of the product. The lower cost flash molds generally sacrifice the ability to tightly control density in favor of reduced tooling costs. These are commonly used in the manufacture of lower cost brake linings. Additionally, semi-positive molds are used less frequently, but are expected to offer an intermediate position at an intermediate cost.

Without regard to the type of compression molding of the various tooling designs, the molding process is critical for manufacturing high-quality tribological composites. If the process is not finely tuned, the potential for defective composites can lead to excessive cost and poor yield.

Issues with Molding Friction Materials

Several factors can arise that affect the compression molding process. However, one of the most prevalent causes of defects during molding can be attributed to trapped vapor. Vapor is created during the polymerization phase because of the chemical reactions of the resins, and it can also be present as a contaminant. Phenolic resins are the most likely source of generated vapor. Unlike resole resins, novolak resins can also produce ammonia and small amounts of formaldehyde vapor in addition to moisture. To prevent trapping these vapors in the composite, the molds are often opened or vented for a short period of time during the curing cycles to liberate gaseous species. They are then quickly closed again to ensure that the resin does not cure without the presence of increased pressure. The process of venting is usually repeated a number of times. The processes must be developed based on the reaction kinetics of the composition and the phenolic resin.

It should be noted that Baekeland experienced a number of issues during the early development of phenolic resin-based composites and finally succeeded in the production of commercial phenoplasts (e.g., Bakelite) by using a similar process of cycling venting and increased pressure [5-31].

Sintered Metallic Materials

A sintered metallic material (also known as a powdered metal or metallic friction material) is, at an elementary level, a compression-molded article. The primary difference is that metallic friction materials develop a metallic matrix by fusion of metal particles, which contrasts with the formation of a polymeric matrix provided by the action of powdered resins under heat and pressure.

Perhaps the earliest example of a friction material produced by sintering metals together was patented in 1930 by Boegehold and Short [5-32]. Their process began by coating a friction fabric with a fluxing agent, and then cladding the fabric with a thin layer of metal. Their patent discloses that the coating and cladding processes can vary widely, but the final product was formed by fusing the metallic-coated fibers to form a friction material based on an asbestos fiber reinforced metal matrix.

Conventional Sintered Metallic Materials

A more conventional method uses powdered metals, such as the alloy of copper and iron. The process patented by Samuel Wellman [5-33] is an excellent example that began by mixing metallic powders with non-metallic ingredients. The non-metallic constituents were considered functional additives for providing specific wear or frictional responses.

There are a number of published compositions, but a typical composition might contain copper, lead, tin, graphite, iron powder, or any mixture with an excess concentration of a fusible metal. The ratios of the material are unique and will be discussed in other sections of this book, but the primary metal ingredient should be present in the highest concentration and will ultimately form the matrix.

In Wellman's example, copper was the matrix-forming metal powder employed and the process begins with compaction of the mixture under pressure to form a briquette, which is an analogous process to preforming in dry-mix friction material manufacture. In the sintered metallic friction material process, the compaction phase uses significantly higher pressures (e.g., 21,750 PSI [approximately 150 MPa]).

The compacted material is then heated in a non-oxidizing atmosphere to temperatures of 1300°F (approximately 700°C) or higher, such that the temperature is high enough to allow the particles to soften, flow, and fuse together. The actual process temperatures will vary depending on the specific melting properties of the metals, and process development of this portion of the manufacturing process is often done empirically. It must be noted that the fusion temperature is an absolutely critical factor to ensure that the final composite has the necessary structure to balance the required strength with the associated tribological performance characteristics.

Sintered metallic friction materials have found widespread use in a host of brake and clutch applications operating in both wet and dry conditions. Sintered materials are specifically targeted toward the higher energy applications that demand the combination of high strength and high thermal capacity. One common market segment for sintered metallic materials that is expected to see growth in the future is for components in the automotive transmission and other powertrain devices.

For instance, Kawamura and Murata [5-34] employed a plasma deposition method to deposit an atomized layer of metallic material on the surface of a synchronizer ring. The process fused the particles together and also fused them to the ring to form a rigid sintered attachment. The final product was a metallic carrier ring with a thin layer of metallic friction material that was highly porous and allowed for efficient fluid transport. This technology may see further advancements and could be an area of great growth as the demands for high-energy materials increase.

Carbon-Carbon Friction Materials

A perfect example of an absorptive material with low mass that is completely suited for high-energy tribological applications is a carbon fiber reinforced carbon matrix material, commonly known as carbon-carbon material. This unique form of composite offers a combination of light weight, high strength, and excellent thermal properties, but is offset by the high cost of manufacture and poor suitability for low energy applications. For these reasons, carbon-carbon materials are typically limited to aerospace, defense, and some limited production for high-performance automotive applications such as those targeted to the racing industry.

There are three basic methods of manufacture [5-35] used in the production of carbon-carbon friction materials. The first method involves impregnation of a carbon fiber preform with a thermosetting carbon precursor material, such as a phenolic resin. After impregnation, the precursor is molded under heat and pressure. After molding, the body is carbonized at elevated temperatures and the process of impregnation and carbonization is repeated until the product achieves the targeted density level. Other carbonaceous materials such as pitch can be used in place of the phenolic resin.

The second method involves deposition of carbon onto the fiber by chemical vapor deposition (CVD) or chemical vapor infiltration (CVI). The deposition process is repeated many times and forms a structure comprised of layers of deposited carbon on the surface of the fibers. It is this layer of carbon that is grown until it achieves the desired density and forms the matrix of the composite.

The third method involves aspects of both processes. In this case, the preform can be coated and carbonized followed by repetitive treatment with CVD or CVI to develop the density of the carbon body. As can be imagined, the number of potential variations on these few themes could lead to carbon-carbon materials processed in a host of different ways.

Regardless of the method used, the final phase of the operation consists of graphitization of the matrix by heating to temperatures of approximately 4500°F (2500°C) until the matrix converts into a highly purified form of carbon. The final carbon-carbon composite is a highly graphitized carbon matrix reinforced with high tensile carbon fiber, typically in laminated fabric form but can also be made with randomly dispersed short fiber.

Since carbon-carbon materials are essentially pure carbon, the coefficient of sliding friction is low under ambient conditions. The low coefficient of friction has been attributed to the lubricating nature of graphite, which is discussed in more detail in later sections of this book and involves the presence of low molecular, weight adsorbed species at the surface. During braking, the heat generated from sliding quickly causes the graphite surfaces to lose these low molecular materials and rapidly generates a pronounced rise in the friction level. This can occur with a few microseconds of use during racing, or during landing of a commercial airliner, and low friction levels are typically not a significant concern.

Carbon-carbon friction materials are also very unique in the manner in which they wear at elevated temperatures. Unlike other friction materials, the wear mechanism is an exaggerated form of chemical wear, specifically thermal oxidation that converts carbon into gaseous oxides because of reactions with oxygen present in the atmosphere. To improve the oxidation resistance of carbon-carbon materials, a number of patents teach the use of additives that can be added as a modifier to the carbonaceous precursor impregnating solution [5-36]. Other patents teach the addition of materials to the process during the fabrication phase, such as boron [5-38]. Each of these can be an effective means of enhancing the oxidation resistance of the carbon matrix.

The manufacturing process used to produce advanced carbon-carbon tribological composites is an expensive process that requires a great deal of time (days or weeks rather than hours). The process also requires a heavy expenditure of energy to repeatedly carbonize and graphitize the carbon matrix. However, after the long wait and the high cost of production, the final product possesses a unique structure that is incredibly strong (even at elevated temperatures), lightweight, and capable of withstanding extremes of operation that would literally disintegrate other material types.

Ceramic Friction Materials

True ceramic materials are comprised of fiber reinforced ceramic matrices. This class of materials has emerged as a niche product for operation in extreme environments that exceed the capabilities of carbon-carbon materials. Ceramic friction materials vary greatly in concept. They range from a ceramic reinforced metallic matrix composite to a ceramic fiber reinforced ceramic matrix.

Gazza and Lindley [5-38] patented their process involving a compaction phase to form a porous body from ceramic powder, which is similar to the briquetting techniques used in sintered metallic processes. The ceramic briquette is then heated under a vacuum and infiltrated with molten metal such as aluminum, which has a melting point of approximately 1220°F (approximately 660°C). The vacuum infiltration method formed a metallic matrix throughout the network of interstitial spaces surrounding the ceramic particles and resulted in a product with an exceptional blend of properties.

It should be noted that there are no typical manufacturing methods for ceramic materials and many techniques are listed in the patent literature of the world. Many of these use high-temperature furnaces capable of firing the ceramic particles to sinter (or vitrify) them together to form a rigid matrix. This method is perhaps one of the oldest types, but advances in synthetic silicon based chemistry have pushed the production of ceramic based materials into other areas that can be processed at lower temperatures.

For example, workers at Starfire Systems [5-39] developed a non-corrosive silicon carbide precursor compound known as TMTSH that can be used to infiltrate and coat materials by using vapor deposition methods. This allows for the development of a nearly stoichiometric layer of silicon carbide to be formed on the surface of various substrate materials and offers the potential to be formed into a number of geometrically complex structures.

The number of potential applications for this type of compound to be used in the development of an effective and highly durable friction material is immense. In spite of the huge potential, ceramic materials are still very limited in their commercial success and high-cost hurdles might be inhibiting further commercialization attempts. Until advances can be made in the processing, the use of these products is expected to remain low.

The Unanswered Question

Despite a long history, many researchers will find a singularly unanswered question in their search for answers: *How does the process affect the final product?* Generally, there is little in the published literature that provides conclusive data. This can be because of the intrinsic need for competitive secrecy, but can also be attributed to differences in composition that convolute the effects of process variables. For example, the rheological behavior of phenolic resins is complicated by temperature and pressure effects during molding, but the kinetics of the polymerization can also be affected by humidity, pH, and the potentially catalytic effects of certain non-ferrous metals.

Typically, the synergistic nature of friction material composition and processes are well-guarded secrets known only to their respective manufacturers. It is sincerely hoped that more work will be done in this area to reveal the relationships between process, materials, and final product performance.

References

5-1. Nicholson, G. 1995. *Facts about Friction*. Winchester, VA: Gedoran Publishing.

5-2. Merrill, G. 1904. *The Non-metallic Minerals: Their Occurrence and Uses*. New York: John Wiley and Sons.

5-3. Halstead, R. 1949. "Brake Linings of Various Types and Their Manufacture." *Asbestos* 10:.3-8.

5-4. Nanfeldt, E .1922. Process for making asbestos yarn. United States Patent No. 1437438, filed January 20, 1920, and issued December 5, 1922.

5-5. Heany, J. 1926. Process for manufacturing yarn. United States Patent No. 1585611, filed January 19, 1926, and issued May 18, 1926.

5-6. Heany, J.1926. Coreless yarn and process for making the same. United States Patent No. 1585617, filed January 4, 1924, and issued May 18, 1926.

5-7. Winckler, P. 1987. Clutch with pyrolytic carbon friction material. United States Patent No. 4700823, filed March 28, 1980, and issued October 20, 1987.

5-8. Winckler, P. 2000. High density, low porosity carbon composite clutch material. United States Patent No. 6132877, filed March 9, 1999, and issued October 17, 2000.

5-9. Bauer, M. et al. 2004. Fiber-reinforced ceramic composites. United States Patent No. 6818288, filed December 23, 2002, and issued November 16, 2004.

5-10. Halstead, R. 1949. "Brake Linings of Various Types and Their Manufacture." *Asbestos* 9: 8-13.

5-11. Eisenhardt, F. 1930. Brake lining. United States Patent No. 1771749, filed May 24, 1924, and issued July 29, 1930.

5-12. Brockway, C. 1933. Apparatus for forming brake lining. United States Patent No. 1909653, filed May 5, 1931, and issued May 16, 1933.

5-13. Smith, H. 1933 Machine for making friction elements. United States Patent No. 1920023, filed October 27, 1930, and issued July 25, 1933.

5-14. Smith, H. 1934. Machine for making friction elements. United States Patent No. 1961176, filed November 10, 1932, and issued June 5, 1934.

5-15. Halstead, R. 1949. "Brake Linings of Various Types and Their Manufacture." *Asbestos* 8: 6-12.

5-16. Novak, I. 1930. Friction element and process of making the same. United States Patent No. 1785701, filed June 20, 1928, and issued December 16, 1930.

5-17. Cofek, H. 1952. Friction material and method of making same. United States Patent No. 2586150, filed December 21, 1948, and issued February 19, 1952.

5-18. Kulis, S. et al. 2005. Noise attenuating friction assembly. United States Patent Application No. 2005/0082124, filed October 15, 2004.

5-19. Chappel, L., 1944. Method for making friction materials. United States Patent No. 2362450, filed September 16, 1940, and issued November 14, 1944.

5-20. Simpson, S. 1919. Process of making friction facings. United States Patent No. 1301612, filed May 22, 1918, and issued April 22, 1919.

5-21. Steck, R. et al. 1958. Friction material. United States Patent No. 2852368, filed August 13, 1954, and issued September 16, 1958.

5-22. Seigle, W. 1921. Brake or clutch band lining. United States Patent No. 1395743, filed June 28, 1920, and issued November 1, 1921.

5-23. Russell, T. 1930. Method of making brake bands and the like. United States Patent No. 1785391, filed November 26, 1926, and issued December 16, 1930.

5-24. Werner, T. 1918. Friction-shoe and method of making the same. United States Patent No. 1262887, filed December 8, 1916, and issued April 16, 1918.

5-25. Howard, W. 1926. Process and apparatus for saturating fibrous materials. United States Patent No. 1598376, filed February 25, 1921, and issued August 31, 1926.

5-26. Pinkhasovich, A. 1997. Reinforcing filler, friction polymer composite material and a brake friction element. Canadian Patent No. 2250023, filed February 21, 1997, and issued August 28, 1997.

5-27. Kawabata, M. et al.2008. Wet type friction material. United States Patent No. 7326742, filed October 16, 2003, and issued February 5, 2008.

5-28. Nanfeldt, W. 1939. Friction material. United States Patent No. 2155020, filed July 3, 1935, and issued April 18, 1939.

5-29. Rosner, A. 1935. Friction material. United States Patent No. 2009207, filed October 27, 1930, and issued July 23, 1935.

5-30. Carswell, T. 1947. *Phenoplasts: Their Structure, Properties, and Chemical Technology*. New York: Interscience Publishers.

5-31. Schwartz, S., and S. Goodman. 1982. *Plastics, materials, and processes*. New York: Van Nostrand Reinhold Company.

5-32. Boegehold, A., and C. Short. 1930. Friction elemental. United States Patent No. 1757408, filed February 23, 1927, and issued May 6, 1930.

5-33. Wellman, S. 1939. Composite machine element and method of making same. United States Patent No. 2178527, filed May 28, 1938, and issued October 31, 1939.

5-34. Kawamura, O., and S. Murata. 2001. Synchronizer ring. United States Patent No. 5337872, field June 19, 1992, and issued July 17, 2001.

5-35. Kosuda, H., and K. Niijima. 1982. Carbon fiber reinforced carbon friction element and method of making. United States Patent No. 4339021, filed July 18, 1980, and issued July 13, 1982.

5-36. Johnson, N., and J. Nickerson. 1959. Oxidation resistant carbon and graphite bodies. United States Patent No. 2868672, filed March 27, 1956, and issued January 13, 1959.

5-37. Ehrenreich, L. 1972. Reinforced carbon and graphite articles. United States Patent No. 3672936, filed April 18, 1968, and issued June 27, 1972.

5-38. Gazza, G., and M. Lindley. Ceramic-metal systems by infiltration. United States Patent No. 3864154, filed November 9, 1972, and issued February 4, 1975.

5-39. Shen, Q., and L. MacDonald. Silicon carbide precursor. United States Patent No. 6730802, filed July 9, 2002, and issued May 4, 2004.

Chapter 6
Types of Friction Materials

The word of man is the most durable of all material.

—Arthur Schopenhauer

DURING THE PAST ONE HUNDRED YEARS, friction material composites have been classified in various ways. Many of the classification methods and the nomenclature describing the materials are the result of an often arbitrary decision-making process. Generally, friction materials have been classified by their raw material constituents. However, some were categorized by their specific form of manufacture, and some were designated by both material composition and manufacturing method. As such, the categories and classifications that occur in the prevailing literature are generally imprecise.

To explore these categories, a number of example compositions will be described from the available literature. These examples should not be viewed as limiting examples, but rather should be considered an illustration of the unique styles of tribological composites that have been developed for commercial use throughout the past century.

It should also be acknowledged that attempts to classify many of the previously published friction material compositions generate an equally large number of exclusions and exceptions to these rules. A few of these exceptions will be discussed as a contrasting perspective to the conventional type of material. The disparity between them will hopefully spark new ideas for growth that will, with any luck, improve our classification methods and associated nomenclature.

The Complexity of Classification

Asbestos was the most prominent ingredient in friction materials for nearly a century and asbestos-based friction materials have been produced in a myriad of ways for applications in the industrial, automotive, and aerospace markets. Because of the extensive reliance on asbestos, the friction material industry became virtually synonymous with asbestos. During the latter half of the last century, several European countries introduced legislation limiting the exposure to airborne asbestos fiber. This led to a frantic search for effective replacement options. It was this period of heightened research activity that led to the emergence of many of the friction material classifications that have survived to this day.

Generally, an asbestos-based friction material was produced with a large amount of asbestos fiber and bound together with an organic matrix such as rubber, asphalt, resin, or drying oils. It was the selection of the binder that earned asbestos-based friction materials the label "organic linings." The term organic was meant to restrict this class to materials with an organic polymeric matrix.

Classes of Organic Linings

Jacko [3-42], who studied the friction heat affected layer (FHAL) of an asbestos-based organic friction material, used the term organic to describe an asbestos-based brake lining. However, in a later work [6-1] he also introduced a separate classification method where a Class A organic material was contrasted with a Class B organic material.

These subdivisions were based on two exclusive sets of customer expectations regarding friction, and wear attributes that differentiated two forms of organic lining. More precisely, it was stated that the class B organic material had been designed with the philosophy of meeting the distinctive needs of European drivers and was contrasted with the Class A organic material, which had been designed to meet the needs of drivers in the United States.

The Class A organic material, which was designed to last longer and be low in noise, did not have the same expectations for frictional performance as the Class B organic material. The class B organic material, on the other hand, was designed to generate a higher level of friction with enhanced frictional stability at elevated temperatures, but sacrificed wear resistance relative to the class A materials. Class B materials were also more likely to generate higher levels of noise during operation than the lower friction class A types.

It is interesting to note that Jacko compared both classes of organic friction material to a semi-metallic material, which was also bound in an organic matrix of phenolic resin and was free of asbestos content. However, the term organic apparently did not apply to semi-metallic compositions.

Organic Versus Semi-Metallic

The work of Libsch and Rhee [6-2] examined micro-structural changes in metals during testing of semi-metallic disc brake pads under low temperature conditions. In this study, the authors continued the use of the terms organic and semi-metallic as separate material classifications and noted that organic linings (i.e., resin-bound asbestos composites) wore more rapidly—as much as five times faster—than the resin-bound semi-metallic material. Regrettably, Libsch and Rhee did not provide specific examples of these types of compositions, which might have enabled further clarification between organic and semi-metallic material types.

Asbestos-Based Friction Materials

In addition to issues with the naming conventions, the composition of an asbestos-based friction material varied widely. Headson [6-3] described an asbestos-based brake lining for use on band brake systems as early as 1918. His composition began by combining 30% rubber, 39% barytes (barium sulfate), 5% organic plant derived fiber, 2% hydrated lime, 2% litharge (lead oxide), 14% lead powder, and 8% sulfur. This rubber mixture was then solvated by mixing with gasoline to form a plastic mass, which was subsequently combined with asbestos and a non-ferrous metal. The final composition was roughly 68% asbestos, 2% copper chips, and 20% rubber mixture.

Other examples used organic drying oils or synthetic phenolic resins instead of rubber to bind the asbestos compositions together. Some others used a combination of both materials [6-4] where a drying oil modified phenolic resin is the binder in simple composites comprised mostly of asbestos, which could be as high as 65% by weight. Compositions such as these also contained small amounts of asphalt, barytes, and particles derived from the polymerization of cashew nut shell liquid. The mixtures were typically blended in a suitable solvent and preformed into sheets of material. The sheets or slabs were dried and molded to form curved drum lining segments.

A Contrasting Perspective

Elerath [6-5] produced a composition containing 42% powdered lead, 42% powdered zinc, 3.4% powdered asbestos, 1.3% graphite, 1.7% iron oxide, and trace amounts of asphalt and carbon black. The binder was a solvated phenolic resin. This mixture was molded to form an organically bound friction lining that contained asbestos, and could technically be described as an organic composite. However, because of the high metal content and low asbestos concentration, Elerath's composition does not fit neatly within the designation of an organic material as easily as the other materials previously described.

There are hundreds of examples in the literature of asbestos-based friction materials. Because of the widespread association of asbestos with organic brake linings, the term organic is not widely used today.

Semi-Metallic Materials

A semi-metallic material, also known as a "semi-met," may be more precisely defined as a resin-bonded metallic (RBM) composition. Often these products contain at least 50% by weight of ferrous metals including iron powder, steel fiber, or combinations thereof. Despite the fact that the term resin-bonded metallic is more precise and differentiates it from sintered metallic materials, the use of the term semi-metallic is prolific in the literature and continues to survive because of its long history of use.

A functional definition for a semi-metallic friction material is limited to friction materials containing a large portion (at least 50% by weight) of ferrous metal ingredients. Furthermore, the semi-metallic friction material is bound together in an organic matrix of resin, rubber, drying oil, or a combination thereof, and excludes metallic and inorganic binder systems. Likewise, for delineation purposes, if the ferrous metal concentration is below 50% by weight, the product will not be considered a true semi-metallic.

Historical Development of Semi-Metallic Materials

Because of the need to find a suitable replacement for asbestos fibers, the first historical evidence of a semi-metallic friction material appeared in 1939 when Rudolf Oswald, a native of Saxony, patented the use of steel or aluminum wool [6-6] in place of asbestos in friction materials. Oswald reported that metallic wool (such as steel wool) improved thermal transport, which, in turn, reduced thermal degradation of the organic binder and improved the wear resistance of the material.

It is interesting to note that Oswald's stated improvements compare remarkably well to the enhancements reported in the more modern work of Libsch and Rhee [6-2]. Despite the benefits, Oswald's patented materials apparently did not find wide acceptance and, when the asbestos supply improved, friction material producers quickly switched back to asbestos.

Conventional Semi-Metallic Materials

A more conventional approach to making a semi-metallic friction material began when Biggs [6-7] developed an iron powder based friction material that relied on the green strength provided by a specific size of ductile iron. The ductile iron was produced by hydrogen reduction and yielded sponge iron powder bound in a resin matrix.

One noteworthy composition was composed of 63.8% by weight iron powder, 23% graphite, 9% phenolic resin, and 0.2% stearic acid, which was added to the mixture as an internal mold release agent. The mixture was compacted under approximately 18,000 PSI (approximately 125 MPa) and molded at a temperature of approximately 356°F (180°C). Many modern semi-metallic formulations are not as simple as Biggs' earliest work, but a high concentration of ferrous metals in combination with a solid lubricant, such as graphite, is a recurring theme.

Aldrich and Deane [6-8] developed a semi-metallic particle composed of approximately 50% iron powder, 20% graphite, and 15% ceramic particles bound together with phenolic resin and molded under heat and pressure to form a molded article. The composite article was pulverized and added to organic friction materials as a friction enhancing additive stated to improve friction performance and impart wear resistance to the organic (asbestos) friction linings.

Rhee and Kwolek [6-9] used a different design philosophy based on volumetric compositions and patented a sponge iron-based friction material with 20% by volume iron powder, 10% ceramic particles (such as magnesium oxide), 28% graphite, 1% non-ferrous metal (such as zinc powder), 10% steel fiber, and 6% rubber particles. This mixture was blended with phenolic resin and molded using standard compression molding techniques.

Modern Semi-Metallic Friction Materials

Modern semi-metallic friction materials still rely on the benefits of combining steel wool and iron powder together to produce a friction material that is highly wear-resistant. However, the enhanced wear benefits of semi-metallic materials are generally outweighed by other performance concerns.

One area of concern involves thermal degradation of the phenolic resin as a leading cause for a rapid drop in friction at elevated temperatures [6-10]. This drop in friction, known as fade, is usually experienced during mountainous descents but can also occur during operation at higher speeds on flat terrain. Because of the high thermal absorption rates, semi-metallic materials are generally capable of returning to a high level of friction, but fade is a major concern during initial use.

Another concern associated with the use of semi-metallic compositions is friction variation. Chan and Stachowiak [6-11] stated that steel fibers introduce a higher degree of friction variation, which was attributed to abrasion of beneficial third body layers present at the friction interface. This leads to vibration that manifests itself in the form of audible brake noise. Additionally, it is pointed out that steel fibers are prone to corrode and may cause excessive wear of the brake linings in highly humid environments.

The Future of Semi-Metallic Friction Materials

Today semi-metallic compositions continue to be comprised of mixtures with only a few raw materials (mostly steel fiber and iron powder). Even though these products are easily manufactured by conventional molding methods, semi-metallic friction materials are losing favor with vehicle designers and are not as widely accepted on many of the popular passenger car applications. This decline in use is anticipated to be a natural progression based on advancements in other tribological composite types such as non-asbestos organic materials.

Carbon-Metallic Materials

A specialized form of semi-metallic material is the carbon-metallic friction material. Aldrich [6-12] developed this type of semi-metallic for use in higher energy automotive applications. One composition used a mixture comprised of 20% by volume phenolic resin, 25% iron powder, and 10% steel fiber. Unlike semi-metallic linings, the carbon additive was present at a concentration of 45% and is specifically stated to be non-graphitic in nature. This salient difference produces a composite with improved friction performance over a wider range of operating temperatures, which can be manufactured using conventional compression molding techniques.

Myers [6-13] developed an unconventional form of the carbon-metallic friction material by using a combination of processing techniques. His hybrid approach begins by mixing the components with resin, and molding using compression molding techniques. After the molding operation was complete, the product was then sintered in an inert atmosphere for roughly 2.5 hours at a temperature of approximately 1562°F (850°C) to fuse the iron powders together, forming a metallic matrix and carbonizing the resin to form a secondary matrix throughout the body of the product. The benefits reported were enhancement in friction performance and improved structural integrity at elevated temperatures.

Today carbon metallic materials occupy a small niche segment of the automotive friction materials market, which is generally limited to various forms of competitive driving. It is expected that carbon metallic friction materials will always be a low-volume fringe product, and many companies have elected not to pursue them for future development.

Low Metallic Materials

Low metallic materials, commonly known as "low steel" or "low mets," are a group of materials that contain small concentrations of steel fiber and do not fit neatly into the semi-metallic classification. The small amount of ferrous metals also prevents their inclusion in other areas of classification such as the non-asbestos organic group. To aid in clarifying what constitutes a low steel material, the literature provides several examples for consideration.

Kim and Jang [6-14] defined this type of material as being composed of small amounts of steel fiber, typically limited to a range of approximately 5-15%. Low metallic materials were further defined by their end use where, as a general rule, they are typically found on high-performance automotive braking systems on European vehicle applications. This suggests that low steel friction materials might be the modern asbestos-free equivalent of the Class B organic material mentioned by Aldrich.

An Example of a Low Metallic Friction Material

Moraw and Paul [6-15] developed a friction material that is an ideal example of the low metallic friction material concept. The inventors employed a blend of rubber and resin to impart flexibility to their composite. It was mentioned that this technique of blending binder ingredients allowed for an effective and practical means of fine-tuning the stiffness of the composite to practically any level desired.

One example composition from the patent by Moraw and Paul is shown in Table 6.1.

Table 6.1 Example Low Metallic Friction Composition from Moraw and Paul

Raw Material	Concentration (wt%)
Aramid fiber	3.5
Mineral fiber	6.5
Steel fiber	24.0
Mica	13.0
Talc	2.0
Calcium hydroxide	6.0
Calcium phosphate	5.0
Iron oxide	4.0
Magnesium oxide	7.0
Graphite	1.5
CNSL powder	5.0
Plasticizer	1.0
NBR rubber	13.0
Sulfur	0.9
Zinc oxide	0.7
Accelerator	0.4
Carbon black	1.0
Resin	5.5

This example shows the use of heat-resistant mineral fiber and aramid fiber with a modest amount of steel fiber. It should be noted that the steel fiber was not simply present as a fibrous reinforcing additive, but was listed as a beneficial ingredient that improved friction stability at elevated temperatures. There is also a large concentration of inorganic materials that impart thermal resistance to the composite, and mica was added specifically to enhance fade performance.

It has been expressed several times [6-10, 6-16, 6-17] that friction materials containing excessive amounts of organic constituents (such as phenolic resin and cashew dust) will generally suffer in fade performance. As such, organic additives are somewhat limited in this composition, probably in an attempt to improve fade resistance, which is a critical performance attribute by many European drivers.

Another Low Metallic Material

Chester [6-18] also used a blend of resin and NBR rubber as the binder system in another low metallic material. His strategy shared aspects of the work of Moraw and Paul (such as the inclusion of mica and steel fiber) but tends to be free of organic filler materials (like CNSL dust) and used a high concentration of carbon. A formulation from his work is shown in Table 6.2.

Barytes, which is a very common mineral filler material in the friction material industry, is present at the largest concentration and occupies over one-fourth of the total friction surface in this example. The composition also contains copper at a modest concentration and a small amount of antimony sulfide, both of which were simply stated to be present as friction and wear modifying agents.

The examples from Chester and Moraw/Paul share a design approach that is evident in many of the low metallic compositions found in the literature. However, despite the enhanced performance and wide acceptance in European countries, the low metallic material is limited in use in other regions around the world. The desired performance level of low metallic materials represents a design philosophy that is uniquely distinct from the desires of casual drivers in many parts of the world that seem to prefer materials that

Table 6.2 Example Low Metallic Friction Composition from Chester

Raw Material	Concentration (vol%)
NBR rubber	6.0
Sulfur	3.2
Phenolic resin	14.9
Steel fiber	10.0
Mica	3.5
Powder coke	14.0
Anorthite	7.0
Aluminum oxide	0.5
Barytes	25.5
Graphite	9.0
Antimony sulfide	2.0
Copper	4.5

generate less wheel dust and are less prone to wear. This presents yet another example of how a single material does not fit all applications in all markets.

Non-Asbestos Organic (NAO) Materials

The non-asbestos organic (NAO) category is essentially a catch-all group that does not fit into the other categories. As such, the diversity of products in this group logically extends to include woven materials, paper based products, rubber linings, and virtually all types of molded articles that do not make use of asbestos as an ingredient.

In an effort to narrow the scope of this group, the NAO material category will be restricted to products that do not contain ferrous metals. To further limit the scope, the manufacturing process will also be limited to extruded, rolled, and molded materials and will exclude those materials that are woven or produced as paper type materials, which are discussed separately.

The NAO category began when producers were not able to obtain asbestos because of shortages in supply. As mentioned earlier, one solution that has been reported by Oswald was to use metallic fibers in place of asbestos. However, other options included the use of other fibrous materials such as fiberglass.

Fiberglass NAOs

While working at General Motors, Rasmussen [6-19] developed a friction material composed of fiberglass fabric bound together with phenolic resin. The product could be saturated, cured, drilled, and riveted to form a conventional drum brake lining. Likewise, glass fiber was also favored in some early asbestos-free clutch facings.

Peters and Barnett [6-20] used a combination of non-ferrous wire and glass filaments to wind a clutch facing. The process began with a continuous strand of fiberglass, comprised of filaments approximately 12 microns in diameter, gathered together in bundles of 800 and helically wound with the wire to form a yarn. The yarn was then wrapped around a mandrel to form a clutch facing and subsequently bound together with a rubber cement.

Despite the success with glass fiber friction materials, it was quickly conceded that no single fiber was a suitable replacement for asbestos. Consequently, the number of potential fiber choices grew rapidly as many producers began to explore blends of fibers. There are hundreds of examples in the available patent literature, but our discussion will be limited to a few noteworthy types that illustrate the more salient points of using a fibrous blend to replace asbestos.

NAOs with Blends of Fiber

One of the more interesting methods of replacing asbestos with a blend of fibers was reported by the design team of Parker and Grele [6-21]. These gentlemen developed a non-asbestos organic brake lining based on high-strength aramid fibers and incorporated a heat-resistant ceramic fiber as a significant portion of the composition. The composition uses 27 parts by volume phenolic resin, 24 parts aramid fiber, and 16 parts ceramic fiber. The remaining portion of the composition was disclosed as friction and wear modifiers, but no specific details were provided. It is interesting to note that the blend of a soft organic fiber with a harder ceramic fiber is a simple ratio (2:3).

Shibata and coworkers [6-22] used a blend of carbon fiber and para-aramid fiber to replace asbestos. Their findings suggest that an optimum ratio of carbon fiber to para-aramid fiber was 1.67:1 with an exemplary composition shown in Table 6.3.

During the discussion of the composition, Shibata's group made several interesting remarks about the different ingredients. For instance, the copper-zinc alloy was intended to impart specific

Table 6.3 Example NAO Friction Composition - Shibata et al.

Raw Material	Concentration (vol%)
Fibrillated para-aramid fiber	5.3
Chopped para-aramid fiber	2.7
Carbon fiber	6.0
Copper powder	6.0
Copper-zinc alloy	9.0
Graphite	5.0
Molybdenum disulfide	8.0
Silicon dioxide	5.4
Barytes	12.6
Melamine dust	16.0
Phenolic resin	23.0
pH controlling agent	1.0

friction attributes to the final product. This is a well-documented property that has also been listed by Halstead [6-23], Batchelor [6-24], and the team of Jacko and Rhee [6-25].

Additionally, Shibata's group divulged a great deal about their reasoning for the selection of the different fibers. Specifically, para-aramid fiber was selected over meta-aramid fiber because of an increased thermal resistance. The para-aramid fibers improved durability of the final product and aided in processing by preventing the carbon fibers from segregating during handling of the mixture. Furthermore, it was noted that carbon fiber was employed to reduce the tendency of the friction material to attack the brake disc and also imparted durability to the material.

NAOs and Alkali Metal Titanate Compounds

The non-asbestos organic friction material discussion would be incomplete if it did not include coverage of alkali metal titanate whiskers as a potential replacement for asbestos. This synthetic friction material additive is one of several types first produced in 1958 by Geir and Salzberg [6-26] as an asbestos replacement. The product was chemically a form of potassium titanate produced as a whisker type fiber with an aspect ratio of approximately 1000:1. The potassium titanate whiskers were designed to have many of the desirable properties of asbestos, including a high degree of thermal resistance. They also exhibited similar morphological characteristics to that of crocidolite asbestos.

A few years later, E. I. DuPont obtained a registered trademark [6-27] for potassium titanate crystals in the form of whisker fibers with the word mark of Fybex. Fybex was targeted as a whisker fiber for reinforcing plastics. It became part of a comprehensive study by Mansfield and colleagues [6-28, 6-29] at the National Aeronautics and Space Administration (NASA) and was intended to be a replacement for asbestos in brake linings.

It was reported that potassium titanate whiskers could be an effective partial replacement for asbestos. They had a synergistic relationship to the phenolic resin concentration, which gave enhanced wear resistance to the final product when used in the correct balance. A full replacement of asbestos was not possible within the constraints of this experiment.

Modern Alkali Titanate Compounds

Today the use of blending strategies often includes mixtures of aramid fibers and alkali metal titanate compounds. These two materials have virtually dominated the landscape of non-asbestos organic friction material development and, in a recent patent application, Phipps and Vyletel [6-30] examined both fibrous and non-fibrous forms. Their comparison presented findings between a fibrous form of potassium titanate and a non-fibrous potassium lithium titanate powder.

The potassium lithium titanate powder provided many of the same benefits of the whisker form. However, it is suggested that the benefits gained by using the non-fibrous form are apparently impacted by the concentration of other mineral fillers such as barytes, which was limited to approximately 25% of the total composition. This suggests that a synergistic relationship exists, but no specific details were implicitly provided.

Ono and colleagues at Hitachi Chemical Company [6-31] designed a friction material using a similar blend of aramid and potassium titanate fibers in a matrix of a silicone-modified phenolic resin. The combination of the raw materials and the selection of the silicon-modified binder gave a final tribological composite with reduced potential for noise generation, especially after the vehicle had been parked for prolonged periods in a high humidity environment.

Final Comments on NAO Materials

To close this section, one final formulation example is taken from the work of Chiba [6-32] as shown in Table 6.4.

Chiba used a blending strategy where the total organic content of the composition is 40% (resin, aramid fiber, CNSL dust, and rubber) and is contrasted to 12% abrasives (vermiculite, mica, iron oxide, and ceramic fiber). The ratio of the organic to abrasives is 3.33, which was considered to be essential for the development of an acceptable NAO friction material.

Non-Asbestos Non-Metallic (NANM) Materials

The non-asbestos non-metallic (NANM) class of materials is viewed as a subdivision of the NAO class of materials. This type is unique in that it does not contain any metallic additives. This broad definition allows for the inclusion of a large number of material types such as carbon-carbon, paper, or woven materials. However, because these

Table 6.4 Example NAO Friction from Chiba

Raw Material	Concentration (vol%)
Phenolic resin	20.0
Aramid fiber	5.0
Polymerized CNSL dust	12.0
Rubber powder	3.0
Vermiculite	3.0
Mica	3.0
Iron oxide	3.0
Ceramic fibers	3.0
Potassium titanate	10.0
Calcium hydroxide	2.0
Barium sulfate	23.0
Graphite	8.0
Copper	5.0

specific types of materials are best discussed as separate entities, we will limit the NANM class of friction materials to a few examples that are produced by more conventional means.

Air-Laid Non-Woven NANM

One of the few examples of a true NANM friction material was described by Bortz [6-33]. This material is produced by combining an air-laid arrangement of heat-resistant fibers with needle punching to produce a three-dimensional felt that is strong in all directions and can easily be saturated with a heat-resistant binder such as solvated phenolic resins or polyimides. The selection of the binder, combined with the novel method of manufacture, produced a friction controlling medium with exceptional performance, durability, and strength.

An additional benefit to this process is that it allows the use of longer that generally do not process easily in a conventional wet paper process, but give the final composite a significant increase in strength over conventional paper-based friction materials based on shorter fiber arrays. Furthermore, this type of non-asbestos non-metallic friction material can be produced without the sacrifices associated with using low temperature fibrous materials (i.e., cotton) that are typically included as a means to enhance web formation or processing aids.

A Molded NANM

A different method for the manufacture of NANM materials uses the more conventional compression molding process. Ogawa and coworkers at Toho Rayon [6-34] patented a friction material based on mixtures of carbon fibers, phosphate fiber, calcium carbonate, and phenolic resin. The four components were blended and molded under heat and pressure to form a rigid friction material that was targeted for use in clutch applications.

Issues with NANM Materials

NANM materials are not typically used in dry friction applications. This limitation is because of the absence of conductive metallic elements that prevents the composite from transferring excessive thermal energy away from the interface. NANM friction materials are therefore very likely to develop severe thermal gradients during use, a fact that has been noted in the discussion of the compositional effects of commercial vehicle brake linings [6-35] and can lead to a number of concerns.

As a result, the NANM class of friction materials is most often limited to operating within a fluid cooled environment like the typical wet friction applications. There is, of course, a very noteworthy exception in the form of carbon-carbon materials, but these are discussed as a separate entry in this book.

Ceramic Materials

True ceramic friction materials have a history that extends back to the earliest days of the industry. These materials have always held the promise of extremely high heat resistance that would endow a friction composite with exceptional life under rigorous conditions. Traditionally the concept of ceramic materials required the same process as the production of ceramic articles such as earthenware. This entails heating the mixture to a temperature sufficient for the materials to fuse and form a glassy matrix (i.e., vitrify).

Vitreous Ceramic Friction Articles

DeLukacsevics [6-36] developed a ceramic friction material in 1931 based on pulverized lava (or other suitable vitrifiable material) that can be fired at a temperature of between 1400–2300°F (approximately 800–1300°C) to yield a solid glassy friction article. This tribological composite was targeted for use on railway applications.

A few years later, DeLukacsevics [6-37] improved on his concept by using a mixture of metal oxides and mineral silicates in combination with a graphite containing clay to produce a ceramic material with improved friction performance. Again, the process involved firing the mixture at elevated temperatures to vitrify the clay to form an amorphous glass matrix containing the various constituents.

It can be imagined that a hard ceramic body would be potentially noisy during operation. In 1932, Nowalk [6-38] developed a solution to this problem. He used a fired porous ceramic body and impregnation process to impregnate the ceramic body with a lubricant mixture. The lubricant mixture was produced by heating the ceramic body and treating it with a mixture of lead oxide and borax (sodium borate) in a calcium stearate base. The end result was a friction lining saturated with lubricant that did not squeak during use.

Ceramic Binder Compositions

The next major step in the progress of making ceramic friction materials was the use of a ceramic binder that could be molded or cast. Eisenhardt [6-39] developed a simple asbestos mixture using 33 parts by weight asbestos, 16 parts clay, 8 parts graphite, 3 parts zinc oxide, 10 parts rubber, and 30 parts sodium silicate. This composition was mixed together and molded to form annular friction rings suited for clutches, or arcuate segments for use in early cam actuated drum brakes.

Today there is still interest in sodium silicate binder solutions (e.g., waterglass). However, there are also a number of commercially available advanced pre-ceramic polymer systems that can be converted into a rigid and very hard ceramic matrix.

Strasser and Atmur [6-40] used an aluminum silicate pre-ceramic polymer and a unique set of processing conditions to produce a friction material and the back plate in a single molding operation. The back plate portion used woven fiber to produce a fiber-reinforced ceramic matrix composite (FRCMC). The friction surface was a mixture of chopped fibers and selected filler materials with the same pre-ceramic polymeric matrix.

The mixtures were compacted and pyrolyzed to convert the pre-polymer into a ceramic matrix. The product could then be saturated with additional pre-polymer, and the pyrolysis step repeated to reach the desired density for the final product. This type of material is a true ceramic material and should be contrasted to the ceramic brake pads that are commonly sold for general purpose use.

Confusion about Ceramic Materials

One point of debate in the industry is based on the term and constitution of a ceramic friction material. To expand on this point of dispute, consider the work of Taylor and colleagues [6-41]. Their patent teaches the use of ceramic powder in the production of a brake lining suited for use on high-speed rail applications. They disclosed a composition comprised of 6% phenolic resin, 4% mineral fiber, 52% coarse ceramic powder, 1% synthetic graphite, 9% flake graphite, 20% fine ceramic powders, and 18% metallic fiber as a structural reinforcement. Despite the fact that the sum of the components is 110% and that the product is listed as having approximately 76% ceramic material, the binder is still organic and susceptible to the thermal degradation issues of any carbon based organic binder system. Therefore, this material is not a true ceramic friction material, but should be classified as a low metallic friction composition if the metallic fiber is ferrous, or as a non-asbestos organic friction material in the case of non-ferrous metals.

Many of today's automotive friction materials labeled and sold as ceramic materials are in fact a resin bonded composite with a high concentration of ceramic ingredients and should not be considered true ceramic materials.

At a functional level, a true ceramic material is defined as a reinforced ceramic matrix material and can include a host of ingredients to provide specific performance attributes. Because of their high cost and prolonged manufacturing process, these materials see limited production and are used in the most extreme operational environments.

Carbon-Carbon Materials

Carbon-carbon materials are a member of the NANM class of materials with a carbonized matrix reinforced with carbon fiber. The concept of carbonizing a polymeric matrix around a structure of fibers is the most common approach. It dates back to the work of Taylor and Komp [6-42] who carbonized a molded compound containing an organic binder at a temperature of approximately 800°F (425°C). This carbonized mixture was then pulverized, added to the original product, and molded to form a friction material capable of absorbing high energy levels during the landing of an aircraft. This product is not a true carbon-carbon material, but modern carbon-carbon friction materials emerged a few years later and can typically be exemplified by the work of Marin [6-43].

Marin wound carbon fibers around a mandrel forming a loose preform in the general shape of the final product. The preform is saturated with a carbon precursor (such as coal tar pitch) and carbonized in a non-oxidizing atmosphere. By using cycles of saturation and carbonization, the carbon level can be increased and ultimately the product is heat treated to graphitize the structure. The final product is deemed acceptable when the density level meets previously defined expectations.

A more recent example by Kosuda and Niijima [5-35] gives an excellent discussion of the manufacturing process for graphitized carbon-carbon friction materials. Much like Marin, this work focuses on the achievement of the high-density carbon structures and involves a cyclic process of carbonization and impregnation that is repeated many times to progressively develop a solid carbon matrix. This process requires a heavy investment in time and money, which limits carbon-carbon friction materials to extreme tribosystems requiring light weight and high strength such as aerospace and high-performance automotive applications.

Sintered Metallic Materials

Sintered metallic materials are metallic matrix compounds with tribological additives present. They can be used in both dry and wet friction applications. Generally, sintered metallic friction materials are of two families: copper-based and iron-based sintered products. However, alloys and other metals can also be used.

An early example by Boegehold and Short [5-32] illustrates a compressed and sintered metallic coated fabric to form a fibrous structure within a metallic matrix. This may be considered the first true sintered metallic friction material; however, a more conventional method of manufacture can be demonstrated by the copper-based friction products developed by Fisher [6-44].

Examples of Conventional Sintered Metallic Materials

One of Fisher's compositions was comprised of 5% zinc, 5% lead, 10% tin, 5% graphite, 3% silica, and the remainder copper. The addition of soft metals (e.g., lead, and hard components) such as silica are essential ingredients in this type of tribological composite. This design philosophy has survived a number of decades. For instance, Valentine [6-45] also used mixtures of zinc, tin, lead, and copper in his friction material composition. Wellman [6-46] later introduced a product with copper as the base metal, but also included the addition of iron powder to the mix. Wellman's example composition was 5.31% tin, 9.3% lead, 6.62% iron, 7.08% graphite, 4.43% silica, with the remaining portion a precipitated copper powder.

Growth of the Industry

By 1945 the powdered metal friction material business had grown significantly in the United States. Batchelor [6-47] points out that sintered metallic friction materials were commonly known as full metallic at this time. The term was apparently an effort to distinguish them from resin-bonded metallic compositions. Batchelor also states that the term full metallic was misleading because many of these materials contained a significant portion of non-metallic materials (such as graphite) and friction augmenting agents (such as alundum, silica, and carborundum).

Despite the naming issues, sintered metallic materials continued to grow in use globally and expanded into areas including aircraft [6-48], agricultural [6-49], and automotive tribological systems [4-5]. Many of these applications involved braking systems, but sintered metallic materials have been extensively used in the field of wet friction.

Iron-Based Sintered Metallic Materials

Genkin and colleagues [6-50] reported the development of an iron-based sintered metallic friction material capable of use in both wet and dry applications. Their unique composition was predominately iron at 61% by weight of the final product, with copper, tin, lead, ceramic particles, and graphite. Additionally, the composition incorporated zinc sulfide, and described the importance of metal sulfides in the development of a stable iron sulfide surface during use.

Today there are a number of iron-based sintered metallic friction materials in production. Many share the same conceptual design by incorporating various alloying agents, minerals, lubricants, and fluxing materials, many of which are intended to improve processing. Iron-based materials are used in higher energy applications that exceed the limits of copper materials and are expected to be a key growth opportunity in the future.

Final Words on Sintered Metallic Materials

A final example by Kondoh and Takano [6-51] described the use of a composite powder produced from an alloy of copper. This powder includes the zinc and nickel in very precise combinations to provide a unique friction surface for transmission components. It is specifically noted that certain extreme pressure additives present in liquid lubricated systems can liberate hydrogen sulfide that attacks certain non-ferrous metals and promotes chemical wear. The alloy described by Kondoh and Takano is not susceptible to this form of attack and provides a novel solution to the issue.

Sintered metallic friction materials are typically based on iron or copper or the alloys thereof. They are able to withstand rigorous duty cycles and extremes of temperature, but because of their heavy concentration of metals, their costs are generally tied to the metal markets and prices can shift rapidly. In addition, the production methods are energy intensive, requiring several hours to successfully sinter and control the grain structure of the final product. This can effectively limit future growth as the price of energy and metals continues to climb.

Paper-Based Materials

As stated previously, friction materials produced on a conventional papermaking process typically involve a secondary saturation with a solvated polymeric binder. Because these materials are often limited to operation in wet friction applications, the materials are generally formed in thin sheets and the surfaces machined to include a series of grooves, slots, or channels that allow fluid transport during use. In addition, the surfaces can also be ground to insure flatness and to reveal a continuous network of internal porosity that becomes a reservoir for fluid storage during use.

Asbestos-Based Paper Materials

The first examples of paper friction materials were recorded by workers at Raybestos [6-52] as early as 1919. Simpson began by grinding asbestos to liberate the fibers and adding them to water in a beater to form a pulp. This was fed onto a paper machine and formed into a felt that was then dried, compressed, and saturated with a solvated mixture of asphalt in drying oil. The saturated product was dried, cured, and ground to produce friction facings.

Along the same lines, we find the work of Stanley [6-53] who used asbestos felt, adding small amounts of graphite for lubrication and allowing the addition of a drying agent to accelerate the oxidation of the drying oil portion of the binder.

In other cases, asbestos paper was simply a precursor raw material that allowed for manufacture of the final product. Simpson [6-54] used strips of saturated asbestos paper that were wrapped around a mandrel under constant tension to form a tube. This tube was then cured, and thin sections were sliced off to form annular friction discs.

Heany [5-5] used asbestos paper to produce yarn by wrapping thin strips around a wire core. The wire core asbestos filaments were then twisted to form wire-reinforced yarn ideally suited for weaving friction materials.

Achtmeyer [6-55] demonstrated another of the many uses for asbestos-based friction paper. Several sheets of asbestos paper were oriented radially with respect to the direction of flow in the paper-making machine and saturated as a laminated structure. It was noted that the paper machine aligned the fibers in the direction of the screen, and preferential fiber alignment resulted in a paper that was prone to crack in the direction of the aligned fibers. By radially arranging sheets of paper; the strength of the final product was significantly improved.

Modern Paper Friction Materials

Today, a modern friction paper can contain a wealth of advanced fibrous materials in various compositions, sizes, and lengths. Modern friction papers are typically produced with a small amount of cellulose or other pulp type fiber, along with functional fillers that impart favorable characteristics such as enhanced compressibility or lubrication.

Additionally, these products usually contain a porosity inducing agent that allows for enhanced retention of fluid and improves the fluid transport. It is an essential design criterion that these materials have efficient fluid transport to provide the necessary friction output during use. One of the many ways to achieve this result involves the use of blends of different fibers [6-56].

Another method relies on the use of porosity enhancing materials. Arledter and colleagues [6-57] developed a paper containing 45% cotton, 45% diatomaceous earth, and 10% asbestos fiber. The high concentration of diatomaceous earth was specified as a critical component that improved resin penetration during saturation, reduced shrinkage of the paper, and imparted enhanced friction characteristics that were not obtained with other conventional mineral fillers.

Since the elimination of asbestos, there are a number of advances that have improved paper friction materials. Much of the advancement has come about because of new synthetic fibers and binder solutions that improve saturation while enhancing friction and wear. One recent patent [6-58] disclosed the use of a paper based friction material saturated with a resol-type phenolic resin mixture with a silane coupling agent to improve adhesion of the binder to the fibers.

The work of Kitahara and coworkers [6-59] used a mixture of naturally occurring and synthetic fibers to produce a friction paper. They also saturated the final product with a mixture of novolac phenolic resin and polydimethylsiloxane silicone resins. The combination of the resins and the unique selection of the fibrous materials resulted in enhanced friction while providing the necessary comfort during use.

There are a number of compositions, methods, and techniques that are currently emerging in the area of paper making. Kearsey [6-60] developed a high porosity heat-resistant dual layer friction paper where the layers are bonded together with a suitable adhesive and formed into friction surfaces for use in synchronizer rings or other transmission components. This is a method that combines many of the benefits of a dual-layer material and, therefore, might present a variety of options for the designers.

Woven Materials

Woven friction materials are the oldest material type and have seen a number of improvements throughout the years, but today there are only a few producers of woven friction materials remaining and their products are generally limited in use. Attempts to improve and advance the state of the art in this area tend to focus on the use of woven carbon fiber.

Winckler [5-7] developed a clutch using a pyrolytic carbon fabric coated with elemental carbon by chemical vapor deposition to form a friction facing substantially composed of carbon. This method is similar to those employed in the carbon-carbon friction material industry and, like those processes, requires an extensive amount of processing time. One specific instance listed a deposition time of

160 hours to achieve the desired density of carbon. In later patents [6-6, 5-8], Winckler demonstrated that the saturation process can use thermoplastic resins in addition to the chemical vapor deposition methods.

Lisowsky [6-62] took the weaving process one step further by saturating a single layer of carbon fabric with a thermosetting polymeric material. After saturation, the coated fabric was then molded under heat and pressure to form a surface with a plurality of grooves. Much like paper friction materials, the grooves were necessary to insure adequate fluid flow and cooling during use. After molding, the product could be treated again with additional saturant if necessary.

References

6-1. Jacko, M. et al. 1984. "Automotive Friction Materials Evolution during the Past Decade." *Wear* 100: 503-515.

6-2. Libsch, T., and S. Rhee. 1978. "Microstructural Changes in Semi-metallic Disc Brake Pads Created by Low Temperature Dynamometer Testing." *Wear* 46: 203-212.

6-3. Headson, F. 1918. Brake-band, brake-shoe, and lining there for. United States Patent No. 1264924, filed May 12, 1913, and issued May 7, 1918.

6-4. Lucid D. 1945. Method of making friction linings. United States Patent No. 2379166, filed November 4, 1943, and issued June 26, 1945.

6-5. Elerath, A. 1938. Frictional composition. United States Patent No. 2110571, filed April 20, 1936, and issued March 8, 1938.

6-6. Oswald, R. 1939. An improvement in or relating to friction linings. Great Britain Patent No. 505922, filed December 6, 1937, and issued May 19, 1939.

6-7. Biggs, R. 1965. Resin bonded friction composition product and method. United States Patent No. 3210303, filed September 26, 1960, and issued October 5, 1965.

6-8. Aldrich, F., and T. Deane. 1969. Modified organic base friction material. United States Patent No. 3434998, filed September 13, 1965, and issued March 25, 1969.

6-9. Rhee, S., and J. Kwolek. 1974. Sponge iron friction material. United States Patent No. 3835118, filed May 14, 1973, and issued September 10, 1974.

6-10. Ho, S. et al. 2005. "Effect of Fiber Addition on Mechanical and Tribological Properties of a Copper/Phenolic-Based Friction Material." *Wear* 258: 861-869.

6-11 Chan, D., and G. Stachowiak. 2004. "Review of Automotive Brake Friction Materials," Proceedings of the Institution of Mechanical Engineers, Part D: *Journal of Automobile Engineering* 218: 956-966.

6-12. Aldrich, F. 1977. Carbon metallic friction composition. United States Patent 4051097, filed April 12, 1976, and issued September 27, 1977.

6-13. Myers, R. 1984. Fused metallic friction materials. United States Patent No. 4438004, filed December 6, 1982, and issued March 20, 1984.

6-14. Kim, S., and H. Jang. 2000. "Friction Characteristics of Non-asbestos Organic (NAO) and Low-Steel Friction Materials: The Comparative Study." *KSTLE International Journal* 1: 1-7.

6-15. Moraw, K., and H. G. Paul. 1983. Asbestos-free friction material. United States Patent No. 4373038, filed October 1, 1981, and issued February 8, 1983.

6-16. Handa, Y., and T. Kato.1996. "Effects of Cu Powder, BaSO4 and Cashew Dust on the Wear and Friction Characteristics of Automotive Brake Pads." *Tribology Transaction* 39: 346-353.

6-17. Lee, J., and H. Jang. 1998. "Study of Anti-fading Phenomena during Automotive Braking." *KSTLE International Journal* 14: 70-78 (Korean).

6-18. Chester, J. 1981. Friction materials. United States Patent No. 4273699, filed July 31, 1978, and issued June 16, 1981.

6-19. Rasmussen, O. 1939. Brake lining material glass fabric. United States Patent No. 2158337, filed September 13, 1937, and issued May 16, 1939.

6-20. Peters, W., and I. Barnett. 1973. Friction material and method of manufacture. United States Patent No. 3756910, filed June 24, 1971, and issued September 4, 1973.

6-21. Parker, E., and B. Grele. Friction materials and their manufacture. United States Patent No. 4775705, filed December 19, 1986, and issued October 4, 1988.

6-22. Shibata, K. et al. 1991. Friction material. United States Patent No. 5004497, filed September 12, 1989, and issued April 2, 1991.

6-23. Halstead, R. 1944. "The Fundamentals of Asbestos Friction Materials." *Paper Trade Journal*: 118: 30-33.

6-24. Batchelor, C. 1996. "Friction Materials." In *Kirk-Othmer Encyclopedia of Chemical Technology*, 2nd ed. New York: John Wiley and Sons.

6-25. Jacko, M., and S. Rhee S. 1983. "Brake Linings and Clutch Facings." In *Encyclopedia of Composite Materials and Components*, edited by M. Grayson. New York: John Wiley and Sons.

6-26. Gier, T., and P. Salzberg. 1958. An inorganic flexible fibrous material consisting of the asbestos-like form of an alkali metal titanate and its preparation." United States Patent No. 2833620, filed September 24, 1954, and issued May 6, 1958.

6-27. Fybex. Registered trademark of E. I. Du Pont De Nemours and Company, 1007 Market Street, Wilmington, Delaware, Registration No. 0959631, May 29, 1973 (Trademark expired).

6-28. Mansfield, J. et al. 1976. "An Improved Automotive Brake Lining Using Fibrous Potassium Titanate." TN-D8363. Washington, DC: National Aeronautics and Space Administration.

6-29. Halberstadt, M.,1976. "Fabrication and Test of Experimental Automotive Friction Materials." CR-137852. Washington, DC: National Aeronautics and Space Administration.

6-30. Phipps, M., and G. Vyletel. 2007. Friction material. United States Patent Application No. 2007/0219289, filed March 14, 2007.

6-31. Ono, M. et al. 2008. Friction material composition and friction material using the same. United States Patent No. 7338987, filed May 21, 2004, and issued March 4, 2008.

6-32. Chiba, M. 2003. Non-asbestos friction material. United States Patent No. 6656240, filed February 20, 2002, and issued December 2, 2003.

6-33. Bortz, D. 1999. Friction controlling devices and methods of their manufacture. United States Patent 5989375, filed June 6, 1995, and issued November 23, 1999.

6-34. Ogawa, H. et al. 1989. Friction material. United States Patent No. 4861809, filed May 27, 1988, and issued August 29, 1989.

6-35. Longley, J. and R. Gardner. 1988. "Some Compositional Effects in the Static and Dynamic Properties of Commercial Vehicle Disc Brakes." *Institution of Mechanical Engineers C453/88*, pp.33-38.

6-36. DeLukacsevics, C. 1931. Lava earthernware or mineral friction element or lining for railway and other brakes in general. United States Patent No. 1823591, filed December 14, 1925, and issued September 15, 1931.

6-37. DeLukacsevics, D. 1933. Friction element. United States Patent No. 1899239, filed December 15, 1928, and issued February 28, 1933.

6-38. Nowalk, R. 1932. Friction equalizing material and use thereof. United States Patent No. 1845096, filed October 6, 1925, and issued February 16, 1932.

6-39. Eisenhardt, F. 1937. Brake. United States Patent No. 2083989, filed March 14, 1935, and issued June 15, 1937.

6-40. Strasser, T., and S. Atmur. 2000. Integrated fiber reinforced ceramic matrix composite brake pad and back plate. United States Patent No. 6062351, filed July 12, 1999, and issued May 16, 2000.

6-41. Taylor, A. et al. 1998. Friction pads for use in disk brakes. United States Patent No. 5725077, filed March 6, 1996, and issued March 10, 1998.

6-42. Taylor, C., and M. Komp. 1959. Friction composition comprising carbonized elements held in a matrix of uncarbonized material and method for preparing same. United States Patent No. 2885379, filed July 15, 1955, and issued May 5, 1959.

6-43. Marin, G. 1973. Reinforced carbon bodies. United States Patent No. 3712428, filed June 22, 1970, and issued January 23, 1973.

6-44. Fisher, J. 1937. Friction article and method of producing same. United States Patent No. 2072070, filed July 24, 1933, and issued February 23, 1937.

6-45. Valentine, I. 1937. Vehicle brake. United States Patent No. 2089080, filed November 7, 1935, and issued August 3, 1937.

6-46. Wellman, S. 1941. Sintered friction article and method of making same. United States Patent No. 2239134, filed January 31, 1940, and issued April 22, 1941.

6-47. Batchelor, C. 1945. "Metal Powder Friction Materials." *Metals and Alloys* 21: 991-993.

6-48. Blau, P. 2001. "Compositions, Functions, and Testing of Friction Brake Materials and Their Additives." ORNL/TM-2001/64. Oak Ridge, TN: Oak Ridge National Laboratories.

6-49. Kryachek, V., and E. Shvedkov. 1969. "Sintered Friction Materials Based on Aluminum Bronze. Part I: Effects of Some Production Technology Factors on Properties Of Materials." *Poroshkovaya Metallurgiya* 9: 60-65 (Russian).

6-50. Genkin, V. et al. 1982. Sintered iron-based friction material. United States Patent No. 4311524, filed April 1, 1980, and issued January 19, 1982.

6-51. Kondoh, K., and Y. Takano. 1999. Sintered friction material, composite copper alloy powder used therefor and manufacturing method thereof. United States Patent No. 5972070, filed August 22, 1997, and issued October 26, 1999.

6-52. Simpson, S. 1919. Process of making friction facings. United States Patent No. 1301612, filed May 22, 1918, and issued April 22, 1919.

6-53. Stanley, F. 1922. Friction facing. United States Patent No. 1414383, filed September 24, 1919, and issued May 2, 1922.

6-54. Simpson, S. Clutch facing and process of making the same. United States Patent No. 1348755, filed February 6, 1919, and issued August 3, 1920.

6-55. Achtmeyer, W. 1922. Disc clutch facing and process of making same. United States Patent No. 1429266, filed March 10, 1921, and issued September 19, 1922.

6-56. Steck, R. et al. 1955. Production of friction materials. United States Patent No. 2702770, filed August 14, 1951, and issued February 22, 1955.

6-57. Arledter, H. et al. 1966. Friction member with friction material. United States Patent No. 3270846, filed March 15, 1957, and issued September 6, 1966.

6-58. Mori, M. et al. 2009. Wet friction material. United States Patent No. 7491664, filed March 24, 2006, and issued February 17, 2009.

6-59. Kitahara, S. et al. 2001. Wet friction material and process for producing the same. United States Patent No. 6331358, filed March 3, 1999, and issued December 18, 2001.

6-60. Kearsey, A. 1996. Friction lining. United States Patent No. 5585166, filed February 24, 1995, and issued December 17, 1996.

6-61. Winckler, P. 1997. Carbon-based friction material for automotive continuous slip service. United States Patent No. 5662993, filed September 8, 1995, and issued September 2, 1997.

6-62. Lisowsky, B. 1999. Grooved friction material, method of making same, and wet friction member using grooved friction material. United States Patent No. 5858511, filed March 11, 1997, and issued January 12, 1999.

Chapter 7
The Five Elements of a Friction Material

The whole is more than the sum of its parts.
—Aristotle

EARLY FRICTION MATERIALS WERE SIMPLE woven fabric composites impregnated with a readily available organic binder material. The fiber, typically asbestos based yarn, provided the necessary structural reinforcement and influenced the tribological characteristics during use. The impregnating agent was the matrix forming material and also aided in imparting favorable frictional responses. For this reason, many early binders were considered frictioning agents [7-1, 7-2].

The simplistic approach of a two-component friction material stands in stark contrast to the immense complexity of modern tribological composites. Many of today's composites contain a wide variety of ingredients in various sizes, shapes, hardness levels, and chemical compositions. The compositional makeup is also highly variable. According to Drava [7-3], a friction material can contain 15–18 materials selected from a list of more than 800 different raw materials. However, variations in the compositions because of philosophical differences around the globe can also limit compositions to mixtures of approximately 10 ingredients [3-42, 6-32].

In all cases, the need for a blend of materials in the composition is essential to achieve the desired balance of friction, wear, and ancillary requirements imposed by customer expectations. This level of complexity makes the development process difficult and often involves a series of empirical evaluations to predict the impact of a given additive on the various aspects of the composite's wear, friction, and noise performance.

In an effort to make the compositions more manageable, the designers generally classify a given raw material by its functional properties. The assignment is strictly comparative and might be arbitrary regarding the specific nature of the comparisons performed. Designers use their own methods of classifying the materials. However, for the purpose of space, and considering the breadth and depth of every potential raw material that can be used in a friction material, the ingredients will be grouped into five classes as a basic form of categorization. Therefore, the five elements of a friction material are defined as: binders, fibers, abrasive materials, lubricating additives, and filler materials.

Admittedly these classifications are not perfect, and many raw materials can play several different roles depending on operational parameters such as load, speed, and temperature. Additionally, there can also be environmental factors present during use that convolute attempts to effectively classify a given additive. To make matters worse, many composites might contain only raw materials from two or three groups and still others might contain curatives, processing aids, and plasticizers that do not fit neatly within our proposed hierarchy of classification. As such, a material that has multiple functions will be discussed under a primary functional designation and secondary attributes will be discussed as necessary.

Validation Is Required

The reader is encouraged to validate these categories of raw materials within their own compositions. It is assumed that each organization will have developed a unique line of friction products based on completely different combinatorial approaches from these listed. Furthermore, the processes disclosed in the methods of manufacture will likely be different, albeit subtly so, from those used, which serves to strengthen the need for verification.

The Binder

The binder becomes the matrix of a friction material and is generally defined by the process. For example, a sintered metallic matrix will possess a metallic matrix that has been formed by manufacturing methods substantially different from those used in the manufacture of a product based on an organic polymeric matrix such as rubber. These processes can be further contrasted with other vastly different forms of composite processing (such as those involved in the formation of a carbon matrix produced by chemical vapor deposition). Each process will therefore dictate the choice of binder.

In many cases, the selection of the binder material is usually based on historical precedence. A process that uses dry mixes and phenolic resin will usually continue to produce this type of product. Therefore, changes to the binder system are not encouraged because of the necessity of procuring new equipment associated with processing different binder types.

In other cases, the process of selection can be the result of cost-driven selection protocols that are combined with the changes in local manufacturing processes such as a need for cycle time reductions. This paradigm shift can force the selection of different binder technologies and drive designers to seek alternative methods and materials.

Binder Selection Is Critical

The matrix is the single most important aspect of the composite. Care must be taken during the selection process to ensure that the binder is compatible with the process and also to provide the final product with the required strength. Additionally, the binder, which can be only a few weight percent of the total product, can have profound implications on the performance of the product.

Georgievskii [4-51] determined that the binder, which included phenolic resins in his study, will determine the mechanical frictional properties and the wear resistance of the composite by influencing the mechanical characteristics of the friction surface.

In a separate work, Georgievskii [7-4] also demonstrated that the binder was dominant in controlling the friction behavior at temperatures below the final processing conditions. This can be explained by the observation that friction levels at temperatures below the curing temperature, specifically the final cure or baking temperature, were controlled by the properties of the polymeric matrix.

If the heat generated exceeded the curing temperature during operation, then the effect of the mineral fillers began to emerge as a contributing factor and overshadowed the binder's contribution to friction. Georgievskii further defined a maximum concentration of about 20% as the upper limit for polymeric matrix materials. Above this concentration, little or no benefit was shown, which indicates that the manufacturing process has an effect on performance and that concentration is also critically important.

Binder Concentration Is Critical

Tewari and Bijwe [7-5] also illustrated how the final product can be affected by the concentration of the polymeric binder. This effect was seen as a marked reduction in strength at low binder concentrations and a pronounced increase in the wear rates. The increase in wear rate was attributed to poor bonding between the reinforcing fibers and the selected matrix material, but also suggested that binder adhesion will affect the performance of all types of ingredients including fillers, abrasives, and the like.

The earlier work of Spurr [7-6] also found that excessive binder concentrations were not acceptable and can lead to poor friction performance at elevated temperatures. Although Spurr did not specify a limit, the overriding conclusion was acknowledgement of a critical binder concentration limit that exists for each product. This limit must be carefully determined during the development process.

These examples can be augmented by a host of other examples including the work of Kim [7-7], and Iwata [7-8]. Both research groups pointed out the influence of binder concentration and the effect of processing, especially when using a phenolic resin binder system. Molding conditions and subsequent heat-treating processes were shown to affect aspects of the surface hardness and porosity of a phenolic bound composite. This in turn affected friction and wear performance, which again clearly showed the link between tribological performance of the product because of the matrix and its processing parameters.

Fibers

The definition of a fiber is not a simple matter. Skinner [7-9] searched through 50 different scientific and technical resources only to conclude that no precise scientific definition for a fiber existed. In a review of the legal literature surrounding occupational exposure to asbestos, it was noted that the National Institute of Occupational Safety and Health had defined a fiber as having a length of five microns and an aspect ratio of at least 1:3. For our purposes, we will treat this as a functional definition and loosely classify a material that meets this requirement as a fibrous material.

Fibers can be naturally occurring organic fiber such as cotton, or naturally occurring inorganic fiber such as asbestos. In other cases, fibers can be synthetically produced from organic polymers such as para-aramid. Yet another subclass of fibers can be defined as synthetic inorganic fibers, which can include a host of extruded, spun, and monocrystalline whisker forms produced from mixtures of metal oxides.

In addition, metals can be processed to form fiber-like materials such as steel wool, which provides a unique blend of mechanical and tribological properties to the composite. Because of the wide range of potential candidates, the process of fiber selection is both an interesting and challenging practice.

The effect of fiber reinforcement on the strength of a composite is well-documented. For this reason, the majority of the discussion will be focused on aspects beyond the enhancement of mechanical strength. We have already seen that fiber de-bonding [7-6] can lead to excessive wear. This represents the fundamental nature of fiber reinforcement and presupposes that the adhesion between the fiber and the matrix is strong enough to ensure a robust product of high strength.

Aside from structural reinforcement, fibers also have a pronounced effect on the tribological behavior as well. One early example by Batchelor [6-24] describes how commercial grades of chrysotile asbestos ranged from grade 3 fibers to grade 7. The grade 3 fibers were the longest and most suitable for use in weaving operations, whereas the grade 7 fibers had a very short grade best suited for molding, paper making, and extrusion operations. The thermal benefits of asbestos fiber were well-known at this time, but Batchelor added that asbestos fibers were incorporated to impart strength as well as influence the friction characteristics of the final product. It is interesting to note that he does not explicitly define the specific benefits.

Metallic Fibers

In contrast to inorganic asbestos fibers, metallic fibers have emerged as the dominant fiber in the semi-metallic class of materials. These fibers are not true fibers per se, but rather are produced by drawing serrated knives along the surface of a metallic rod to form thin filaments of metal. Metallic fibers can also be in the form of finely chopped wire. They typically have a circular cross section resulting from the wire drawing process. Regardless of the method used in production, metallic fibers are added to friction compositions to improve many performance aspects including enhancing structural reinforcement, improving wear resistance, and increasing the thermal transport characteristic of the lining.

Eriksson [7-10] found that metallic fibers at the surface of a friction material composite aided in the formation of load carrying plateaus and provided beneficial wear enhancement as well as increases in the real area. The creation mechanism was described as an accumulation of wear debris particles lodged behind the fiber. This acted as a dam to restrict particle flow at the surface and subsequently trapped more particles. As the process continued, these particles began to accrete together to form a plateau that would grow in size and remain until the fiber was worn away.

Because the formation process of the plateau affected the real area of contact, the pressure distribution across the surface of the friction material was altered, which impacted frictional performance as well. This process can also be expected to occur with any fiber that is thermally stable at the operating temperatures developed during use. It is only one of the many aspects of the fiber selection process.

Fiber Blends

There are many different fibers that have been explored during the last one-hundred years, with asbestos being the primary fiber of choice. Like the highly reinforced composites of the past, today's modern engineered tribological composites also rely heavily on fibers but tend to optimize the benefits by using selective blending strategies. It is therefore commonplace to see a blend of organic, inorganic, and metallic fibers operating in conjunction in a given composite. The use of fibrous mixtures, which were at one time called fiber cocktails, is an area that is expected to remain rich in development. It isrecommended fiber blends be explored extensively to achieve new levels of performance and strength.

Abrasives

It is also difficult to define precisely what constitutes an abrasive material. A general rule is that abrasives are usually harder than the mating material and have the fundamental purpose of generating material removal from the surface during sliding. Typically, these include powdered forms of naturally occurring minerals such as garnet, quartz, diamond, and corundum, but, can also include synthetic materials such as aluminum oxide, silicon carbide, and cubic boron nitride.

Scales of Hardness

The relative differences in the hardness of these materials provide a convenient method for defining their abrasive ability, but this technique is not without its failings. For instance, iron oxide can be considered an abrasive when used against a cast iron surface, but would not be notably abrasive against a much harder ceramic surface such as silicon carbide. As such, the classification of abrasives is often a matter of comparison based on measures of the relative hardness of the material.

The Mohs Scale

A commonly cited scale for relative hardness of minerals is the Mohs scale. The Mohs scale of mineral hardness uses the concept of scratch resistance to classify geologic samples. It was created by the German mineralogist Friedrich Mohs [7-11] and is based on ten "standard" materials. These materials are ranked from softest to hardest and include: talc, gypsum, calcite, fluorite, apatite, feldspar (orthoclase), quartz, topaz, corundum (sapphire), and diamond.

To classify a material such as cast iron by using the Mohs scale, the researcher would simply scratch a sample of each of the standards against the surface of the material with unknown hardness and observe to see if a visible scratch was produced. The hardness would then be between the two standard hardness values where scratching was observed to happen.

It should be noted that this scale is not linear and is not as precise as other methods for measuring hardness. Therefore, the Mohs scale should be used only as a rough guide during the abrasive selection process. Other scales such as Knoop or Vickers are more definitive and should be used when possible. Aside from hardness, abrasive materials can have a pronounced response in a tribological composite based on the concentration and size of the particles.

The Effect of Abrasives in Friction Materials

Crosa and colleagues [7-12] studied the effect of abrasive materials in friction material composites. This study focused on three aspects of abrasive selection: different grain sizes, different abrasive materials of the same size, and concentration dependencies of a given abrasive of a given size within the composition. Their work compared aluminum oxide and chromium oxide in a simple semi-metallic friction formulation.

The results of this study indicated that changes in the grain size and shape were important. It cited the differences between spherical chromium oxide particles, which gave less friction generation than pyramidal aluminum oxide particles. It also suggested that at higher temperatures, the resin would begin to degrade and the abrasive grains were able to penetrate the matrix when under an applied load. This particle regression mechanism effectively reduced the plowing action of the particles and thus decreased the observed frictional responses.

It was also noted that possible improvements in load-carrying capacity of the materials and reductions in pad wear can be gained by using abrasive materials. However, their addition to a composite would generally produce an accompanying increase in the rate of wear of the disc.

The Concentration Dependence of Abrasives

Fan [7-13] also used aluminum oxide additions to enhance a semi-metallic formulation. Aluminum oxide was reported to increase abrasion but also imparted a higher degree of friction stability. The use of abrasives materials did not provide simple solutions to increasing friction output and wear. For example, at a concentration of 3.4% by volume, the composition was shown to have the least amount of wear. However, when the concentration was increased to the level of 5.6%, the product was shown to provide the most acceptable friction levels. This balance of properties is another example of how the selection process can often involve a sacrifice of one characteristic to achieve another.

The Size Limits of Abrasive Particles

The selection of an abrasive constituent must also consider the dynamic nature of operation. As a case in point, work by Avient [7-14] examined the abrasion of metals and found finer particle sizes tend to generate the least amount of wear. While this may seem intuitive, Avient also reported that finer sizes were more likely to clog with wear debris, which would reduce the abrasion by growing localized regions of packed wear debris to accommodate the load. From his work came the suggestion that an optimum size for abrasive materials might be limited to particles approximately 70 microns in size. Larger particles apparently gave little benefit.

The size limit concept was also given further confirmation by the work of Sin [7-15]. As the size of the abrasive particle increased, the wear rates increased rapidly up to critical size. Once the limit was reached, the wear rates would not increase dramatically. Sin concluded that the critical size limit in his study was for the particles smaller than 80 microns in size. A key point from Sin's work was that large particle sizes (larger grit sizes) tended to produce more extensive subsurface damage than the finer grades, which could lead to delamination wear.

Sin's approach was based on the assertion that the plowing component of friction is strongly dependent on particle geometry. Sin explained the effect of grit size in terms of relative bluntness of the particles, whereby particle wear processes during motion serve to limit the depth of penetration, which lessen their ability to plow through the surface layers.

To have continual material removal processes, an abrasive grain should splinter or fracture to reveal fresh cutting edges during use. On the other hand, a particle that polishes down to a smooth surface does not plow as effectively and, with prolonged use, can be expected to show a drop in friction with reduced wear rates.

Even though a typical friction material composition might contain only a few percent of an abrasive material, the selection process is one that must not be casually disregarded. In addition, abrasive constituents can be present as contaminants and many mineral species, which are moderately soft in comparison to cast iron, can contain hard impurities that can shift the concentration of abrasive materials to an undesirable level. It is therefore essential that the designer understand the composition of each raw material, and the potential impurities that can be contained to prevent unwanted abrasion.

Lubricants

Solid lubricants have been a part of the friction material industry since the earliest days and are generally intended as a counterpoint to the addition of abrasive materials. Fundamentally, their purpose is to control friction and wear. There are a number of solid lubricants that have been used in the friction material industry that typically include mixtures of graphite, in at least one of its forms, and a reactive metal sulfide. The selection process of a lubricant additive therefore involves an understanding of the advantages and disadvantages that each form of solid lubricant can have on the performance of the final composite.

Advantages and Disadvantages

Campbell [7-16] listed the advantages and disadvantages associated with solid lubricants. The advantages of solid lubricants include better stability at extreme temperatures where liquids are subject to vaporize and typically more effectiveness in chemically reactive environments. This allows for low weight simple designs that do not require complicated cooling systems with pumps, tubing, and heat exchangers. Additionally, fewer seals are required with solid lubricants, and many devices gain mechanical stability that enables high-speed operation.

One of the disadvantages of solid lubricants is a higher friction coefficient when compared to the ideal hydrodynamic lubrication of liquid lubricated systems. Additionally, solid lubricants operate in a regime where sliding contact between surfaces will produce measurable amounts of wear, especially in dry film coatings where the solid lubricants have a defined thickness and therefore have a definable life expectancy. Furthermore, solid lubricants have no means to cool the system and thus can experience operating temperatures higher than desired.

Despite the negatives associated with solid lubricants, they are the only tenable solution to the lubrication of a composite that operates in the dry friction regime of automotive brakes. As mentioned, the friction materials industry commonly uses blends of materials to offset the disadvantages and amplify the advantages.

An Example of an Effective Lubricant Blend

Centers [7-17] studied the role of several metal sulfides as solid lubricants and suggested that trace amounts of molybdenum oxide in a molybdenum disulfide compact can improve the performance of the lubricant system. The proposed mechanism was the formation of a vitreous binder that retains the lubricant at the surface. The retention of lubricant promotes a smooth tribological surface and limits the interaction between asperities, thus resulting in favorable friction and wear characteristics.

The U.S. Army Corp of Engineers [7-18] states that a successful lubricant will form a tenacious film that is strongly adhered to the base metal. Additionally, an effective lubricant should promote regeneration of the film during use. These specific attributes explain why molybdenum disulfide, which readily forms a highly tenacious film on specific metals, is an effective lubricant. Graphite, on the other hand, is an effective and commonly used lubricant but operates in a different way and provides different properties.

Graphite Lubrication Mechanism

Molybdenum disulfide in an intrinsically lubricating solid, whereas graphite provides effective lubrication only in the presence of other adjuvants. Savage [7-19] demonstrated this by using several techniques of vacuum and atmosphere cycling to study the wear behavior of graphite brushes for electrical contacts. Under a vacuum, the wear rates for graphite were quite high and the wear debris was a highly activated form of carbon. When the atmosphere was admitted, the friction level decreased rapidly and the wear rates were dramatically reduced. This was demonstrated repeatedly by the cyclic use of evacuation and admission of selected gases.

In a later study, Deacon and Goodman [7-20] evaluated several lamellar lubricant candidates. It was believed that the lubricating ability was a physical manifestation of weakly bonded basal planes that would shear easily under an imposed stress in a similar manner to molybdenum disulfide and graphite, which share this structural characteristic.

Based on this study, the friction of graphite was shown to exhibit a transition as the temperatures were increased to approximately 750–930°F (400–500°C). At this point, the adsorbed contaminant species were driven off and the friction level increased as shown by Savage. Upon cooling, the contaminants adsorbed onto the surfaces and the friction level again decreased. It should be noted that the conversion of molybdenum disulfide into the oxide was also responsible for a similar transition to higher friction levels that occurred at approximately 662°F (350°C), but this reaction is not reversible.

Film Formation and Blends of Lubricant Additives

As mentioned previously, the lubricant should form a tenacious film [7-18, 7-21] that allows for constant regeneration and protection of the surfaces. Because of the effects of the environment and temperature on the stability of the third body layer formed during sliding, the most successful method for accomplishing a stable transfer film has been to use lubricant blends. For example, mixtures of antimony sulfide and graphite have been used as "anti-friction" agents since 1889 [3-19]. Centers [7-17] demonstrated that antimony oxide would impart favorable reductions in wear when used in combination with graphite. Elemental lead, which was a commonly used lubricant [7-22, 4-5], was also blended with graphite [6-24] in an attempt to provide for effective lubrication across a wider range of operational conditions. Each of these techniques was shown to give improvements in their respective operational regimes.

Taking the blend concept one step further, Longley and Gardner [6-35] studied the compositional effects of commercial vehicle disc brakes and concluded that lubricants, which were used in relatively small amounts, would perform at an optimum level within a given temperature range. To extend the performance across a wider range of temperatures, the use of blends was an absolute necessity. One of the listed examples demonstrates that graphite is an effective low temperature lubricant. However, at elevated temperatures, the addition of elemental copper was necessary to ensure effective lubrication.

Issues with Lubricant Use

Longley and Gardner also warned against the use of excessive concentrations that can induce poor friction performance under high pressures, which was given the name "pressure sensitivity." As operating pressures increase, the understanding of how these materials perform under increased loading is a key selection criterion that must be considered during the design phase.

The effect of pressure is analogous to liquid lubrication mechanisms, and the use of solid lubricants can at a simplistic level be viewed as an attempt to control the formation and rheology of the third body layer during operation. This has been well documented by Cho and colleagues [7-23] where blends of materials aid in the formation of the transfer film that was shown to greatly affect the behavior of the product during subsequent use (as listed in previous sections of this book where this topic was discussed).

Another issue with solid lubricant use has been reported by the team of Lee and Jang [7-24], who illustrated that blends of solid lubricants have significant effects on the performance and wear of automotive friction materials, but also highlighted some of the potential pitfalls to the use of solid lubricants. When examined in certain situations, the blends of solid lubricants were found to give higher wear rates than expected. This was also manifested as non-uniform wear of the brake disc, which suggests that an optimum level might exist for each unique blend.

Lubricant Mixtures Should Be Optimized

To illustrate the need for lubricant optimization, one case in point is the work of Kim and colleagues [7-25]. This group explored the complementary effects of solid lubricants in friction materials and discovered that blends of graphite with antimony sulfide had an optimum performance when blended at a ratio of 2:1. Variations in the concentration were therefore deemed to be unacceptable outside of this ratio.

In summary, the world of knowledge regarding solid lubricants is growing each day. There are a number of sources in the available literature indicating that key performance attributes can be obtained by using specific ratios and combinations of certain mixtures. This recurring theme is due to the number of synergies that exist and should be further explored. It is hoped that advances in the use of lubricant blends will push the development of friction materials forward to a new level of performance.

Fillers

The ubiquitous filler category is the catch-all receptacle for those materials that do not easily fit within the other groups. For instance, according to Blau [6-48], a filler material can be metallic, organic, or inorganic in nature and can be used to "maintain the overall composition of the friction material." Blau also stated that fillers can have other functions as well. This allows for a wide range of materials to be selected and classified as filler materials. It further suggests that filler materials might be functional by imparting specific attributes to the composite.

On the other hand, Gopal [7-26] stated that fillers are low-cost materials that are intended to cut the cost of the composite. In essence, they are present only to occupy space. This perspective was also maintained by Longley and Gardner who stated that fillers are generally low-cost additives that are added in higher concentrations, which implies they have little impact on the final product.

Filler Effects Can Be Masked

In reality, a filler material can be used as a low-cost additive; however, filler materials do provide specific and definable attributes. Generally, some of the attributes can be overshadowed by other ingredients present and therefore seem to be innocuous. The addition of very finely divided magnesium oxide powder to a friction material can serve as a wear reducing filler if the role of an abrasive material is already being played by a harder material such as aluminum oxide. In other words, the highly abrasive nature of the aluminum oxide addition overshadows the abrasive component provided by the magnesium oxide and results in magnesium oxide being classified as a filler material.

Fillers Affect the Structure of the Composite

In other cases, filler materials can provide structural effects. The rubber industry has widely accepted that filler materials can be non-reinforcing, semi-reinforcing, or reinforcing. Ohm [7-27] lists these attributes as the result of the selected particle size where particles greater than approximately 10 microns can be harmful to the physical properties because of the development of localized areas of stress. These particles are considered non-reinforcing.

Materials between 1 and 10 microns neither harm nor help the measured properties of the vulcanized compound and therefore are considered inert diluents that reduce the amount of higher cost elastomer. Semi-reinforcing fillers, which provide slight improvements in strength and modulus properties, have particle sizes of 0.1 to 1 micron. True reinforcing fillers, which significantly improve the physical properties, have particle sizes of 0.01 to 0.1 micron. Outside of the rubber industry, there are also a number of sources that demonstrate that polymeric friction material composites can also be influenced by the addition of selected fillers.

Pogosian [7-28] recorded that the adhesive strength and effectiveness of the third body layer can be improved by the addition of fillers that have high surface energies. Particles that are small and have surface energies will increase the cohesive strength of the particles, enhance the adhesion of the particles to the mating surface, and promote the formation of a tenacious third body layer, which will in turn provide an enhanced friction response during use.

Abrasive Fillers

The abrasive aspect of filler materials is also a critical concern that should be addressed during the selection process. Gahr [7-29] demonstrated that fillers that are highly abrasive will tend to give elevated wear rates. This was reinforced by earlier work by Batchelor who compared the effect of barytes, a commonly used powdered form of naturally occurring barium sulfate, with the effect of rottenstone, a form of decomposed limestone rich in silica. The noted outcome was that barytes was inert relative to the friction augmenting behavior of the rottenstone. Similar results by Spurr add weight by demonstrating a correlation between the hardness of the mineral filler and higher friction output, which was accompanied by higher wear rates, suggesting that fillers can provide an abrasive component.

The effect that filler materials have on wear performance is often a combination of many factors including size, hardness, and particle retention. Gahr reported that filler sizes should be evaluated to determine if an optimum size is available. If the filler particles are too large or too small, the wear rates of the composite can be affected. This can also be seen along the lines of the reinforcing filler concept previously described.

Fillers Affect the Physical Properties

Aside from elastomeric mixture, the physical properties of rigid composites are also affected by the addition of fillers and will thus affect the tribological performance of the product. One example studied the effect of ceramic filler materials in a phenolic resin matrix [7-30] and found that the physical properties of the composite and the wear rates were dependent on the type of filler used.

Tanaka [7-31] also studied friction materials with a phenolic resin matrix and determined that one of the major causes of wear was by detachment of the filler particles. This supports that filler materials must be compatible with the polymeric matrix to provide many of the benefits that are possible with the addition of a filler material.

Support of this was also provided by Bely and colleagues [3-22] who listed one of their criteria for obtaining a quality friction composite was based on the degree of adhesion between the filler and the polymer. If the polymer was capable of forming a strong bond between the materials, then an ideal composite was produced. In other words, the filler material can provide benefit only when it is compatible and rigidly fixed in the matrix.

Optimization of Filler Materials

Like many aspects of the raw material selection process, a filler material must also be used in the correct concentration. Bartenev and Lavrentev [7-32] disclosed that the hardness of a rubber composite was generally affected by the type and concentration of filler. The increase in stiffness [7-33] and hardness that occurs in polymeric systems is commonly seen with the addition of mineral fillers and is the result of carefully determining the required concentration [7-34] of optimally sized filler.

Reactive Fillers

Aside from size, hardness, and concentration effects, the filler material must also be examined to determine if chemical reactions will take place during sliding. Mizutani and co-workers [7-35] studied the wear of resin bonded asbestos composites by the use of x-ray analysis. It was concluded that select filler materials had a pronounced effect on the thermal degradation behavior of asbestos fibers. It was further suggested that the degradation process could be altered by correctly selecting the best filler material.

Oehl and Paul [3-12] also expanded on the use of functional fillers in their discussion of wear behavior and suggested the use of endothermic filler materials in the composition. The addition of an endothermic filler material is intended to undergo phase transformations as the braking element is heated by sliding. The phase transitions will then absorb the energy, thus cooling the interface during use. It was noted that fusible elements that can react with oxygen to form abrasive oxides (such as aluminum) should be avoided.

Additionally, it has been noted that if the filler dispersion was present as an aggregate, the friction variation observed during use was likely to increase. In other words, a highly dispersed system would give better friction stability, which could lead to reduction in the noise level during use.

One cautionary point that was also made regarding filler materials is the thermal expansion rate of the selected filler. If the rate of expansion does not match that of the polymer, stresses would be produced as heat was generated from sliding. These points of stress could develop into micro-fissures that would lead to the generation of larger cracks and ultimately would generate particles of wear debris. This should be avoided whenever possible, preferably by selecting polymers and materials that have similar rates of thermal expansion.

Organic Filler Materials

To this point, the topic has focused mainly on inorganic fillers that are generally minerals of various types. However, many friction material compositions also rely on the benefits of softer organic filler materials to regulate the friction level and reduce brake noise [7-30]. These particles (typically polymeric materials such as rubber crumb or CNSL particles) provide a large macro-asperity that aids in making the surfaces more conformable, which in turn increases the area of contact and helps dissipate energy more effectively.

In conclusion, virtually all materials can be classified as filler and as such the filler category does not truly exist. For the sake of consistency, we will use this area to discuss materials that are not conventionally used as binders, lubricants, fibers, or abrasives.

References

7-1. Seigle, W. 1922. Method of modifying oils and the product thereof. United States Patent No. 1436362, filed June 28, 1920, and issued November 21, 1922.

7-2. Nanfeldt, W. 1934. Method of making brake linings containing friction compounds. United States Patent No. 1959686, filed April 13, 1929, and issued May 22, 1934.

7-3. Drava, G. et al. 1996. "Application of Chemometrics to the Production of Friction Materials: Analysis of Previous Data and Search of New Formulations." *Chemometrics and Intelligent Laboratory System* 32: 245-255.

7-4. Georgievskii, G. 1959. "Peculiarities in the Design of Heat-Resisting Friction Materials: Increase in the Efficiency of Brakes and Improvement of Properties of Friction Materials." USSR: Izdatel'stvo (Russian).

7-5. Tewari, U., and J. Bijwe. 1993. "Recent Developments in Tribology of Fibre Reinforced Composites with Thermoplastic and Thermosetting Matrices." In *Advances in Composite Tribology*, edited by K. Friedrich. Amsterdam: Elsevier.

7-6. Spurr, R. 1972. "Fillers in Friction Materials." *Wear* 22: 367-372.

7-7. Kim, S. et al. 2002. "Manufacturing Parameters Affecting Physical Properties and Tribological Behavior of Brake Linings." *KSTLE International Journal* 18: 127-132 (Korean).

7-8. Iwata, K., and H. Asano. 1987. Friction material. European Patent No. 208326, filed July 10, 1986, and issued January 14, 1987.

7-9. Skinner, H. et al. 1988. *Asbestos and Other Fibrous Materials*. New York: Oxford University Press.

7-10. Eriksson, M. et al. 2001. "Wear and Contact Conditions of Brake Pads: Dynamical in situ Studies of Pad on Glass." *Wear* 249: 272-278.

7-11. Winkler, E. 1973. *Stone: Properties, Durability in Man's Environment*. New York: Springer-Verlag.

7-12. Crosa, G. et al. 1992. "Study on the Effect of Abrasives in Friction Materials Composites." Proceedings of Symposium A4 on Composite Materials of the International Conference on Advanced Materials. Elsevier, Amsterdam.

7-13. Fan, Y. et al. 2008. "Role of Al2O3 in Semi-Metallic Friction Materials and Its Effects on Friction and Wear Performance." *Tribology Transaction* 51: 771-778.

7-14. Avient, B. et al. 1960. "An Experimental Study of Friction and Wear During Abrasion of Metals." Proceedings of the Royal Society of London, A258, pp.159-180.

7-15. Sin, H. et al. 1979. "Abrasive Wear Mechanisms and the Grit Size Effect." Wear 55: 163-190.

7-16. Campbell, M. et al. 1966. "Solid Lubricants." SP-5059. Washington, DC: National Aeronautics and Space Administration.

7-17. Centers, P. 1988. "The Role of Oxide and Sulfide Additions in Solid Lubricant Compacts." *Tribology Transactions* 31: 149-156.

7-18. U.S. Army Corps of Engineers. 1999. "Lubricants and Hydraulic Fluids." USACE Manual 1110-2-1424. Washington, DC: U.S. Army Corps of Engineers.

7-19. Savage, R. 1948. "Graphite Lubrication." *Journal of Applied Physics* 19: 1-10.

7-20. Deacon, R., and J. Goodman. 1958. "Lubrication by Lamellar Solids." Proceedings of the Royal Society of London, A 243, pp.464-482.

7-21. Bowden, F. 1951. "The Influence of Surface Films on the Friction, Adhesion, and Surface Damage of Solids." *Annals New York Academy of Sciences* 53: 805-823.

7-22. Seabury, R., and L. Murray. 1934. Brake Lining Composition. United States Patent No. 1969041, filed May 29, 1930, and issued August 7, 1934.

7-23. Cho, M. et al. 2001. "The Role of Raw Material Ingredients of Brake Linings on the Formation of Transfer Film and Friction Characteristics." Technical Paper No. 2001-01-3130. Warrendale, PA: Society of Automotive Engineers.

7-24. Lee, J., and H. Jang. 1998. "Effect of Different Solid Lubricants in the Automotive Friction Material On Friction Characteristics." *KSTLE International Journal* 14: 17-23 (Korean).

7-25. Kim, S., and H. Jang. "Effects of the Relative Amounts of Graphite and Antimony Trisulfide (Sb2S3) on Brake Performance of Non-asbestos Organic (NAO) Type Brake Linings." *KSTLE International Journal* 17: 351-357 (Korean).

7-26. Gopal, P. et al. 1994. "Fade and Wear Characteristics of a Glass Fiber Reinforced Phenolic Friction Material." *Wear* 174: 119-127.

7-27. Ohm, R. 1990. *The Vanderbilt Rubber Handbook,* 13th ed. Nowalk, CT: R. T. Vanderbilt Company, Inc.

7-28. Pogosian, A. 1984. "Friction Transfer Film Thickness Calculation and Film Effectiveness Conditions." *Trenie i Iznos* 5: 204-211 (Russian).

7-29. Gahr, K, H. 1987. *Microstructure and Wear of Materials.* Amsterdam: Elsevier.

7-30. Kim, H., and S. Huh. 1998. "Role of Ceramic Fillers in the Friction and Wear Behaviour Of Phenolics Resin Composites." *Journal of Material Science* 17: 501-502.

7-31. Tanaka, K. et al. 1973. "Fundamental Studies on the Brake Friction of Resin-Based Friction Materials." *Wear* 23: 349-365.

7-32. Bartenev, G., and V. Lavrentev. 1981. *Friction and Wear of Polymers*. Amsterdam: Elsevier.

7-33. Bhattacharya, S. 1986. *Metal-Filled Polymers: Properties and Applications*. New York: Marcel Dekker.

7-34. Cramer, W., and I. Silver. 1951. "Acoustical Properties of Rubber as a Function of Chemical Composition." NAVORD Report # 1778. White Oak, MD: U.S. Naval Ordinance Laboratory.

7-35. Mizutani, Y. et al.1973. "X-ray Study of Friction And Wear of Resin-Bonded Asbestos." *Wear* 23: 387-392.

Chapter 8
Early Organic Binder Systems

I was reading a book...'the history of glue' - I could not put it down.

—Tom Vine

BINDERS ARE THE GLUE THAT UNITES the individual components together and forms the structure of the composite. Commonly used binder materials in the manufacture of early tribological composites were readily available organic materials such as asphalt, drying oils, rubber compounds, and naturally occurring resins (such as those derived from cashew nut shell liquid).

Many of the binders were used as simple dispersions in a suitable solvent or for the saturation and coating of early woven friction fabrics. Advances in manufacturing soon led to blending of binder materials based on the supposition that no single material was capable of meeting the desired design requirements. Blends and the addition of additives were effective in improving areas of weakness.

Even though the use of blends provided an optimum balance of properties, it also typically introduced a series of compromises that had to be considered. Therefore, sacrifices were often made.

Binder Selection

The selection of a binder for friction materials is based on design requirements for a matrix that is capable of resisting intense mechanical forces, yet remains compliant enough to prevent brittle failure or premature fatigue. A suitable binder material must not only be tough, but must also form a heat-resistant matrix and yet remain low in cost. Often these criteria are in opposition and require some degree of compromise.

Tribologically speaking, the binder must also be a material that provides a modest effect to the coefficient of friction and exhibits resistance to wear. The matrix material should not overpower the performance attributes of the other ingredients in the composition, but should serve to augment their contributions to the total performance of the final product.

Finally, the binder material must be compatible with a wide range of materials that differ in chemistry and morphology. The necessity of compatibility insures the final composite will perform as desired. In short, binders are the single most important aspect of the composite and must be selected with the utmost attention to detail.

Binders and the Manufacturing Process

The selection of binder materials has typically been intimately tied to the manufacturing process. During the early days of the industry, the first composite friction materials were typified by woven fabric materials saturated with an organic binder solution. Other types of friction materials were based on non-woven methods, but also required impregnation with a solvated binder. Furthermore, polymer science was still in its infancy during this time and the selection of the binder was limited to naturally occurring materials with the desired solubility. Therefore, the selected binders were required to be compatible with the solvent and also to coat the fibers in the fabric easily. Both of these factors were effective limitations for considering other binder options. As manufacturers began to explore alternative processes involving rubber binder systems, which led to the development of calendaring, sheeter, and extrusion processes, the binder selection choices increased to include natural and synthetic elastomers. Like saturation, early versions of these processes relied heavily on organic solvents to improve mixing.

Other manufacturing methods required a different approach and ultimately reached a point where common manufacturing methods were based on dry mix formulations. This also allowed for the introduction of various types of molding techniques and a drive toward modern methods including the use of powdered thermosetting resins.

An excellent diagram, which illustrates the numerous manufacturing methods used to produce many types of early commercial friction materials, was recorded by Halstead in 1944 [6-23] and has been included for reference in Fig. 8.1

Asphalt and Bitumen

From a historical perspective, the progression of organic matrix forming materials begins with asphaltic bitumen. Chemically, asphalt, bitumen, and pitch are similar products. They are all derived from petroleum and are extremely complicated mixtures of hydrocarbon compounds. The composition of each individual material type is incredibly variable. The early patent literature

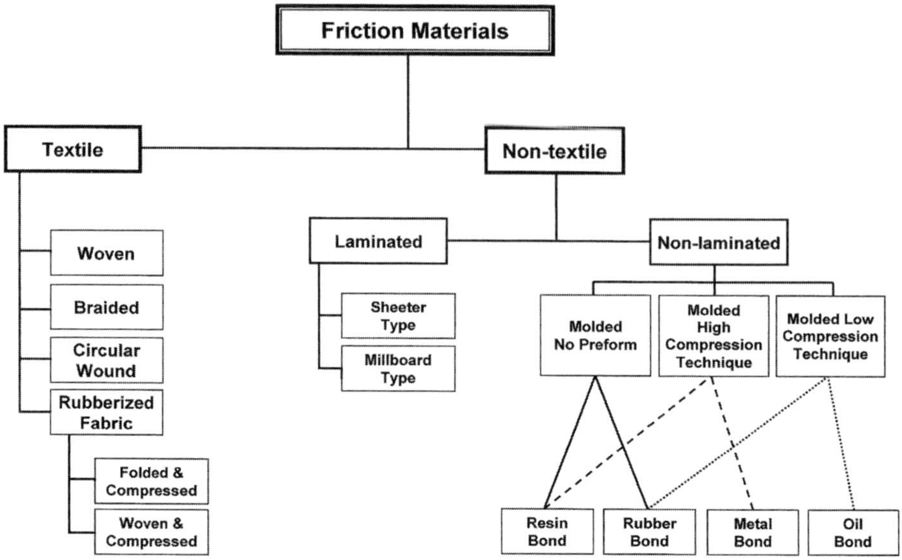

Fig. 8.1 Commercial types of friction materials from Halstead.

is filled with a host of examples that discuss the incorporation of various forms of asphaltic pitch, bitumen, or other suitable pyrobitumenous materials into friction materials.

The terminology used is often interchangeable, but according to Abraham [8-1] each of these types of materials is unique in their composition and properties. For our purpose, which is mostly historical, the terms asphalt and bitumen will encompass all members of this class of materials under an umbrella of naturally occurring petroleum-derived solids.

These products are thermoplastic and by their very nature soften and melt at elevated temperatures. In addition, asphaltic compounds have a degree of solubility in certain organic solvents such as gasoline. This allows these materials to saturate and coat the fibers of a woven fabric making it impermeable to water. Additionally, the local availability, relative ease of procurement, and modest heat resistance made asphaltic compounds ideal for use in early friction material composites.

Asphalt as a Saturant

An early example of a bituminous material used as the binder during production of tribological composite dates back to the work of Frood and Holmes [1-2]. Their friction material was comprised of layers of fabric bound together with a "cementitious substance." The cement was made from a heated mixture of resin pitch, a plant-derived resinous material now known as dammar gum, and recycled "India rubber." The fabric, after saturation, was arranged in layers and compressed together to form them into a block of material. The laminated block was subsequently machined into the final geometry of the brake block.

Asphalt and bitumen were also beneficial for improving the friction output of the composite. Bonner [8-2] used a woven asbestos fabric that had been "frictioned" with an asphaltic binder. Bonner's method involved heating the asphalt with a stream of oxygen bubbling through it. This oxidation process elevated the boiling point and improved the thermal resistance of the binder, which was expected to provide better performance at higher temperatures. The fabric layers were then impregnated and stitched together to form a product as shown in Fig. 8-2.

Bonner points out that asphalt and bitumen binder materials provided a number of benefits that were considered "valuable properties for the present purposes." These included the ability of the lining to yield and bend, which was a "very convenient property" during the assembly process. The softness of the lining also allowed the lining to conform to variations in the surface and enabled better engagement of the brake during use. Because of the thermoplastic nature of asphalt, a tertiary benefit took place during use because of melting as asphaltic binders would lubricate the surfaces.

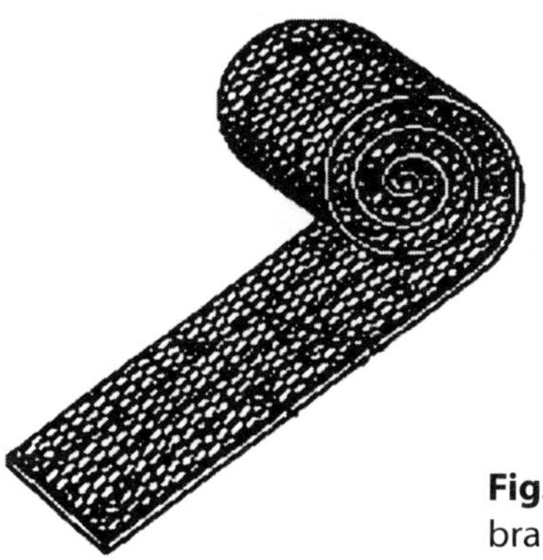

Fig. 8.2 Stitch-bonded asbestos brake liners saturated with asphalt.

Gilsonite

One of the most successful and longest running forms of commercially available asphaltic materials is Gilsonite. Gilsonite was a registered trademark of the American Gilsonite Company [8-3]. Its commercial use dates back to 1886. Chemically, Gilsonite is a naturally occurring hydrocarbon resin known as uintaite found in the Uinta Mountains of northeastern Utah.

Physically, uintaite is a brittle black solid that is soluble in many organic solvents and can be easily crushed into granular powder. One of the first uses of uintaite in a friction material was recorded in 1911 by the Asbestos Brake Company [8-4]. Their process involved dissolving the hydrocarbon resin in a solvent, such as naphtha, to produce a saturant for asbestos fabric. The saturated material was then allowed to sit until the solvent evaporated, leaving the uintaite dispersed throughout the material, forming a hydrocarbon matrix. The importance of uintaite was the development of a higher degree of friction in the asbestos-based linings and the provision of water resistance to the final product. This practice continued with uintaite as the primary binder for a number of years.

In solid form, uintaite can be added to a friction material as an organic filler material. More recent examples from the patent literature [8-5, 8-6] list the addition of Gilsonite as organic filler, but do not specifically mention the benefits gained by its use in these compositions.

Modern Use of Asphalt

Asphaltic compounds have virtually been eliminated from consideration as the primary matrix for use in clutches and brake linings. This can be attributed to changes in the manufacturing process and because of the low thermal resistance of naturally occurring thermoplastics. The softening of the matrix at elevated temperatures limits their functionality at the higher operating temperatures of today's automotive world.

There may still be a potential use for asphalt and bitumen in low-energy friction applications for industrial markets or possibly as a carbon precursor in the production of carbon-carbon composites. Generally, however, other chemical compounds have replaced asphaltic bitumen completely.

Drying Oils

Drying oils are another example of a classical misnomer that is prevalent in the friction materials industry. A drying oil is typically obtained from plant sources and will harden with time to form a tough film in the presence of oxygen. The "drying action" is not a form of evaporation [8-7] but is a chemical reaction, namely oxidation, which occurs between atmospheric oxygen and the double bonds present in the molecular backbone. The oxidation reactions produce ether linkages between the molecular chains and, in the presence of certain metallic compounds that were given the common name dryers, the oxidation process can be accelerated. A simple reaction diagram showing this process is provided in Fig. 8.3.

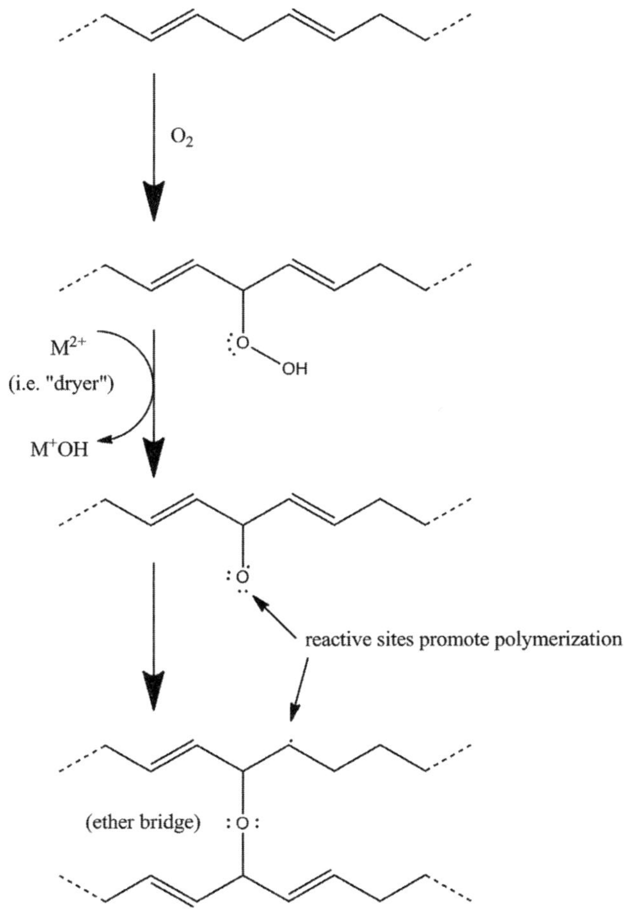

Fig. 8.3 Oxidation of drying oils.

The drying reaction begins when oxygen reacts with the molecular chain forming a hydroperoxide group. In the presence of a metallic ion, the hydroxyl ion will be removed from the hydroperoxide, forming a highly reactive oxygen species. This will react with an adjacent carbon-carbon double bond from a different molecule, forming an ether linkage that ties two molecules together with a loose ladder-type structure (similar in concept to the process of vulcanization). As the process continues, the molecular weight will increase with a concurrent increase in viscosity and the oil will ultimately become a solid material.

Iodine Number and Drying Oil

The susceptibility of a drying oil to oxidation is based on the unique chemistry of these naturally occurring plant oils. The plant oils are rich in mixtures of glycerol esters of fatty acids and often have high levels of polyunsaturated fatty acids such as linolenic or linoleic acid. The amount of unsaturation present in these various molecular compounds can be measured by their ability to react with iodine, which has lead to the development of the iodine number as an analytical method used as a quality control indicator [8-8].

This test allows for a basic classification of oils based on their reactivity. Generally, oils with iodine numbers in excess of 130 are considered drying oils. Oils with iodine number of 115–130 are semi-drying and those with iodine numbers less than 115 are typically non-drying. In many cases, the oils can vary within a range of iodine values based on the natural variation in the type and concentration of the unsaturated molecules present.

Drying Oils in Friction

The producers of friction materials used drying oils in much the same manner as asphalt. Oils were mixed with an appropriate solvent and used to impregnate the woven fabric. The oils would harden upon drying, but remain flexible enough to curve onto a brake shoe or a brake band. To enhance the properties of the oil, many producers used different means for pre-oxidizing the oil—a process that has become known as bodying or blowing oil.

Seigle [8-9] "bodied" oil by heating with a flow of oxygen bubbling through. Seigle's process was similar to that of Bonner [8-2] in many aspects and his "bodying process" was also intended to enhance the thermal properties of the oil prior to saturation.

The use of bodied or blown oils continued throughout much of the industry for many years and continues today. However, even with drying oils that had been pre-oxidized to fully reach the desired strength, the oils require additional drying after saturation. This step is necessary to ensure that the friction facings are cured sufficiently, which was most often determined by the product reaching the desired hardness level. The simplest curing method was to bake the product, which usually involved long, slow bake cycles.

As an example, Raybestos-Manhattan [5-21] developed a paper-based friction material comprised of 60% asbestos with small amounts of cellulose pulp. Cellulose pulp had been added to aid in the formation of the web during the papermaking process and was considered a processing aid. After the paper was formed and dried, it was saturated with a drying oil and cured in a process that required 30 hours to reach the desired level of hardness. Other examples record cure times of as little as 8 to 15 hours [8-10] and others show cure times approaching 48 hours [8-11].

The duration of the drying process was lengthy in many cases because the oils reacted with oxygen molecules closest to the surface and formed a film that covered the body of the composite. This film served to reduce the permeability of the composite and effectively inhibited further oxidation of the oil located near the center of the friction lining. Since the final brake or clutch facing required a thermally resistant material, complete oxidation of the oil was necessary and often resulted in long cure cycles.

This was a significant technical hurdle that had to be overcome. In an effort to improve the saturation and the drying processes, organizations used a number of methods including vacuum saturation techniques [5-25]. The most common method involved the addition of drying agents (i.e., oxidation catalysts) to the saturant to enhance the depth and degree of oxidation.

Drying Agents (aka Dryers)

Drying agents, commonly known as dryers, were chemicals added to promote oxidation of the oil. Many of these additives were based on compounds of nickel [8-9], lead [6-53], or cobalt [8-12]. Often metallic dryers were metallic oxides added during various stages of the process. They were considered to be an absolutely essential ingredient for reducing drying times to a more manageable level.

In many of the references, the use of drying agents was limited to a concentration of about 5% by weight of the oil and varied depending on the process and the type of oil.

Sulfurized Drying Oils

One interesting diversion from the classical approach of using a drying oil with a drying agent was the development of the sulfurized oil. The first example of this can be found in 1928 by Laughlin [8-13] who used a complicated mixture of rubber, linseed oil, rosin, and wax to produce a brake lining composition. During the various stages of the process, sulfur was added and heated until the odor of sulfur dioxide was no longer detected. This was an indication that a chemical reaction between sulfur and sites of unsaturation along the molecular backbone of the oil had taken place. Sulfur formed a cross-linked structure with the oil molecules that increased the viscosity of the oil and prevented oxidative hardening.

Another example of the sulfurized oil concept was developed by workers at World Bestos [8-12]. Nanfeldt records that brake linings were manufactured with frictioning compounds to make the lining "sticky" during operation, which provided resistance to sliding motion. His version of the sulfurized oil was produced by heating tung oil in the presence of elemental sulfur to a temperature of 392°F (200°C) until the solution darkened. The concentration of the sulfur was limited to about 5% by weight of the mixture and was expected to produce sulfur-based cross links between the chains in a manner similar to the vulcanization of rubber. Like Laughlin, the sulfurized oil frictioning agent was also reported to prevent oxidative hardening of the linings, allowing them to remain flexible for a longer period of time.

Common Drying Oils

As mentioned, oils with higher iodine values were usually considered better drying oils. In theory, any oil with sufficient sites that are unsaturated (i.e., a large iodine number) can be used. However, the majority of the literature citations seem to prefer tung oil and linseed oil, both of which are highly unsaturated plant-derived oils.

Tung Oil

Tung oil is extracted from the seed of the tung tree (Vernicia fordii) which is indigenous to southern China and is therefore also known as China Wood oil. It was widely used and is cited in many of the early literature sources. It has one of the highest iodine values and is considered one of the best drying oils available. Tung oil is also a potential physical property modifier for phenolic resins.

Linseed Oil

Linseed oil is extracted from the seeds of Linum usitatissimum and is also known as flax seed oil. It is also a very common choice for friction material producers and has a high iodine value. Linseed oil is still in use today, but typically only as a modifier to phenolic resins. The use of linseed oil as a primary binder system in friction material manufacture has been virtually eliminated from consideration because of poor thermal resistance.

Other Oils

Other oils including perilla oil, oiticica oil, and fish oils (such as menhaden oil) have all been used in the production of friction materials [8-14]. However, as the commercial availability of synthetic resins and elastomeric compounds increased, the use of mixtures including phenolic resins, rubber compounds, and drying oils become more common in practice. In addition to blending oils with resins and elastomers, it was also common to find drying oils used with asphalt.

Modern Use of Drying Oils

Drying oils are rarely used as the primary binder for the production of modern friction material composites. Oils are still used, but at very low concentrations. They are necessary to modify the physical properties by acting as a plasticizer. Drying oils are most commonly found in flexible linings such as those on band brakes or drum brakes. These are usually blended with more thermally resistant phenolic resins. They were at one time known as "bending resins" to facilitate bending of warm slabs of friction material in the heat and curve process [5-29].

Aside from phenolic resins, drying oils can also be mixed with rubber compounds to supply selected performance characteristics. However, the oil and elastomer must be extensively tested beforehand to determine compatibility between the constituents.

Despite the rich history and diversity in the use of drying oils, they are expected to see a gradual decline in consumption because of poor thermal resistance. In areas where imparting flexibility is desired, the use of oils has also diminished and, because of the use of more thermally stable elastomeric materials, drying oils been removed from future consideration.

Rubber-Based Binders

Rubber-based friction materials were first introduced in 1898 when McTaggart [8-15] developed a spoon brake using a simple rubber friction material as the primary friction generating material. The rubber compound was forced into contact against the tire of a bicycle resulting in the necessary retardation and braking. From this point onward, rubber and rubber-modified binder systems have seen continual service in the friction material industry and remain in widespread use in modern friction materials.

Rubberized Fabrics

Rubber as a saturant for fabric was the method used by Frood and Holmes [1-2], Bonner [8-16], and Abert [8-17]. These are only a few of the number of examples from prior art where rubber was the matrix surrounding a fabric of woven asbestos. In most cases, these methods share many similarities in manufacturing and in performance.

One difference of merit was the work of Abert, which illustrates a unique method using a woven asbestos fabric saturated with a rubber solution and bonded to a strong heat-resistant carrier produced out of millboard asbestos to form an early dual layer friction composite. The benefit of the compliant woven layer on a rigid heat-resistant layer was that it formed an ideal friction material for early annular brake and clutch designs by improving the engagement properties. The blend of materials was also well-suited for use in the higher energy applications operating during this time.

Rubber Binders in Other Manufacturing Processes

Watson and Redfield [8-18] reported the development of a hard rubber friction compound using smoked rubber as the base. The conventional vulcanization agents of the time such as zinc oxide and sulfur were added to this compound. An additional ingredient was Jeffersite, a form of vermiculite added to improve wear. Cotton fibers were also added to impart strength. This combination of materials was milled for one hour, allowed to sit for twenty-four hours, and then processed into the friction material by calendaring. Along the same lines as calendaring were a number of illustrations regarding the sheeter process and rolling of continuous strips of friction material, which was also known as continuous molding. All of these were reliant on rubber compounds as the primary matrix material.

The Rolling Mill

The rolling process was patented by workers at Raybestos-Manhattan [8-19] and made use of combinations of rubber and asbestos fiber. A generalized composition was composed of a mixture of rubber and asbestos with sulfur and additives, such as litharge, barytes, and magnesium carbonate. These were mixed in solvent to form a blend with the consistency of thick dough. This was fed into the nips of the roller and formed into continuous strips of lining in a number of widths. The commercial product was partially vulcanized and supplied in coils of a given length. This was sold to repair shops, freeing shop owners from having to carry an extensive inventory of replacement brake linings.

Modern Rubber Binder Systems

Today, rubber-based friction material compounds are produced in many of the same ways discussed previously, while others have followed a parallel path to that of the tire industry. These products are calendared, folded and compressed, extruded, and molded. To enhance the products, the rubber component can be modified with other polymeric materials or it can actually be a modifier for a different organic matrix material.

A modern example by Pinkhasovich [5-26] presented information about the development of a friction material fabric from fibers of glass, basalt, and copper. The fibers were woven into a fabric and saturated with a rubber modified phenolic resin to produce a friction element for use in brakes. This process relied on technology that is more than a century old. However, advances in fiber science and polymer technology have resulted in a new product within the class of rubber bonded woven materials.

In the rail industry, traditional friction materials have historically been cast iron shoes, but the development of composite shoes has been growing progressively for a number of decades. These are generally made with a rubber-based binder system and fall within the molded friction material group. One example was patented by Ferodo [8-20] and made use of high hysteresis rubber compounds.

This specific binder system employed a blend of SBR rubber compounds with differing amounts of bound styrene. A small amount of a resole phenolic resin was added to this blend to impart rigidity. The remainder of the composition was composed of blends of metallic and inorganic fibers, mineral fillers, and friction modifiers. The final mixture was molded to form a brake shoe with an exceptional set of performance benefits compared to the existing shoe brake materials. This brake show derived its unique enhancements from the blending of elastomeric compounds.

Applicable to High Static Friction Uses

Another aspect of rubber-based friction materials is their ability to generate a high static coefficient of friction that is greatly sought after in holding or parking brakes. Siemon and Meier [8-21] developed a very high friction formulation using an elastomer compound that was ideally suited for static braking applications. The composition was a blend of W and WHV grades of Neoprene rubber. This was reinforced with carbon black and employed a curing system that relied on sulfur and zinc oxide. The final product was able to supply extremely high levels of static friction for use in the defense industry.

Future Application of Rubber Binder Systems

Despite the benefits of elastomeric friction materials, their use is limited to a select number of applications. For the most part, rubber-bound friction materials in automotive applications have virtually disappeared and have been replaced by the more thermally resistant phenolic resins. However, rubber remains prevalent as an ingredient in friction materials, especially in modern disc brake systems that typically use them in the form of granulated powder. Granulated powder, known as rubber crumb or tire peel, can be obtained as a recycled material produced by grinding tire or other rubber compounds. It is discussed further in chapter 19 of this book.

Rubber is also used as a modifier in phenolic resin systems to toughen the matrix. It can be chemically reacted with resin, or simply mixed into the compound by the designer. Each approach has its own drawbacks. As a modifier, the various forms of rubber powder are expected to have a significant role in friction material development during the next few years and should see the potential for sustained growth.

Cashew Nut Shell Liquid (CNSL)

Cashew nut shell liquid (CNSL) is a viscous amber fluid obtained from the cashew tree (Anacardium occidentale). The tree bears fruit, known as an apple, which is unique because the seed (or nut) is located outside of the fruit. The seed has an outer leathery skin and a hard inner shell. Inside the shell is the edible portion that is highly prized for its rich flavor. Between the skin and the hard inner shell is a porous honeycomb structure that contains CNSL.

CNSL is considered to be inedible because of its toxicity [8-22], but is considered highly valuable because of its moderately high iodine value. This is the result of several unsaturated meta-substituted phenolic compounds present in the fluid.

The naturally occurring phenolic compounds are generally extracted by heating [8-23] until the liquid is expelled from the leathery skin and collected. More recent sophisticated extraction methods include solvents and the use of supercritical fluids [8-24, 8-25]. These methods have increased the extraction yields to very promising levels. Many examples are available in the prior art.

Chemistry of CNSL

Raw CNSL has been chemically examined and analyzed [8-26]. CNSL is typically stated to be a mixture of compounds containing cardol, anacardic acid, cardanol, and 2-methylcardol. The specific constituents in raw CNSL and variants of structural isomers of the meta side chain were reported by Rodrigues' group [8-27] and are shown in Fig. 8.4.

Each of these constituents will vary in concentration depending on the source and extraction method. In addition, each compound has a meta-substituted side chain that is fifteen carbon atoms in length and possesses several sites of unsaturation that can also vary. The large side chain and presence of the unsaturated bonds give CNSL a distinctive chemical behavior that is ideal when used in friction materials.

Uses of CNSL

Harvey [8-28], who spent a great deal of his life studying CNSL, developed one of its first commercial uses: an electrical insulation based on mixtures of rubber and CNSL. CNSL was added to convey heat resistance to the rubber compound and eventually led to the development of a varnish suited for windings in electrical motor manufacture.

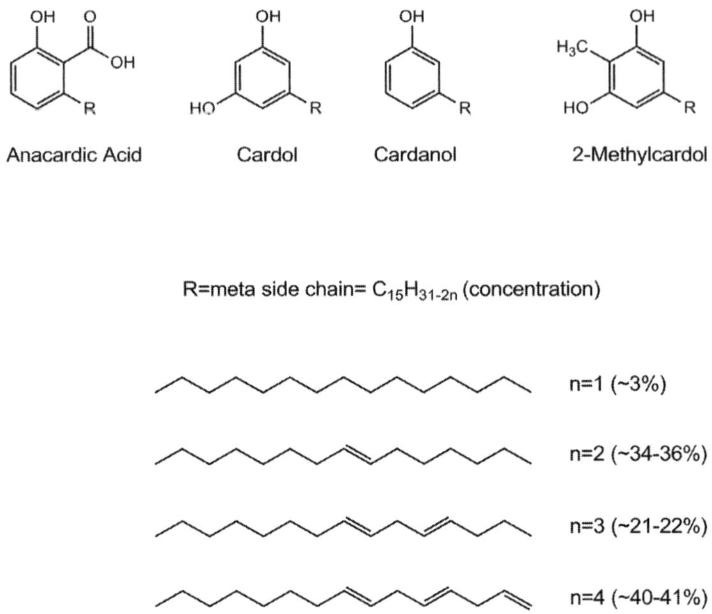

Fig. 8.4 Structure of the constituents of cashew nut shell liquid.

During the following years, Harvey generated a number of patents involving a host of potential applications for further uses of CNSL. One of these products was a granulated form of polymerized CNSL [8-29] produced by reacting CNSL with an aldehyde such as formaldehyde, and baking to form a cured polymer. The cured polymer was subsequently ground into a fine flour or dust, which is now known commonly as CNSL dust, or sometimes as friction dust. Friction dust is organic filler and will be examined more in chapter 19 of this book, but the role of CNSL was vital to the early friction material industry in a number of other ways.

CNSL as a Binder

The concept of reacting CNSL with an aldehyde to form a phenolic resin was also applied to the manufacture of brake linings as the primary binder. Thompson [8-30] mixed CNSL and hexamethylenetetramine (HMTA) to form brake linings by molding the compositions onto a thin strip of expanded metal. The CNSL acted as the phenolic portion of the resin and was condensed to form a flexible polymeric matrix. The action of the formaldehyde and ammonia generated by the decomposition of the HMTA during molding operations converted the thermoplastic CNSL into a pliable thermosetting compound. Within a few short years, polymerized forms of CNSL were being used by a number of other groups including Bruce and Halstead [8-31], as well as in blends.

Many groups were apparently not completely satisfied with the benefits provided by polymerized CNSL and resorted to using polymer blends. Spokes [8-32] used a cresylic acid CNSL reaction product that was miscible with rubber to form an asbestos brake lining. It is interesting to note that the product also contained thermoplastic pitch. This example blends aspects of asphalt, rubber, and CNSL with reactions between phenolic members to form a rather complicated binder system.

Spokes and Keller [8-33] used a combination of polymerized linseed oil, CNSL, phenolic resin, and a copolymer of acrylonitrile and butadiene to form a very complicated binder system. In addition, the product was cured with sulfur similar to vulcanization. This sulfur cure was necessary to ensure that the product achieved the desired results.

Finally, Harvey [8-34] also used a mixture of drying oils and CNSL, but developed an oil-modified friction particle for use as an organic additive. This product had some properties of a soft compliant drying oil and the improved thermal resistance of a phenolic resin.

Modern Use of CNSL

In today's modern tribological composites, CNSL is typically limited to the role of a physical property modifier. However, one interesting use for CNSL was reported by workers at Akebono Brake Industries [8-35]. They employed an organic dust, possibly polymerized CNSL, which had been coated with a highly viscous material in their friction material compositions. The stated improvements given by the use of a coated particle were better tribological performance and improved strength of the product. One of the many recommended highly viscous materials in their list contains CNSL as one of the preferred coating materials, suggesting the use of a liquid CNSL coating on a granulated CNSL particle.

Whether CNSL is used as a primary binder, secondary binder, physical property modifier, coating, or polymerized particle, it is a valuable additive that enhances the structural and tribological properties of many friction materials. It is expected to see continued use in the industry and, with further advancements in polymer science, may lead to many promising avenues for future growth.

References

8-1. Abraham, H. 1938. *Asphalts and Allied Substances*, 4th ed. New York: D. van Nostrand Company.

8-2. Bonner, W. 1911. Brake and brake-liner. United States Patent No. 980481, filed July 28, 1909, and issued January 3, 1911.

8-3. Gilsonite. Registered trademark of the American Gilsonite Company, Salt Lake City, Utah, Registration No. 0144545, July 12, 1921 (Trademark expired).

8-4. Hoff. L., 1911. Asbestos brake-band. United States Patent No. 1012356, filed November 5, 1909, and issued December 19, 1911.

8-5. Johnson, H. 1972. Friction lining composition. United States Patent No. 3684062, filed January 29, 1971, and issued August 15, 1972.

8-6. Adamczak, L. et al. 2000. Friction liner for a dry friction device, a method of making such a liner, and a clutch friction wheel having such a liner. United States Patent No. 6056100, filed March 5, 1998, and issued May 2, 2000.

8-7. Bailey, A. 1985. *Industrial Oil and Fat Products*. New York: Wiley-Interscience.

8-8. ASTM International. 2003. "Standard Test Method for Determination of Iodine Value of Tall Oil Fatty Acids." ASTM D5768-02 (2006). West Conshohocken, PA: ASTM International, PA. doi: 10.1520/C0033-03, www.astm.org.

8-9. Seigle, W. 1921. Method of impregnating porous or bibulous materials and the product thereof. United States Patent No. 1395744, filed June 28, 1920, and issued November 1, 1921.

8-10. Blume, W. 1932. Friction elements and method of making. United States Patent No. 1866287, filed October 2, 1929, and issued July 5, 1932.

8-11. Kirschenbaum, L. 1931.Process of making asbestos friction elements. United States Patent No. 1810714, filed September 10, 1920, and issued June 16, 1931.

8-12. Nanfeldt, W. 1931. Transmission lining. United States Patent No. 1810723, filed November 21, 1928, and issued June 16, 1931.

8-13. Laughlin, R. 1928. Brake lining composition. United States Patent No. 1693394, filed November 3, 1924, and issued November 27, 1928.

8-14. Grazen, F. et al. 1976. Friction particle for brake lining. United States Patent No. 3966670, filed December 23, 1974, and issued June 29, 1976.

8-15. McTaggart, W. 1898. A cycle brake spoon. Great Britain Patent No. 189730771, filed December 30, 1897, and issued November 5, 1898.

8-16. Bonner, W. 1909. Friction member. United States Patent No. 931988, filed January 20, 1909, and issued August 24, 1909.

8-17. Abert, H. 1934. Friction facing. United States Patent No. 1941872, filed December 31, 1931, and issued January 2, 1934.

8-18. Watson, J., and C. Redfield. Friction member. United States Patent No. 1845858, filed July 14, 1930, and issued February 16, 1932.

8-19. Abert, H. et al. 1933. Molded brake lining. United States Patent No. 1932919, filed July 8, 1929, and issued October 31, 1933.

8-20. Loftipour, M.1994. Brake pad friction linings. European Patent No. 578668, filed April 6, 1991, and issued January 19, 1994.

8-21. Siemon, J., and J. Meier. 1992. Very high friction elastomer formulation for use in static braking applications. United States Patent No. 5093388, filed March 28, 1989, and issued May 3, 1992.

8-22. Akinhanmi, T., al. 2008. "Chemical Composition and Physiochemical Properties of Cashew Nut (Anacardium Occidentale) Oil and Cashew Nut Shell Liquid." *Science and Technology/Journal of Agriculture and Food Chemistry* 2: 1-10.

8-23. Hughes, E. 1930. Method for removing oil from the shell of the cashew nut. United States Patent No. 1784628, filed November 29, 1926, and issued December 9, 1930.

8-24. Egoda, M. et al. 2006. "Extraction of Polyphenols from Cashew Nut Shell." *Leonardo Electronic Journal of Practices and Technologies* 9: 107-112.

8-25. Patel, R. et al. 2006. "Extraction of Cashew (Anacardium Occidentale) Nut Shell Liquid Using Supercritical Carbon Dioxide." *Bioresource Technology* 97:847-853.

8-26. Baillie, C., 2005. *Green Composites: Polymer Composites and the Environment*. Cambridge: Woodhead Publishing Ltd.

8-27. Rodrigues, F. et al. 2006. "Antioxidant Activity of Cashew Nut Shell Liquid (CNSL) Derivatives on the Thermal Oxidation of Synthetic cis-1,4-polyisoprene." *Journal of Brazilian Chemistry Society* 17 (2).

8-28. Harvey, M. 1931. Compositions of materials and method for producing them. United States Patent No. 1819416, filed August 8, 1928, and issued August 18, 1931.

8-29. Harvey, M. Compositions of matter and methods and steps of making and using the same. United States Patent No. 2165140, filed October 25, 1935, and issued July 4, 1939.

8-30. Thompson, J. 1931. Method of making friction blocks. United States Patent No. 1821487, filed December 3, 1923, and issued September 1, 1931.

8-31. Bruce, D., and R. Halstead. Friction material. United States Patent No 2227424, filed October 25, 1935, and issued January 7, 1941.

8-32. Spokes, R. 1937. Friction element and bond therefor. United States Patent No 2078617, filed February 8, 1934, and issued April 27, 1937.

8-33. Spokes, R., and E. Keller. 1947. Friction elements bonded with a reaction product of polymerized linseed oil, sulfur, and phenol-aldehyde resin mixed with a butadiene-acrylonitrile copolymer. United States Patent No 2428299, filed November 16, 1942, and issued September 30, 1947.

8-34. Harvey, M. 1943. Compositions of matter and methods and steps of making and using the same. United States Patent No 2327785, filed May 13, 1939, and issued August 24, 1943.

8-35. Saito, N., and H. Kobayashi. 1993. Frictional material reduced in segregation of raw material. Japanese Patent No. 05-156235, filed November 29, 1991, and issued June 22, 1993.

Chapter 9
Phenolic Resins

The atom being for all practical purposes the stable unit of the physical plane is a constantly changing vortex of reactions.
—Kabbalah

THE BACKBONE OF MANY MODERN FRICTION materials is a rigid thermally resistant family of reactive polymers commonly known as phenolic resins. This class of polymers has been universally accepted as the binder of choice by the vast majority of friction material manufacturers around the world. Its prominence is based on a long history of success provided by the innate benefits and desirable attributes of phenolic resins.

In addition, phenolic resins are globally available from a number of commercial suppliers and can be obtained in either liquid or solid form. Furthermore, many of the commercial grades are obtainable over a large range of molecular weight distributions with a multitude of different curing profiles. Phenolic resins are also generally viewed as being moderately easy to process and to provide the finished tribological composite with a rigid polymeric matrix of high compressive strength.

Phenolic Resins are Tunable

Many commercial phenolic resins can be tuned by a number of different modification methods, and either by chemical or physical means, which permits further enhancement of selected properties. Modifiers can be added to convey higher degrees of thermal resistance by promoting higher char yields (e.g., carbonaceous residue after exposure to elevated temperatures). They can also include additives that improve the flexibility of the final product by reducing the brittle nature of amorphous resins in the cured state. In rare cases, both areas of flexibility and thermal resistance can be enhanced simultaneously by the selection of suitable modifiers.

Specific modifications to phenolic resins will be discussed in more detail later, but the addition of a modifying agent can have a number of ancillary effects that must be considered carefully during the design phase of the project. For instance, the cost of a modified phenolic resin is dependent on the type and concentration of any property modifiers included in the resin. Modified phenolic resins are usually more expensive than unmodified phenolic resins. It is commonly anticipated that benefits provided by modification will offset the higher cost by bringing value to the final product; however, many of the highly desired attributes can also bring sacrifices in other areas of product performance.

A chemical modifier can influence processing characteristics of the resin. This can include changes to the rheology of the resin during its molten phase, which can affect physical properties by preventing the resin from effectively coating and binding the constituents together into a solid mass. Other forms of resin modification, such as those intended to improve flexibility, typically alter the compressibility or compliance of the final composite. This physical response has a number of favorable outcomes, especially in the wet clutch market segment, but can also introduce poor wear resistance by allowing the composite to erode rapidly during abrasive wear regimes.

All of the methods employed in the modification of phenolic resins allow for a potentially infinite number of combinations and permutations for the composite designer. The different options provide a wealth of unique opportunities that can be utilized during the design process. Therefore, it is critically important that the designer be equipped with a basic understanding of the history and chemistry of phenolic resins to effectively foresee the impact of potential modifications.

History of Phenolic Resins

Chemists began to explore phenolic resins as early as 1872 when Bayer [9-1] used acid catalyzed reactions between phenol and members of the aldehyde family of chemical compounds. Bayer noted that formaldehyde, the simplest aldehyde, reacted with phenol to form a clear material, but the reaction product was not a crystalline compound.

It should be noted that chemists had little interest in polymers at this time and were only considered successful if they were able to isolate, purify, and classify new chemical substances. In addition, the reaction between phenol and formaldehyde was difficult to control and an attempt to isolate a single pure crystalline substance from the mixture of reaction products was not successful.

Work was not abandoned and Carswell [5-30] tells us that over the next thirty years, various researchers studied reactions of phenol in an effort to produce synthetic forms of shellac. Other efforts involved attempts at molding early phenolic materials to form solid bodies. These attempts were generally considered failures because of poor strength resulting from the high internal porosity generated by entrapped water vapor. The resolution to the manufacture and the molding problem was twofold, and a patent outlining the solution was applied for by Leo Baekeland [9-2] in 1907.

One aspect of Baekeland's answer to the molding problem involved incorporation of filler materials such as asbestos, wood fiber, or barium sulfate into the resin. The second key was the use of molding conditions where the plastic article was subjected to high pressure at elevated temperature. The addition of fillers imparted strength to the resin and the high processing pressure was necessary to force the moisture out of the product. The combination of these insured compaction of the mixture into a rigid non-porous body.

Baekeland's early attempts used molding times generally on the order of one to two hours, but stated that these could be reduced by using higher molding temperatures or chemical accelerants such as zinc chloride. Today the benefits of higher molding temperatures and chemical accelerants are still being used to improve processing and reduce manufacturing costs. Baekeland's efforts soon spawned Bakelite, the first synthetic plastic material. Bakelite has excellent thermal and electrical resistance properties, which quickly became a commercial success in a number of consumer industries. To fully appreciate the widespread success of Bakelite, and to aid in the proper application of phenolic resins in tribological composites, the chemistry of phenolic resins must be understood at a fundamental level. There are a number of excellent resources including Carswell and Gardziella, Pilato and Knop [9-3], that discuss phenolic resins in great detail for those interested in gaining a deeper understanding.

Elementary Phenolic Resin Chemistry

The reactions that produce phenolic resins are best exemplified by illustrations with phenol (the simplest phenolic member) and formaldehyde (the simplest member of the aldehyde family of chemicals). The products of these reactions are classified as either resole or novolak resins depending on the concentration of each of the reagents and type of catalyst employed. Earlier terminology referred to these classes as "one step" and "two step" resins, where the one step resole required only heat to cure, but the two step novolak required additional curing agents and heat to fully cure. Because much of the terminology is based on the unique chemistry of

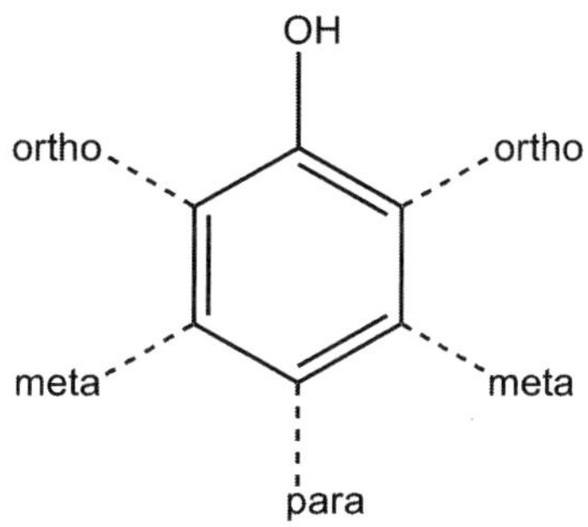

Fig. 9.1 Phenol molecule showing vacant reaction sites.

the phenol molecule, and to aid in understanding the nomenclature generally employed in the literature regarding phenolic resins, a map of the phenol molecule is presented in Fig. 9.1.

Phenol is the common name for hydroxybenzene. Its chemical structure is a benzene ring with a single hydroxyl group. Adjacent to the hydroxyl group are two open sites that are known as the ortho positions. These are symmetrically identical and have an equal probability of reacting to form connections with other molecular species. Directly opposite of the hydroxyl group is the para position. Between the ortho and para on either side of the benzene ring are two additional vacant sites, which are the meta positions. These three distinctive sets of positions have different reaction energy profiles because of subtleties in the electron configuration of the molecule. Phenol will preferentially react at the ortho and para positions. Generally speaking, the meta positions are considered to be deactivated and do not react easily.

Because phenolic resins are the product of phenol and formaldehyde, a simplified reaction diagram is shown in Fig. 9.2.

The phenol molecule is shown with dotted lines representing the available reaction sites and will be used throughout to describe the potential sites for reaction. Formaldehyde (CH_2O) reacts with

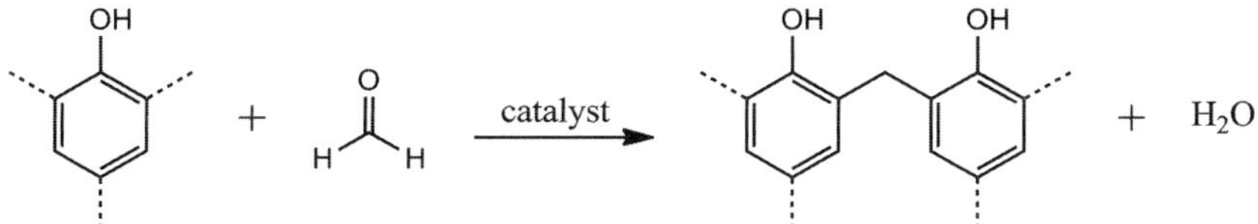

Fig. 9.2 Idealized reaction between phenol and formaldehyde.

phenol in the presence of a catalyst to form methylene bridges that connect phenol atoms via a methylene bridge and produces water (i.e., a condensation reaction). The connection shown in Fig. 9.2 is an ortho-ortho connection, where the connection is formed at the ortho sites of two phenol molecules. In reality, the reaction can form linkages between two phenol molecules that are ortho-ortho linked, ortho-para linked, and para-para linked. Further reactions between the dimeric intermediates will form more bridges, ultimately leading to the formation of a rigid three-dimensional network that gives phenolic resins the desired strength and thermal resistance that makes them ideally suited for use in friction applications.

Resole Resins

Resole resins are produced by reacting phenol with an excess of formaldehyde in basic conditions. The specific mechanism involves the formation of hydroxymethyl groups on the reactive sites of phenol, which leads to the development of resole resins. This was outlined by Gardziella, Pilato, and Knop and is considered to be the most relevant for our discussion.

The resole reaction, which is referred to as the Lederer-Manasse reaction [9-4, 9-5], begins by adding a suitable base such as sodium hydroxide to a reactor charged with phenol. The base abstracts a hydrogen atom from the hydroxyl portion of phenol, forming water and a negatively charged phenolate ion. This negative charge on the oxygen atom of the phenol molecule alters the electron configuration of the benzene ring and allows for formaldehyde to attack the benzene ring, liberating a hydrogen ion and forming a hydroxymethyl group on the ring at the ortho or para position. The phenolate ion then reacts with water, removes a hydrogen atom to convert the phenolate ion into phenol, and reforms basic ions to continue catalysis of subsequent reactions. For reference, the Lederer-Manasse reaction and subsequent conversion to the polymeric stage is shown in Fig. 9.3.

Resole resins are typically sold in liquid form and contain low molecular weight hydroxymethyl phenol species. Because of the reactive nature of resole resins, an understanding of the pH characteristics of raw materials employed in the development of the tribological composite is critical.

If the friction material mixture contains materials that are acidic in nature, or that can react to form acidic materials, the reaction

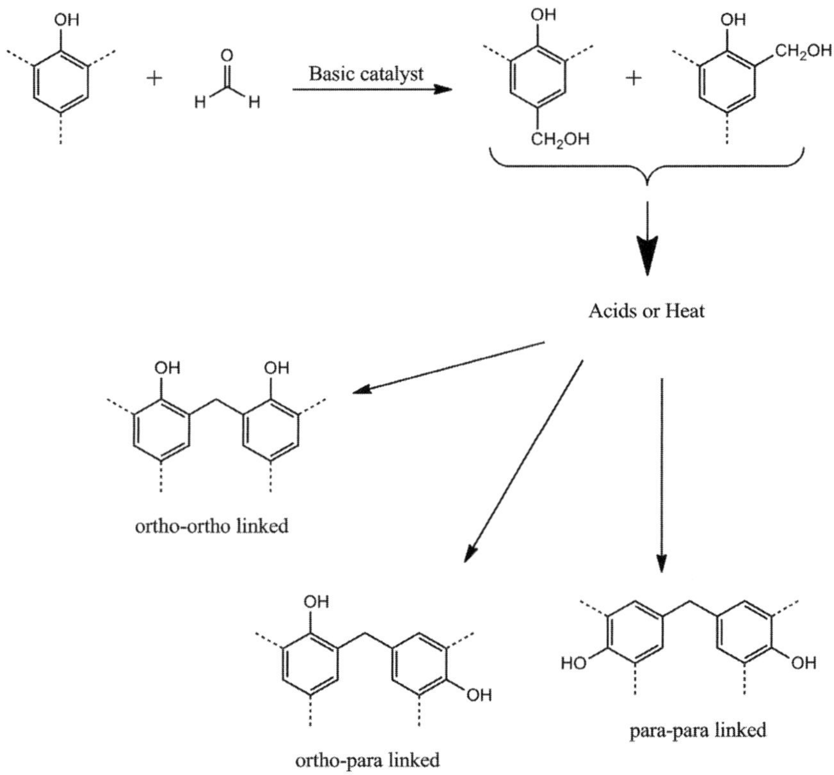

Fig. 9.3 Lederer-Manasse reaction (resole resin).

profile of resole resin may change. If the resin reaction profile changes, the cure rates, which determine the liberation of moisture, are likely to be out of sync with previously established processing conditions. This can lead to issues where the duration and number of necessary venting cycles is not sufficient to ensure gases are liberated effectively. As such, acidic raw materials can react with resole resins during processing and can result in products that are excessively porous, blistered, or cracked.

Additionally, storage conditions and shelf life considerations are another area of critical concern. Resole resins are active thermosetting resins that will react during storage, especially in the presence of heat. Therefore, they should be stored under cold conditions, and attention should be given to the use of basic inventory management techniques that employ a first in, first out approach to ensure that the oldest resins are consumed first.

Even under ideal conditions, molecular motion and the kinetic properties of the molecules will allow resole resins to react. This will produce a slow increase in molecular weight that can result in changes in the resins' solubility. Higher molecular weight species can slowly precipitate out of solution, ultimately solidifying in the bottom of the container. The use of cold storage conditions inhibits thermally activated motion and increases the shelf life dramatically.

Use of Resole Resins in the Friction Industry

Historically, resole resins were sold as alcoholic solutions based on dispersions of resin in methanol or ethanol. These resins have seen continuous service since the earliest days of the friction industry. For example, consider the work of William Achtmeyer [6-55] who patented a design for clutch facings in 1922 that is an excellent illustration for the early use of the resole class of phenolic resins.

Achtmeyer's process began by producing a relatively thick "paper" of asbestos fibers (e.g., millboard). Annular discs of the desired size were cut from the thick sheets and laminated by using phenolic resin as an adhesive. After the lamination process was complete, the disc was saturated with additional phenolic resin and cured in a stepwise fashion to temperatures of approximately 572°F (300°C). The final product was reported to possess excellent friction and wear characteristics.

The use of solvated resole resins in modern friction materials continues to thrive. However, they appear to remain limited to use as saturants in the paper friction material sector and are often modified with other materials. For instance, Mori and colleagues [6-58] taught the use of a base catalyzed resole resin in conjunction with an aminosilane coupling agent. The resole resin, in a methanol solution, along with the aminosilane coupling agent was controlled to a specific dilution value to saturate the friction material. Their paper friction material, which was based on a mixture of cellulose fiber, aramid fiber, and selected heat-resistant porous mineral filler materials, was shown to generate a positive friction response relative to velocity and to exhibit ideal compression values, both of which were highly desired in this system. It was further noted that ancillary improvements in fatigue resistance over a conventional non-modified resin was found with this type of modified resole.

Today, resole resins are produced with precise control over the reactant proportions as the reactions are usually stopped when the resin has reached the desired range of molecular weight. The resins are either dispensed as an aqueous resole, or the water is driven off by vacuum distillation to increase the viscosity. When the viscosity reaches the target level, a solvent such as ethanol can be added to produce solvated forms.

Spray-Dried Resole Resins

Resole resins, though commonly found in liquid form, can also be produced in a solid form by spray drying. Spray drying consists of spraying the liquid solution to form an atomized mist of fine droplets that are mixed with hot air, which evaporates the water. Spray-dried resole resins have not found wide use in friction composites, but research in other areas may allow for their introduction and acceptance for future use.

A quick review of the patent literature shows that Parker and Grele [9-6] used a phenolic resin dispersed in an aqueous solution to produce a novel brake lining comprised of very high concentrations of fibrous materials. Their slurry process, similar in some ways to a paper making process, would have required a resin that was compatible with excess concentrations of water. An aqueous resole would be perfectly matched for this type of wet process.

Walisser [9-7] developed a method of spray drying resole resins that allowed for ease in handling and could be used in dry mix applications. This was further expanded by Chiu [9-8] who reported a method of producing boron-modified resole resins that could also be manufactured in solid form by spray drying. It is interesting to note that boron-modified resole resins are typically expected to have higher thermal resistance than conventional unmodified phenolic resins.

Resole resins occupy a large portion of the phenolic resin marketplace. Improvements in the development of resole resins by physical property modification are enabling them to provide unique physical and chemical properties to the final tribological composite. Another area that is expected to experience growth is that of water dispersible forms of resole resins, which alleviate concerns with volatile organic solvents and eliminate the storage and handling of flammable carrier solvents.

Novolak Resins

Unlike resole resins, novolak resins are produced from an acid catalyzed reaction where phenol is used in excess relative to the concentration of formaldehyde. Early novolak resin producers employed strong mineral acids such as sulfuric acid. Today, the use of strong mineral acids has declined and the novolak resin industry generally prefers to use solid organic acids (such as oxalic acid), which are typically less reactive and do not require the same level of handling precautions associated with the use of strongly acidic solutions.

A key point, which is a result of the low formaldehyde concentration, is that novolak resins are not thermosetting, but are actually thermoplastic in nature. The thermoplastic nature of novolak resins is due to limitations in the number of methylene bridges formed during the reaction. A permanent rigid network does not form. To convert a novolak resin into a fully thermosetting resin, the resin must be processed with an additional ingredient such as hexamethylenetetramine (HMTA) that will promote the formation of additional linkages to form the rigid three-dimensional polymeric network.

The mechanism behind the formation of the novolak precursor molecules begins with protonation of methylene glycol. The positively charged glycolic ion reacts with phenol in the ortho or para positions, forming an intermediate that reacts with additional phenol present to form Bisphenol F (which is the simplest novolak molecule). An idealized reaction scheme, including the addition of HMTA to develop the final fully cured polymeric network (i.e., resite stage of the cure) is shown in Fig. 9.4.

The novolak precursor molecule shown in Fig. 9.4 is linked by ortho-ortho linkages. As with the resole, the reader is reminded that this reaction yields a final product that is actually comprised of mixtures including bisphenol units formed with para-para and ortho-para linkages. This resin is then mixed with the appropriate curing agents to form the polymer network.

Novolak Production

The production process for novolak resins is also very different from that used in the manufacture of resole resins. After the phenol and aldehyde have reacted forming the pre-polymeric material of the desired molecular weight (typically much higher than for resole

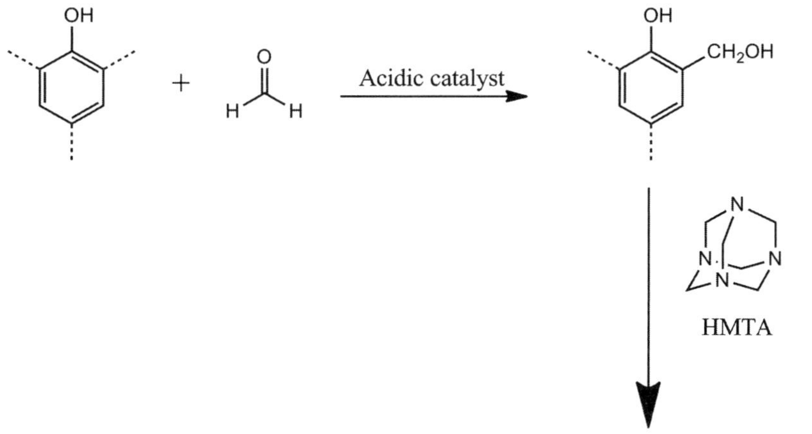

Fig. 9.4 Novolak reaction.

resins), the material is heated to distill off water. Since the polymeric precursor is thermoplastic, the heat can be increased to the reactor and combined with vacuum to reduce the water content. Molten resin is poured from the reactor onto a larger metal belt and cooled into an amorphous glassy solid. At the end of the belt, the solidified resin falls into a storage bin in the form of large flakes were it is stored until needed.

Novolak resins in flake form are usually considered a base resin that can be mixed with different additives and curatives to meet the needs of the consumer. Most novolak resins for use in tribological composites are ground into a fine powder with an appropriate amount of cross-linking agent. In some cases, the same base resin can be ordered with varying amounts of curatives to optimize the performance of the final composite.

Free Phenol Concerns

The presence of free phenol is one area of concern with novolak resins. Free phenol is defined as the concentration of un-reacted phenol remaining in the resin after the reaction is considered to be complete. The free phenol content can range from a very small fraction of a percent to higher levels (sometimes as high as 5–10%). High levels of free phenol were at one time considered to be desirable and exhibited a profound effect on the flow of novolak resins during processing. Free phenol also enabled the use of lower compaction pressures that were inherent in many of the older molding equipment designs. Additionally, high free phenol concentrations can improve wetting of the various ingredients and promote effective binding, providing the composite with high strength.

Despite these benefits, phenol is a toxic mutagen. This classification exulted in lowering of permissible exposure limits and, therefore, many of today's novolak resins are produced with free phenol concentrations below 1%. Resin manufacturers have adapted to these tighter regulations, using other methods or chemical modifiers to enhance flow and wetting characteristics.

Hexamethylenetetramine

Hexamethylenetetramine (HMTA), also known as hexamine or simply hexa, is a heterocyclic molecule and may be considered the ultimate cross-linking agent for novolak resins. It is produced by reacting ammonia and formaldehyde to form a white crystalline solid that can be added to novolak resin during the grinding process. The conversion from flake to powder with HMTA also ensures effective dispersion of the curative in the final powdered resin.

During molding, novolak powder undergoes a phase transition from solid to liquid. HMTA present in the resin begins to dissociate because of the heat present during molding. The dissociation of HMTA liberates formaldehyde and ammonia. Formaldehyde reacts with the phenolic portions of the resin precursor molecules in the presence of the ammonia (the basic catalyst) and forms the necessary methylene bridges, leading to the development of the final rigid solid.

The HMTA content of commercial novolak resins can vary greatly and depends on the desires of the consumer. Typical friction material producers can purchase a number of different resins that vary in both molecular weight and concentration of HMTA. The effect of HMTA at low concentrations, which is typically about 5–7% by weight, generates less methylene bridges than the same base resin with a higher level of HMTA and, therefore, would be expected to give a final resin with different physical attributes.

Aside from physical effects, Rogers and Long [9-9] reported that novolak resins with higher HMTA concentrations did not have as much weight loss as resins with less HMTA. This indicates that higher HMTA levels might be beneficial for the enhancement of brake linings with regard to improving the thermal resistance of the product. Also it must be noted that higher HMTA concentrations liberate more gas during cure and might require that adjustments be made to the manufacturing processes.

Coloring Novolak Resins

HMTA as a curative will produce a polymeric network where the molecular backbone can contain nitrogen atoms present in the connecting bridges. The nitrogen atoms are derived from the decomposition, or partial decomposition, of HMTA. Their presence can form chemical structures that impart cured novolak resins with a brilliant yellow hue. This is a special consideration for companies interested in the final color of their composites and explains why many friction material composites are limited in their available color schemes.

Finally, the chemistry of novolak resins has been studied by Zhang and colleagues in great detail [9-10 to 9-15]. The general conclusion is that novolak reactions are incredibly complex and follow many different reaction paths. For instance, reactions between HMTA and novolak resins can produce a number of intermediates including ethers, benzoxazines, benzylamines, diamines, triazines, amines, stillbenequinones, and imides.

Additionally, the effect of contaminants (such as trace amounts of water), which tend to hydrolyze HMTA, can also have a pronounced effect on the type of reaction intermediate formed. It should also be noted that the effects of HMTA concentration and the temperatures used in the curing process were found to be key process variables that have a significant effect on the chemical composition of the final polymeric structures. Changes in the process can yield a final resin with distinctly different properties and should be avoided unless absolutely necessary.

Thermo-Oxidative Degradation of Phenolic Resins

As kinetic energy profiles, and thermal excursions during vehicle braking events, continue to increase, the primary binder system in a tribological composite is required to survive sustained operation at high temperature. Traditionally, the success of phenolic resin over other materials has been based on its intrinsically superior heat resistance. However, several vehicle platforms have emerged that are exceeding the limits of conventional phenolic resins and, because of the amount of energy being converted, their brake systems promote rapid oxidation of the binder during braking. This can lead to a host of potential concerns such as wear and loss of frictional output in the form of brake fade.

The process of oxidation is a critical aspect in wear processes [9-16, 9-17] and is a form of chemical wear. With polymer systems, the heat of braking generates thermal effects that, when combined with operation in an oxygen-rich atmosphere, lead to thermo-oxidative degradation of the polymeric matrix. This was widely studied by Conley and his colleagues [9-18, 9-19, 9-20, 9-21, 9-22]. Their work provides us with a proposed mechanism for the degradation of phenolic polymer systems, beginning with oxidation of the methylene bridges. Conley's proposed mechanism is shown in Fig. 9.5.

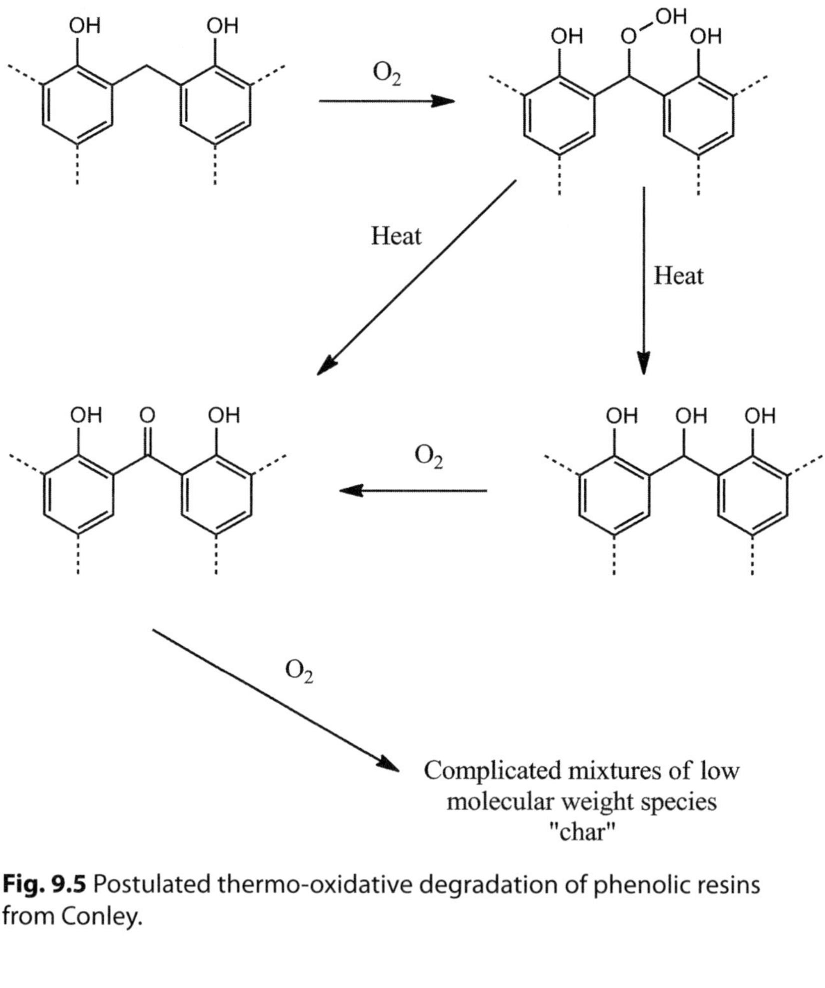

Fig. 9.5 Postulated thermo-oxidative degradation of phenolic resins from Conley.

The essential points of their model begin with the dihydrodiphenylmethane portion of the polymer reacting with oxygen to form a peroxide (or alcohol) at the methylene bridge. This can subsequently convert into a carbonyl functional group forming the structural analog of benzophenone. Additional reactions with oxygen at the carbonyl group results in destruction of the methylene bridges by scission of this carbonyl group resulting in the formation of carbon oxides and produces a carbon-rich char structure.

Lochte and group [9-23] specifically examined the process of char formation by using phenolic resins under a wide range of carefully controlled thermogravimetric studies. It was concluded that the Conley model is correct and the degradation of phenolic resins at elevated temperatures is fundamentally associated with the chemical stability of the dihydroxydiphenylmethane unit.

If this structural unit were chemically modified, the properties of the resin would be changed and oxidation resistance could be improved. Lochte's group also noted that changes in the curing methods can have an effect on the degradation process and suggest that thermooxidative degradation cannot be completely eliminated because the resin itself can act as an oxygen source.

Other materials present in the composite can also affect the oxidation process. Kristkova's group [9-24] examined the influence of metals on the degradation of phenolic resins used in friction material composites. It was determined that pure metallic additives such as copper or iron can have a significant influence on the degradation of phenolic resins. These metals have a catalytic effect where the metals, or their oxides, accelerate the process of oxidation. The significance of this finding may explain why certain compositions outperform others with regard to wear.

References

9-1. Bayer, A. 1872. "Über die Verbindungen der Aldehyde mit den Phenolen." *Berichte der Deutschen Chemischen Gesellschaft* 5: pp.25-56 (German).

9-2. Baekeland, H. 1909. Method of making insoluble products of phenol and formaldehyde. United States Patent No. 942699, filed July 13, 1907, and issued December 7, 1909.

9-3. Gardziella, A., L. Pilato, and A. Knop. 2000. *Phenolic Resins: Chemistry, Applications, Standardization, Safety, and Ecology*, 2nd ed. Berlin: Springer-Verlag.

9-4. Lederer, L. 1894. "Eine neue Synthese von Phenolalkoholen." *Journal für Praktische Chemie* 50: 223-226 (German).

9-5. Manasse, O. 1894. "Uber eine Synthese aromatischer Oxyalkohole." *Berichte der Deutschen Chemischen Gesellschaft* 27: 2409 (German).

9-6. Parker, E., and B. Grele. Brake linings. United States Patent No. 5190991, filed November 1, 1990, and issued May 2, 1993.

9-7. Walisser, W. 2004. Spray dried phenolic resole molding powder with crystalline phenolic compounds. United States Patent No. 6818707, filed January 8, 2002, and issued November 16, 2004.

9-8. Chiu, S. T. 1991. Spray dried phenol-formaldehyde resin compositions. United States Patent No. 5047275, filed January 16, 1991, and issued September 10, 1991.

9-9. Rogers, M., and T. Long. 2003. "Synthetic methods in step-growth polymers." New Jersey: John Wiley and Sons.

9-10. Zhang, X. et al. 1997. "The Chemistry of Novolac Resins – 3. 13C and 15N N.M.R. Studies of Curing with Hexamethylenetetramine." *Polymer* 38: 5835-5845.

9-11. Zhang, X. et al. 1998. "The Chemistry of Novolac Resins – V. Reactions of Benzoxazine Intermediates." Polymer 39: 399-404.

9-12. Zhang, X., and D. Solomon. "The Chemistry of Novolac Resins – VI. Reactions Between Benzoxazine Intermediates and Model Phenols." *Polymer* 39: 405-412.

9-13. Zhang X. et al. 1998. "The Chemistry of Novolac Resins: Part 7. Reactions of Para-hydroxybenzylamine Intermediates." *Polymer* 39: 1957-1966.

9-14. Zhang, X. et al. 1998. "The Chemistry of Novolac Resins: 8. Reactions of Para-hydroxybenzylamines with Model Phenols." *Polymer* 39: 1967-1974.

9-15. Zhang X., and D. Solomon. 1998. "The Chemistry of Novolac Resins: 9. Reaction Pathways Studied via Model Systems of Ortho-hydroxybenzylamine Intermediates and Phenols." *Polymer* 39: 6153-6162.

9-16. Quinn, T. 1983. "Review of Oxidation Wear. Part 1: The Origins of Oxidational Wear." *Tribology International* 16: 257-271.

9-17. Quinn, T. 1983. "Review of Oxidation Wear. Part 2: Recent Developments and Future Trends in Oxidational Wear Research." *Tribology International* 16: 305-315.

9-18. Conley, R., and J. Bieron. "A study of the Oxidative Degradation of Phenol-formaldehyde Polycondensates Using Infrared Spectroscopy." *Journal of Applied Polymer Science* 7: 103-117.

9-19. Conley, R., and J. Bieron. "A Kinetic Study of the Oxidative Degradation of Phenol-formaldehyde Polycondensates Using Infrared Spectroscopy." *Journal of Applied Polymer Science* 7: 171-180.

9-20. Jackson, W. and R. Conley. 1964. "High Temperature Oxidative Degradation of Phenol-formaldehyde Polycondensates." *Journal of Applied Polymer Science* 8: 2163-2193.

9-21. Conley, R. 1965. "Oxidative Degradation of Polybenzyl." *Journal of Applied Polymer Science* 9: 1107-1115.

9-22. Conley, R., 1965. "Oxidative Degradation of Phenol-formaldehyde Polycondensation Initial Degradation Reactions." *Journal of Applied Polymer Science* 9: 1117-1126.

9-23. Lochte, H. et al. 1965. "The Thermo-oxidative Degradation of Phenol-formaldehyde Condensates: Thermogravimetric and Elemental Composition Studies of Char Formation." *Journal of Applied Polymer Science* 9: 2799-2810.

9-24. Kristkova, M. et al. 2004. "Influence of Metals on the Phenol-formaldehyde Resin Degradation in Friction Composites." *Polymer Degradation and Stability* 84: 49-60.

Chapter 10
Modified Phenolic Resins

It is a bad plan that admits of no modification.
—Publilius Syrus

UNMODIFIED PHENOLIC RESINS IMPART rigidity and thermal resistance to tribological composites and are sufficient for general purpose use. However, in certain applications a friction material can be subjected to increased levels of thermal energy. In other cases, the friction composite needs to be conformable to the mating member and therefore less rigidity is required. In these circumstances, a modified phenolic resin is often the only acceptable solution.

There are innumerable examples were phenolic resins have been blended with other materials in an attempt to alter a specific property of the final composite. At other times, the resin manufacturer will introduce a highly modified commercial resin specifically designed for use in a given market segment. These resins are usually submitted to several industries to determine if the new modification has merit in the widest possible scope of use.

Oil Modified Phenolic Resins

Drying oil modified phenolic resins date back to at least 1912, but might have been in practice earlier. The General Bakelite Company [10-1] developed a varnish by blending phenolic resin with selected alcohols and suitable drying oils. This varnish was anticipated for use in the coating of various materials including woven articles.

The varnish concept was also accepted by members of the friction material fraternity. Novak [10-2] described a resinous compound of rosin and other "oleo-resinous" compounds, which included drying oils, to enhance a phenolic resin coating solution for woven friction linings. This compound was stated to beneficially provide higher friction levels.

Daniels and Snell [10-3] used the approach of mixing castor oil with phenolic resin to coat granules of cork. The coated granules were allowed to dry and then molded into segments that formed a friction material. The benefits of this product included the thermal stability of a phenolic resin with the compliance of cork granules and water repellency because of the oil.

Another patent by Novak [10-4] described a cresol-formaldehyde resin blended with oil for use as a saturating resin. Novak's use of cresol, which contains a blend of ortho, para and meta isomers of methyl phenol, is a very subtle example of how modification to resins can occur at the molecular level. The methyl moiety of cresol blocks some of the potential reaction sites and restricts the formation of a very rigid network. As such, cresol resins are typically more flexible than a pure phenol-formaldehyde resin, especially in conjunction with drying oils.

Early Chemical Modification

In 1932 Byck [10-5] combined drying oils and phenol in the presence of a catalyst (e.g., aluminum chloride) to produce a phenolic intermediate. This intermediate was reacted with formaldehyde to produce a phenolic resin with a blend of properties. The final resin was used in the production of molded plastic elements. Later, Martin [10-6] classified the reaction products of phenolic resins with drying oils as a unique class of compounds known as chromans. The use of chromans in friction materials is an area that is essentially devoid of information.

Oil as a "Flexibilizer"

Perhaps the most widely cited benefit for using an oil modified phenolic resin is to improve the flexibility of the final product. As we have seen, oil modification also tends to affect the friction of the linings [8-12]. However, a more recent example by Noguchi [10-7] contrasted the intent of Nanfeldt's design and stated that oil modified phenolic resins lower the coefficient of friction, which suggests a lubricating action.

Confirmation of this contrasting perspective was also provided by Bijwe and colleagues [10-8] who found that linseed oil modified phenolic resins gave poor (i.e., lower than acceptable) friction levels but had the longest life of the resins tested. It was concluded that the beneficial wear results were the result of the friction lining having done less work because of the lowered coefficient of friction.

The frictional effect of using an oil modified phenolic can be dependent on the operating temperatures at which it is tested. At lower temperatures, the oil can provide enhanced friction, analogous to the "frictioning" effect reported by Novak. However, the oil can become an effective lubricant and give poor performance as the operating temperatures increase.

Modern Oil Modified Phenolic Resins

The use of oil modified phenolic resins is still commonplace in certain segments of the friction material industry. They are typically used in low-energy applications that demand a material that is flexible and compliant. Another point of merit regarding the use of oil modified phenolic resins is the ability to provide resistance to water [10-9], which can reduce noise generation in certain automotive braking applications.

Aside from these peripheral benefits, their use is beginning to dwindle because of inherently poor thermal resistance [10-10]. In addition, oil modified phenolic resins are facing stiff competition from more advanced polymer systems that simultaneously impart flexibility, provide water resistance, and enhance thermal characteristics in the final product .

Cashew Nut Shell Liquid (CNSL) Modified Phenolic Resins

Cashew nut shell liquid (CNSL) is comprised of naturally occurring phenolic compounds that have a fifteen carbon side chain at the meta position, which is also unsaturated to varying degrees. Because CNSL contains naturally occurring phenolic compounds, the combination of CNSL and phenolic resins has been explored for many years. However, much of the early literature is convoluted and makes it difficult to understand many of the beneficial attributes of CNSL and CNSL modified phenolic resins.

For instance, Spokes [8-32] developed a binding mixture that employed thermoplastic pitch to aid in the generation of friction during use (i.e., a frictioning agent). In addition to pitch, the binder also contained partially polymerized CNSL that had been reacted with cresylic acid, which is a complicated mixture of cresol isomers derived from coal tar. This mixture of pitch, CNSL, and cresylic acid was miscible when mixed with rubber and formed a rubber cement compound that was consequently blended with asbestos, graphite, and sulfur. The mixture was molded into the final shape of the desired friction material and cured/vulcanized to make a friction material that was free from noise and resistant to heat, moisture, oil, and grease.

From the work of Spokes, it is difficult to ascribe a single specific benefit to the use of CNSL as a modifier to phenolic resins. However, in recent years the benefits of CNSL modification have been more widely studied under more precise conditions. It is generally concluded that the use of CNSL modified phenolic resins in tribological composites is one of the most promising forms of phenolic resin for consideration today.

Modern Use of CNSL Modified Phenolic Resins

Blau [6-48] taught that one advantage of cashew resin, as compared to an unmodified phenolic resin, is the development of a softer material that is more wear-resistant at lower speeds when the brakes are relatively cold. In simple terms, CNSL modified phenolic resins are stated to impart cold wear enhancement to tribological composites.

Longley and Gardner [6-35] reported that a hard brittle friction material would tend to drop in friction at higher speeds. They proposed a remedy involving the addition of CNSL to the composition. It is somewhat uncertain whether the authors intended for the addition of polymerized CNSL as a granular material or the use of CNSL as a binder in its raw liquid form. Regardless of the ambiguity, it can be safely stated that the intended benefit from the addition of the CNSL was to impart a degree of compliance that allowed for better conformity between the sliding surfaces. This was expected to improve energy dissipation at the interface and help stabilize friction levels.

More recently, research has been conducted in other countries and continues to explore CNSL modified resins in the development of advanced phenolic based tribological composites [10-11]. Sergienko [10-12] developed an asbestos-free friction material for use as brake linings and determined that poor adhesion between the resole resin and the filler materials was a significant shortcoming to tribological composites. In addition, the presence of residual stresses inherent in the polymeric matrix after curing was considered another significant shortcoming that must be overcome during the design and development process. One method for surmounting both of these shortcomings was the use of plasticizing agents such as CNSL to improve the properties of the polymeric matrix.

Sergienko continued on to show that the addition of small amounts of CNSL (as little as 4% by weight of the resole resin) was capable of providing significant improvement in non-asbestos organic brake linings, specifically citing an increase in the coefficient of friction of 14% and stabilization of the overall friction level by 6%. The plasticizing effect of the CNSL served to reduce the rigidity of the composite by nearly twofold while making the polymer more heat-resistant.

In short, the use of CNSL as a polymer modifier is effective for enhancing many aspects of the tribological performance of asbestos free friction materials. It has fallen out of favor in recent years, but many of the findings suggest that CNSL modified phenolic resins should be explored by designers who seek to simultaneously improve several facets their products performance.

A Word of Caution about CNSL Modified Phenolic Resins

Phenolic resins that have been modified with CNSL can cause dermatitis and other topical skin reactions in sensitive individuals. The reaction is similar to that of poison ivy exposure that, interestingly enough, is due to the similarity in the chemical compositions of the molecular compounds. It is recommended that all personnel should use the appropriate personal protective equipment when handling CNSL modified phenolic resins.

Rubber Modified Phenolic Resins

The use of rubber and phenolic resin mixtures dates back to the early part of the last century. In general, nearly all attempts to use rubber in combination with phenolic resin were intended to impart flexibility to the final product. However, the role of the resin was also essential in improving other areas.

For instance, an early example from Harvey [8-28] used rubber in combination with CNSL to form a varnish ideally suited for electrical insulation. The blend of rubber and natural phenolic resin was an attempt to improve the heat resistance of the rubber in this varnish. In many cases, the blends of rubber and resin have pronounced synergies that include enhanced thermal properties, but can also yield a number of other benefits.

Rubber Modified Resins in Friction Materials

Regarding brake linings, one of the first examples of the mixing rubber and phenolic resin comes from an unusual source: Colt's Patent Fire Arms Manufacturing Company. In 1932, Whitworth [10-13] developed an asbestos-based brake lining using a solvated mixture of sulfur and rubber in combination with phenolic resin. The mixture was formed into a mass and rolled onto a sheeter machine, forming a laminated material as described previously. Whitworth improved on the design of the laminated sheeter stock by perforating the surface of the friction material with a secondary rolling operation where the perforations provided channels at the braking surface to improve wet friction performance. The perforations allowed liquids to be forced away from the interface and increased the region of direct contact between the lining and the drum. Additionally, the perforations were found to prevent delamination of the laminated layers and to enhance durability.

Chemical Modification with Rubber

Much like the oil processes mentioned earlier, the modification of phenolic resins can be accomplished by simple mixtures or by chemical modification. Chemical modification with an elastomer involves reactions between a phenolic resin and elastomers, forming a unique chemical intermediate that can then be polymerized during processing. The bounds of modification are limited only by the ingenuity of the chemist and availability of ingredients.

To limit this discussion to a reasonable length, the process of chemically modifying phenolic resins with rubber compounds is illustrated by the work of Jeffreys and Williams [10-14]. Their work recorded that prior attempts to incorporate rubber into a phenolic resin involved a labor intensive method whereby rubber and phenolic resin were compounded with extensive mixing.

Jeffreys and Williams went one step further by reacting selected phenolic compounds with an appropriate aldehyde to form a novolak resin. Toward the end of the reaction, when the aldehyde concentration was nearly depleted, latex rubber (a copolymer of acrylonitrile-butadiene) was added to the reactor in concentrations up to 30% by weight. The result was a resin with a highly dispersed elastomer component. This method was also stated to be appropriate for use with other elastomers such as styrene-butadiene and methyl methacrylate.

More recently, McCormick [10-15] examined the use of modified phenolic resins in friction materials and contrasted the concept of chemical modification with physical mixtures. During his discussion, McCormick reiterates that rubber modified systems generally have higher toughness and the dispersion of elastomer particles in the phenolic network, which can be on the order of a few microns in size, is an important factor to be considered. It was concluded that the ability to finely disperse small particles of elastomer throughout a phenolic network can be achieved only by chemical modification, and the use of physical mixtures does not generate the same level of performance. Therefore, simple mixtures do not provide the level of dispersion that chemical modification can achieve, and chemically modified systems were generally regarded as superior in this regard.

NBR Rubber Modified Phenolic Resins

Today the use of rubber compounds as modifiers to phenolic resins is still an area of intense interest and widespread study. Bijwe [10-8] and colleagues like Jeffreys and Williams used acrylonitrile-butadiene modified phenolic resin and found that it showed poor friction at high temperatures (i.e., fade) but recovered well upon cooling. Satapathy and Bijwe [10-16] used the same type of rubber modified resin in a study of organic fibers and their role in the performance of friction materials, but did not specifically describe any of the benefits of this type of resin.

In addition, McCormick found that NBR modified phenolic resins gave higher specific wear rates than other types of elastomer modification. It appears that despite the popularity, NBR modification does not seem to provide a significant improvement over other forms of modified resins. The general conclusion is that NBR modified phenolic resins exhibit a number of conflicting results and more work should be conducted to further expand on this area of much needed research.

Polyvinylbutyral (PVB) Rubber Modified Phenolic Resins

Another common modification to phenolic resins involves the use of polyvinylbutyral (PVB), which is produced by reacting polyvinyl alcohol with butyraldehyde. Verma [10-17] and the team of Tewari and Bijwe [7-5] evaluated a phenolic resin that had been modified with PVB in fiber-reinforced composites. It was generally concluded by both groups that the use of a PVB modified phenolic resin improved the impact resistance of the composite. It was also noted that these resins could provide wear reductions, but could also suppress the friction level to an unacceptable level. It is feasible that the suppressed friction (i.e., less work) would account for the wear reductions observed when using PVB modified phenolic resins. These findings conclude that a systematic study of PVB resins should be conducted to fully appreciate their benefits.

Epoxy Rubber Modified Phenolic Resins

Another favorite modification method is the epoxy modified phenolic resin. The modification process involves reacting phenolic resins (typically novolak resins) with epichlorohydrin in presence

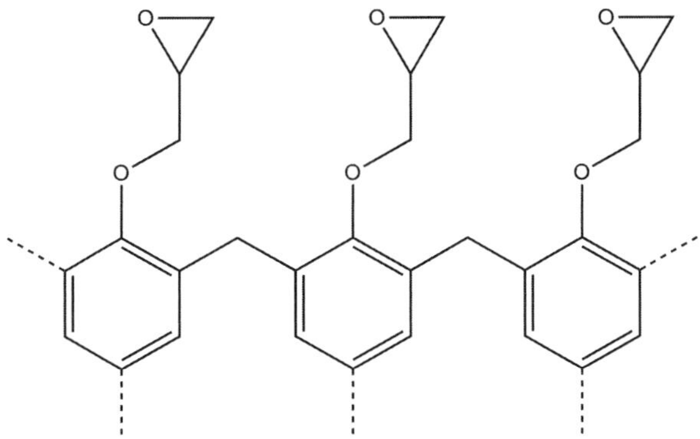

Fig. 10.1 Epoxy modified novolak resin.

of a suitable basic catalyst. This converts the hydroxyl group into a reactive epoxide, which can be illustrated as shown in Fig. 10.1.

Chan and Stachowiak [6-11] discussed an epoxy modified phenolic resin and stated that epoxy compounds would be ideal for use in friction materials. The epoxy addition would not only impart toughness that enhances the structural aspects of the composite, but was also considered a synergistic mixture that provided improved thermal resistance in comparison with either pure phenolic or pure epoxy.

Ganguly and George [10-18] developed an asbestos-free friction material for use as a brake lining that employed an epoxy modified phenolic resin as the primary binder. In this case, the epoxy was also added to toughen the phenolic matrix, but no mention was made of enhanced thermal properties, suggesting that more research is needed regarding the use of epoxy modified phenolic resins.

Issues with Rubber Modified Phenolic Resins

Shibata and co-workers [10-19] stated that rubber modified phenolic resins, as well as cresol modified resins, improved flexibility and could be effective for noise reduction. It was noted that these modifications to the phenolic resin also gave a reduction in the coefficient of friction during operation at higher temperatures and had poor thermal resistance properties, which can be linked with poor wear performance at elevated temperatures.

The general conclusion is that modification of phenolic resins with a conventional elastomer will impart flexibility or conformability to the friction material composite. Additional benefits might include enhanced toughness, but this is usually offset by reduced thermal performance. Therefore, the use of rubber modified phenolic resins has generally been limited to composites designed for lower energy braking applications with moderately low operational temperatures. However, advances in polymer science are continually changing our understanding of the abilities of modern polymeric additives. Two unique forms of elastomer modified phenolic resin have become commercialized that are of special interest: acrylic modified phenolic resins and silicone modified phenolic resins.

Acrylic Modified Phenolic Resins

An acrylic modified phenolic resin is a form of elastomer modified phenolic resin that has become increasingly common over the past few years. The increase in use is due to advances in the manufacturing of modified phenolic resins such as those described by Ema and colleagues [10-20] that have made these resins more readily available. In addition, there are a number of sources that suggest acrylic modified phenolic resins are the best choice for tribological composites.

Acrylic compounds can offer an effective means for imparting flexibility to a polymer and improving the impact resistance of a composite [10-21]. Additionally, Ohm [7-27] discussed the beneficial resilience of acrylic rubber and stated that acrylic rubber leaves a high percentage of char after exposure to elevated temperatures, which can aid in the retention of a stable third body layer at the friction interface.

Perhaps the single largest area of interest concerning acrylic modified phenolic resins lies in the area of noise control. Ozeki [10-22] discussed the use of modified phenolic resins in high-performance friction applications. One of the key benefits was the ability to reduce the transmission rates of vibration (i.e., the resins possessed a higher damping capacity than conventional resins and were more effective at reducing noise).

Imai [10-23] reported on the development of a novel acrylic rubber modified phenolic resin that was specifically designed to retain its favorable noise characteristics at lower temperatures. The specific form of acrylic rubber retained higher degrees of flexibility at reduced temperatures compared to other forms.

Based on the highly desirable char yield and effective noise reduction potential of these products, the use of acrylic modified phenolic resins is expected to see continued growth over the next few years.

Silicone Modified Phenolic Resins

One very promising modification that can be made with phenolic resin involves the use of silicone elastomers. An example of the use of silicone modification to phenolic resins was recorded during the late 1970s as a means of inhibiting the liberation of graphite fiber from aerospace laminates [10-24]. The mechanism was based on the decomposition of the resin, forming a glassy, silica-rich char layer that trapped fibers and prevented dispersal.

Several years later, Chan and Stachowiak [6-11] included silicone binder systems in their review of automotive brake friction materials. It was concluded that silicone based binders would be another ideal system in the design of tribological composites. These resins are expected to be able to impart impact and thermal resistance to the product while also making the surface hydrophobic in nature. This is considered a positive attribute that might inhibit absorption of excessive amounts of moisture, which has been linked with various forms of noise generation concerns [10-25].

In addition, recent patent literature is beginning to show a surge in the use of silicone modified resin systems. In brake linings, the work of Ono and colleagues [10-26] described the development of an enhanced non-asbestos organic formulation bound together with a silicone modified phenolic resin. The silicone modified resin was employed to prevent noise after the brakes had been sitting for a prolonged period of time.

The use of silicone modified resins is also growing rapidly in the area of wet friction applications [6-59]. Kitahara and colleagues described the use of a silicone modified resin and its ability to enhance both friction performance and the durability of the clutch facing during use. The primary enhancement was associated with the thermal resistance and natural flexibility of silicone based elastomers.

In general, the use of silicone modified phenolic resins is a result of a unique combination of physical property effects including thermal resistance, flexibility, water resistance, and noise reductions. The single largest concern with this type of resin is associated with cost, but the key benefits will most likely continue to generate growth in the use of silicone modified phenolic resins during the next few years.

Final Words about Elastomer Modified Phenolic Resins

It should be obvious by now that friction material producers have attempted to use virtually all types of elastomeric additives and mixtures of additives to modify the physical properties of phenolic resin based composites. One final example, which illustrates the complexity of some prior attempts, also presents an alternative perspective. Holmes [10-27] developed a binder based on a mixture of phenol and formaldehyde. To this was added a blend of xylenols, linseed oil, and CNSL. Because of the complexity of the additives, it can be imagined that the number of potentially competing side reactions would be immensely difficult to predict. Furthermore, after the resinous mixture was reacted to the desired level, it was then added to polychloroprene rubber as a "hardener" to make the rubber stiffer.

Thermally Enhanced Phenolic Resins

Despite the thermal resistance provided by conventional phenolic resins, there are applications that operate at excessively high temperatures. Operation at these sustained temperatures promotes thermo-oxidative degradation leading to poor wear and an increased proclivity for structural failure at the surface of the composite. This has created a need for modified phenolic resins that are thermally more resistant and capable of surviving for longer periods of time under these conditions.

As outlined in the previous chapter, the thermo-oxidative degradation mechanism illustrated by Conley and Lochte suggested that structural modification would enhance the ability of the polymer to operate at sustained temperatures for a longer period of time. This can be accomplished by changes to the structure that prevent formation of peroxides at the methylene bridges. Based on this approach, there are several methods that have shown promise and are suitable for friction material composites.

Acetylation of Phenolic Resins

Chemical modification of phenolic fibers was disclosed by Hayes [10-28]. His method converted the hydroxyl portion of phenol into an acetyl group, which inhibits formation of peroxides at the methylene linkages. This process, which is effective for a fibrous material, might be suitable for a non-woven paper type material comprised of (or containing) phenolic fibers. However, it is not ideal for low permeability rigid molded composites.

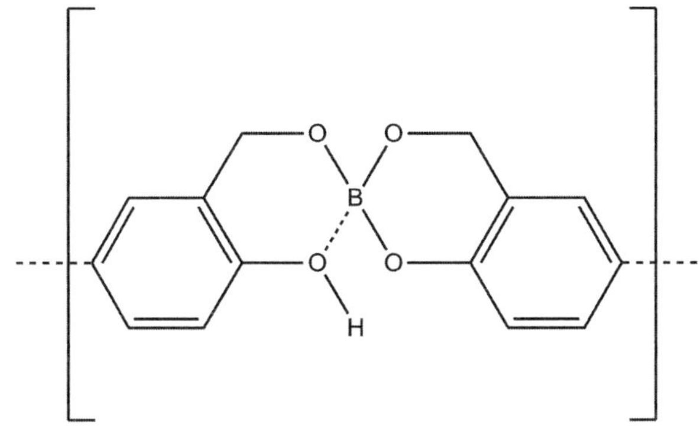

Fig. 10.2 Idealized structure of a boron modified phenolic resin.

Boron Additions

Boron modified resins have increased thermal resistance and were used in friction materials as early as 1938. Bender [10-29] used boric acid as the catalyst during condensation to produce a boron containing resin suitable for saturating friction materials. This resin was reported to give improved friction and wear performance.

Gao and co-workers [10-30, 10-31] recently reported that boron modified phenolic resins condensed by reactions of Bisphenol F with formaldehyde would form a bisphenol with methylol groups at the ortho positions. Additions of boric acid formed six-member cyclic structures by reacting with the methylol groups and the hydroxyl portions of the phenol. The final structure was a polymer with boron atoms effectively blocking the oxygen atoms from further oxidation. The reaction is shown in Fig. 10.2.

Phosphorous Modified Resins

Another common approach for enhancing thermal resistance in phenolic resins uses phosphorous compounds. These compounds react with ortho-ortho linked phenolic resins to form a cyclic molecule that incorporates phosphorous into the structure and blocks labile oxygen atoms from further reaction. Daniels and Shepard [10-32] described this reaction, which is shown in Fig. 10.3, in great detail.

Arylalkyl Modification

A final method involves the addition of a large arylalkyl group to the structure as taught in the patent of Robinson and Brain [10-33]. Their

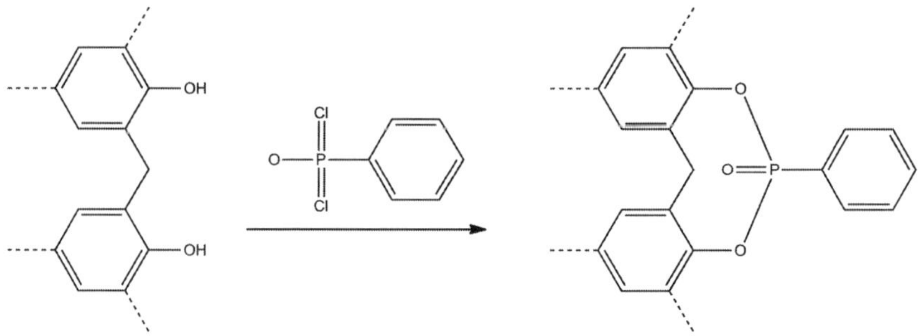

Fig. 10.3 Proposed phosphorous modified phenolic resin structure

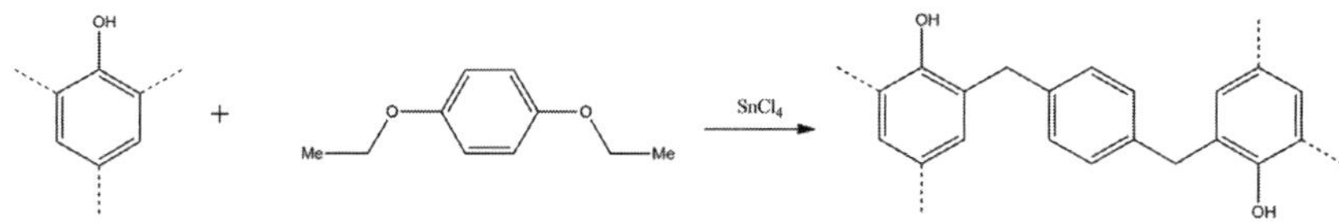

Fig. 10.4 Arylalkyl modified phenolic resin

ideal reaction involved combining phenol and a reactive modifier in the presence of a tin catalyst to form a polymer, where a relatively un-reactive aromatic member is positioned between the phenolic moieties as described in Fig. 10.4.

This resin, which is known as a Xylok [10-34] resin, is produced by reacting phenol with dimethoxy-p-xylene in the presence of catalytic amounts of stannic chloride. The reaction product is a solid resin that can be dissolved in solvents for use in saturation or coating applications. It can also be reacted with additional HMTA to produce a fully thermoset rigid polymeric matrix that is more heat-resistant than an unmodified phenolic resin and is capable of retaining its physical properties up to temperatures of 482°F (250°C).

The Underlying Mechanism

In all cases, whether by addition of phosphorous, boron, or large arylalkyl groups, these methods are focused on blocking the methylene bridge from attack in an effort to inhibit the formation of peroxides. The intent is to provide friction material composites with enhanced thermal resistance and reductions in flammability [10-35]. In many cases, significant improvements to phenolic resins can be found with small changes in the structure of the polymeric network.

Tyberg states that as little as 6% phosphorous additions to phenolic resin can render them self-extinguishing. Blends of 2% nitrogen and 2% phosphorous will also make resins self-extinguishing and can be reasonably expected to occur when HMTA is used in the condensation of a phosphorous modified novolak. Small additions of boron can make phenolic resins inflammable and resins with higher aromatic content, such as those modified with large arylalkyl groups, are more likely to form protective char layers because of their intrinsically higher carbon content. In essence, the amount of modification does not have to be significantly large.

Other Benefits of Thermally Resistant Resins

One of the benefits of thermally resistant polymers is the development of a carbonized char layer. The char layer formed with boron and phosphorous modified resins is important in that it can also affect the development of a carbon-rich third body layer [10-36] that will affect the tribological performance of ensuing operation. Support for this noted enhancement was also provided by Hong and co-workers [10-37] who reported that wear and fade performance at elevated temperatures of friction materials was directly related to the chemical structure of the resin. It was further noted that boron and phosphorous modified phenolic resins were superior to straight (unmodified phenolic) resins. Therefore, even minor structural changes can have substantial impact on the product.

Concerns with Thermally Modified Resins

Of the modification types, the arylalkyl modified phenolic resins appear to be less widely used than boron and phosphorous. Cost may be one potential reason, but there can be other concerns with this type of resin. Bijwe and co-workers found resins modified with alkybenzene functional groups to have the best friction performance of the resins tested. However, the wear performance was poor in comparison to the other types of resin under evaluation.

Additionally, Kim and Jang [3-42] used a form of arylalkyl modified phenolic resin in a study of aramid fiber reinforced friction materials. It was concluded that modified resins were weaker and subject to more abrasion than unmodified resins. This might have been a physical response to the stiffness of the resins as the modified resins were generally softer. However, not everyone found concern with their use.

In contrast, Ozeki [10-22] examined the properties of a xylilene modified phenolic resin and found that it provided excellent thermal resistance. In addition, this type of arylalkyl modified phenolic resin system provided reduced pad wear rates while enhancing the friction levels at elevated temperatures (e.g., fade), which might coincide with the findings of Bijwe's group.

These opposing observations might seem surprising, but insight into this behavior was provided by workers at Honda [10-19]. It was determined during their research on the development of brake friction materials that silicon, phosphorous, and boron modified phenolic resin have increased thermal resistance when compared to unmodified phenolic resins. However, these resins yielded products that were reduced in strength and exhibited higher rates of wear. It was suggested that the processing of these resins was more difficult, indicating that when thermally modified phenolic resins are used to replace an unmodified phenolic resin, the process must be carefully optimized to gain the potential benefits from the use of the enhanced resin.

Benzoxazine Resins

Though technically not a modified phenolic resin, benzoxazine resins are an emerging class of resins that offer significant promise for future products. The resins are produced by reactions involving phenolic compounds, formaldehyde, and a primary amine (such as methylamine) [10-38]. The product is a monomer with an oxazine ring that opens when heated, allowing the monomers to polymerize and forming a phenolic structure as shown in Fig. 10.5.

Benzoxazine monomers polymerize by addition reactions and do not liberate moisture vapor like traditional phenolic resins. Additionally, benzoxazine resins are less likely to shrink during molding and therefore do not impart excessive stress to the composite structure.

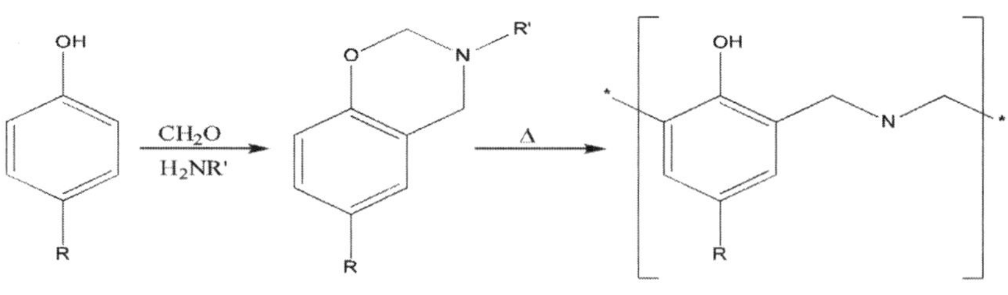

Fig. 10.5 Benzoxazine ring opening reaction

In some cases, the resins can actually expand, resulting in increases to the porosity level of the composite. This increase in porosity can influence the overall strength of the composite as well as the wear behavior and should be examined in depth during the process of evaluating benzoxazine resins.

One of the more interesting aspects of benzoxazine resins is their ability to form a very high char yield [10-39]. This potentially makes them suitable for higher temperature operation for more severe applications. Benzoxazine resins can also be more hydrophobic than conventional phenolic resins [10-40]. It is not known whether this effect is pronounced when compared to an unmodified phenolic resin, but increases in the hydrophobic nature of the primary binder could lead to improvements in the area of noise prevention as suggested by the use of silicones previously.

References

10-1. Baekeland, L. 1912. Varnish. United States Patent No. 1018385, filed March 14, 1911, and issued February 20, 1912.

10-2. Novak, I. 1928. Friction element. United States Patent No. 1672538, filed October 21, 1922, and issued June 5, 1928.

10-3. Daniels, E., and H. Snel. Cork composition. United States Patent No. 1732398, filed July 9, 1924, and issued October 22, 1929.

10-4. Novak, I. 1930. Hardened cellular of fibrous friction elements and method of producing same. United States Patent No. 1766932, filed June 9, 1926, and issued June 24, 1930.

10-5. Byck, L. 1932. Improvement in phenolic condensation products and process of making same. United States Patent No. 1887833, filed September 27, 1922, and issued November 15, 1932.

10-6. Martin, R. *The Chemistry of Phenolic Resins*. New York: John Wiley and Sons.

10-7. Noguchi, K. 1987. Modified liquid resole phenolic resinous composition for friction materials. United States Patent No. 4647631, filed January 13, 1986, and issued March 3, 1987.

10-8. Bijwe, J. et al. 2005. "Influence of Modified Phenolic Resins on the Fade and Recovery Behavior of Friction Materials." *Wear* 259: 1068-1078.

10-9. Powers, P. 1943. *Synthetic Resins and Rubbers*. New York: John Wiley and Sons.

10-10. Suzuki, M., and M. Mori. 2003. Wet friction material. United States Patent No. 6231977, filed August 18, 1999, and issued July 1, 2003.

10-11. Gopal, P. et al. 1995. "Load, Speed, and Temperature Sensitivities of a Carbon-fiber Reinforced Phenolic Friction Material." *Wear* 181-183: 913-921.

10-12. Sergienko, V. 1995. "Asbestos-free Materials for Fabricating Brakes." *Trenie i Iznos* 16:33-136 (Russian).

10-13. Whitworth, H. 1932. Brake lining and method of making. United States Patent No. 1890424, filed December 19, 1930, and issued December 6, 1932.

10-14. Jeffreys, K., and H. Williams. Process for the preparation of modified resins. United States Patent No. 3536783, filed September 28, 1967, and issued October 27, 1970.

10-15. McCormick, B. 2002. "The Interactions of Phenolic Polymers in Friction Material." In *Braking 2002*. London: Professional Engineering Publishing.

10-16. Satapathy, B., and J. Bijwe. 2004. "Performance of Friction Materials Based on Variation in Nature of Organic Fibres: Part I. Fade And Recovery Behavior." *Wear* 257: 573-584.

10-17. Verma, A. et al. 1996. "Effect of Resin Modification on Friction and Wear of Glass Phenolic Composites." *Wear* 193:.193-198.

10-18. Ganguly, A., and R. George. 2008. "Asbestos Free Friction Composition for Brake Linings." *Bulletin of Material Science* 31:19-22.

10-19. Shibata, K. et al. 1993. "Development of Brake Friction Materials." Technical Paper No. 930806. Warrendale, PA: Society of Automotive Engineers.

10-20. Ema, K. et al. 1984. Phenolic resin compositions. United States Patent No. 4435543, filed April 12, 1982, and issued March 6, 1984.

10-21. Deanin, R., and N. Schott. 1974. "Fillers and Reinforcements for Plastics." *Advances in Chemistry Series*, No. 134. Washington, DC: American Chemical Society.

10-22. Ozeki, S. 2001. "Modified Phenolic Resins for High Performance Friction Applications." Technical Paper No. 2001-01-3127. Warrendale, PA: Society of Automotive Engineers.

10-23. Imai, J., 2008. "An Acrylic Rubber Modified Phenolic Resin Which Has Good Flexibility at Low Temperature." Technical Paper No. 2008-01-2539. Warrendale, PA: SAE International.

10-24. Frost, L., and G. Bower. 1979. "Silicone Modified Resins for Graphite Fiber Laminates." CR-159750. Washington, DC: National Aeronautics and Space Administration.

10-25. Hara, Y. et al. 1999. Non-asbestos disc brake pad for automobiles. European Patent # 959262, filed May 18, 1999, and issued November 24, 1999.

10-26. Ono, M. et al. 2008. Friction material composition and friction material using the same. United States Patent No. 7338987, filed May 21, 2004, and issued March 4, 2008.

10-27. Holmes, R. 1953. Polychloroprene-phenol-aldehyde compositions. United States Patent No. 2635088, filed January 16, 1951, and issued April 14, 1953.

10-28. Hayes, J. 1981. "Novoloid Fibers." In *Kirk-Othmer Encyclopedia of Chemical Technology*, 3rd ed., 16: 125-138.

10-29. Bender, H. 1938. Friction material. United States Patent No. 2114121, filed July 27, 1935, and issued April 12, 1938.

10-30. Gao, J. et al. 2001. "Structure and Properties of Boron-containing Bisphenol - A Formaldehyde Resin." *European Polymer Journal* 37: 207-210.

10-31. Gao, J. et al. 2004. "Structure of a Boron-containing bisphenol-F Formaldehyde Resin and Kenetics of Its Thermal Degradation." Polymer Degradation and Stability 83: 71-77.

10-32. Daniels, B., and F. Shepard. 1969. "Inorganic Esters of Novolacs." *Journal of Polymer Science* 6: 2051-2058.

10-33. Robinson, J., and S. Brain. 1981. Method of producing a resin for producing heat stable products, the resin and the products. United States Patent No. 4276408, filed February 11, 1980, and issued June 30, 1981.

10-34. Xylok. Registered Trademark of Albright & Wilson, Ltd., West Midlands, England, Registration No. 0992261, September 3, 1974 (Trademark expired).

10-35. Tyberg, C., 2000. "Void Free Flame Retardant Phenolic Networks: Properties And Processablity." PhD Dissertation. Blacksburg, VA: Virginia Polytechnic Institute.

10-36. Lancaster, J. 1985. "Transfer Lubrication for High Temperatures: A Review." *Journal of Tribology* 107: 437-443.

10-37. Hong, U. et al. 1009. "Wear Mechanism of Multiphase Friction Materials with Different Phenolic Resin Matrices." *Wear* 266:739- 744.

10-38. Ishida, H., and H. Low. 1997. "A Study on the Volumetric Expansion Of Benzoxazine-based Phenolic Resins." *Macromolecules* 30: 1099-1106.

10-39. Shen, S., and H. Ishida. 1996. "Synthesis and Characterization of Polyfunctional Naphthoxazines and Related Polymers." Journal of Applied Polymer Science 61: 1595-1605.

10-40. Wirasate, S. et al. 1998. "Molecular Origin of Unusual Physical and Mechanical Properties in Novel Phenolic Materials Based on Benzoxazine Chemistry." Journal of Applied Polymer Science 70:1299-1306.

Chapter 11
Advanced Organic Binder Systems

To raise new questions, new possibilities, to regard old problems from a new angle, requires creative imagination and marks real advance in science.
—Albert Einstein

TRIBOLOGICAL COMPOSITES OCCASIONALLY experience operational extremes when frictionally generated surface temperatures can reach 5432°F (3000°C) during use [4-53]. In many cases, this extreme temperature range is achieved in only a few exceptional conditions, but phenolic resins begin degradation at temperatures greater than 842°F (450°C), exhibiting reduced strength and higher porosity at the surface [6-11]. To develop a friction material that will survive under these harsh conditions, the composite must be made with a binder capable of sustaining its strength in such situations.

Typical approaches to replacing phenolic resins with more advanced polymer systems have been based on the chemistry of three classes of chemical compounds: polyimides, bismaleimide-triazines (BT), and condensed polynuclear aromatic (COPNA) resins. Each of these advanced polymer systems presents a unique method for achieving this goal, but each also includes its own subtly distinct challenges that must be overcome, especially during the early phases of product development.

Polyimides

Polyimides are a class of thermally stable polymers with molecular structures composed of either long chain polyimides (linear polyimides) or heterocyclic polyimides with large aromatic groups present in the polymer network. The chemistry of polyimides includes a number of potential polymers that can be modified by the selection of individual monomers. This provides a wealth of possibilities for the composite designer. However, despite the numerous forms, polyimide chemistry can be described by the polyamic acid process.

Polyamic Acid Process

The polyamic acid process is a prevalent method of synthesis of polyimide compounds that is based on a two-step reaction process pioneered by workers at E. I. DuPont [11-1] during the 1950s.

This effort led to the development of a number of novel chemical compounds and commercial products. The science of polyimides has advanced considerably and Takekoshi [11-2] provided an extensive discussion regarding the entire class of polyimide polymers.

A typical two-stage polyamic process begins with the reaction of a suitable dianhydride and a selected diamine in an aprotic solvent, which is typically an organic solvent that does not exchange protons with the reactants present in solution. The reaction between the dianhydride and the diamine forms a polyamic acid with a large number of carboxylic acid functional groups.

The second stage of the process involves the liberation of water, partly from the carboxylic acid groups, thus forming the cyclic structure of the final polyimide polymer. This stage is known as the imidization stage and can be performed either chemically or thermally. The end result is a polymer that possesses planar aromatic groups and provides excellent thermal, electrical, and structural properties, which makes it ideally suited for the rigorous operating conditions of many friction materials.

The Chemistry of Polyimides

To examine the chemistry of polyimides a bit more thoroughly, the polyamic process can be illustrated by the synthesis of the first commercially significant Kapton polyimide [11-3]. The process uses a mixture of monomeric pyromellitic dianhydride (PMDA) and 4, 4′-oxydianiline (ODA) in an aprotic solvent, such as N,N-dimethylacetamide (DMAc). The reaction diagram is illustrated in Fig. 11.1.

Polyimides in Tribology

Polyimides have been a subject of keen interest in the field of tribology for a number of years. Fusaro [11-4] examined dry film lubricant systems based on polyimide binders for use in aerospace applications and determined that polyimides provided a low initial coefficient of friction that rose to higher levels during sliding, which has been attributed to the development of a third body layer. Taylor and colleagues [6-41] listed the family of polyimide compounds as one of the preferred binder systems for the development of friction materials, especially for those targeted for use on high-speed railway applications. In this high-energy braking application, the polyimide might have been a primary binder, but it could also have been used as a property modifier for enhancing phenolic resins. It

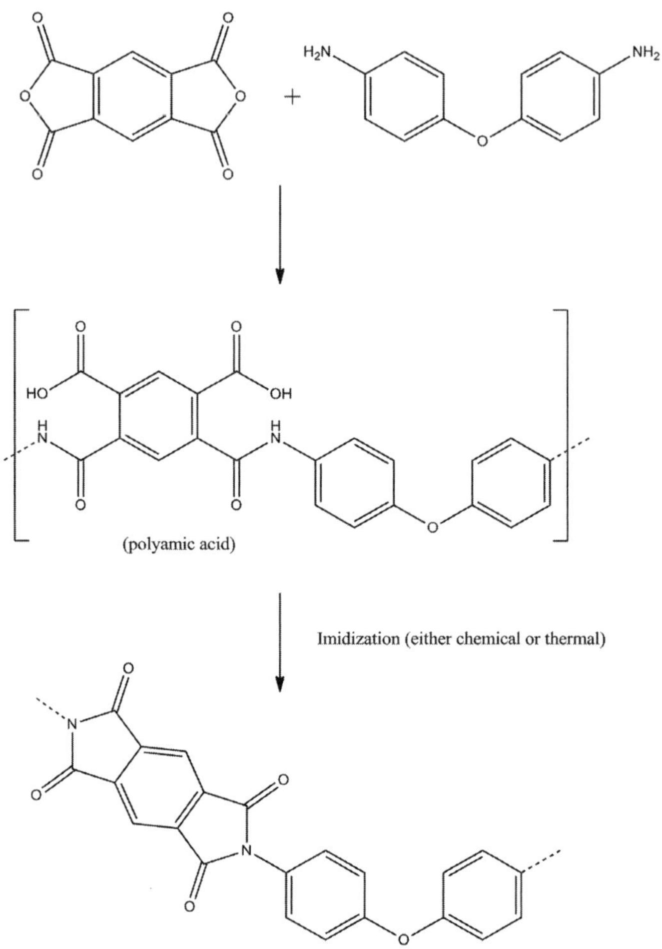

Fig. 11.1 Synthesis of Kapton polyimide.

is interesting to note that their composition also used polyimide compounds in conjunction with other carbonaceous materials such as pitch or asphalt.

Nagahiro and colleagues [11-5] described a specific member of the polyimide class of compounds during the development of a friction material. The compound incorporated a fluoro resin, calcium carbonate, and a synthetic polyimide synthesized by reacting 4,4′-bis(3-aminophenoxy)biphenyl with pyromellitic anhydride. The chemical composition of their unique resin was based on the repeating structure shown in Fig. 11.2.

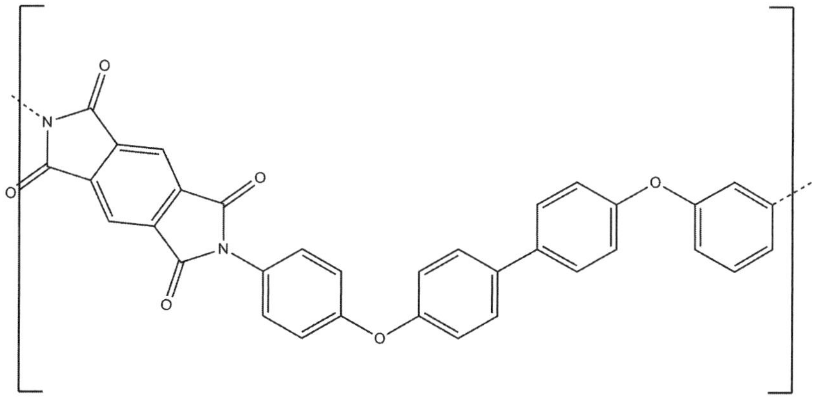

Fig. 11.2 Polyimide structure used by Nagahiro and group.

Results indicate that this original polyimide was a method for improving the abrasion resistance while simultaneously maintaining a stable coefficient of friction under high-temperature operating conditions. In addition to the impressive thermal and mechanical properties, the polyimide matrix also had a significant effect on the friction level of the final product that was further augmented by the performance characteristics of the fluoro additive.

The overriding consideration for selecting a polyimide polymer system as the primary binder in a friction material is often a desire to improve tribological performance and to assure survival at elevated temperatures [11-6]. In addition to the enhanced thermal stability gained from their use, polyimides are also ideally suited for friction applications because the mechanical properties are considerably more abrasion-resistant than other polymer systems. Polyimides have been shown to exhibit stable coefficients of friction over a wide range of operating conditions, including more stable fade performance [6-11].

Issues with Polyimides

Not all studies have found polyimides to beneficial for use in the friction material industry. It has been noted that polyimides are very low in thermal conductivity and do not dissipate heat away from the surface as readily as phenolic resins [6-11]. This insulative attribute can lead to localized thermal excursions at the interface and an increased potential for hot spot formation (i.e., transient thermo-elastic instabilities). This can in turn manifest itself as unacceptable vibrations commonly known as judder.

Jacko [11-7] examined polyimide compounds in a semi-metallic friction material of undefined composition and determined phenolic resins to be superior compared to polyimides. This was a result of the beneficial third body formation properties of phenolic resins that, when tested under rigorous conditions, were found to readily form a coherent third body layer at the friction interface. Polyimide binders were reported to transfer from the friction material composite to the mating metal in large non-uniform masses, which caused higher wear rates. The concern with elevated wear was also reported by Fusaro (11-4) and confirms that polyimide polymer systems have modes of material transfer different from conventional phenolic resin systems.

Polyimides as a Polymer Modifier

Halberstadt [6-29] tested polyimide compounds as a replacement for phenolic resins and found both binders to give similar wear and friction performance during his evaluations. Despite the similarities, it was noted that polyimide polymers have a tendency to be brittle and fracture easily. To offset the brittleness of polyimides, a blend of phenolic resin and polyimide were combined and found to provide a synergistic reduction in wear rates in comparison to either of the individual constituents.

Years later, the concept of polyimide as additive to thermoset resin systems is still being considered. Sasaki and co-workers [11-8] suggested that polyimides can be used to modify phenolic resins for use in a non-asbestos organic friction material.

Lamport's group [11-9] incorporated many aspects of prior attempts (such as Taylor and Halberstadt), but expands on the concept by using a multi-component blend. Lamport's pitch-polyimide-phenolic resin mixture was stated to improve the fracture toughness of a commercial friction material and enhanced thermal stability, which would be reasonably anticipated to improve wear at elevated operating conditions.

The use of polyimide binders is limited to a few industrial and wet friction applications, especially those that demand a friction material with exceptional heat resistance. The limitations in use can also be associated with cost. Polyimide binder systems are significantly more expensive than conventional phenolic resins. Growth in the use of polyimides will continue to be limited until a lower cost manufacturing method has been developed.

Condensed Polynuclear Aromatic (COPNA) Resins

One of the most interesting binders for friction materials is a group of compounds known as condensed polynuclear aromatic (COPNA) resins. COPNA resins were first developed by workers at Fuji Standard Research [11-10] in 1988 and are currently still being explored for use in friction material applications [11-11]. COPNA resins are considered to have significant potential for enhancing the thermal properties of tribological composites.

COPNA resins are based on reactions between large polynuclear aromatic compounds derived from coal tar pitch (such as phenanthrene, anthracene, pyrene, and chrysene), which are polymerized with p-xylylene glycol in the presence of an acid catalyst (such as para-toluenesulfonic acid). The reactions are time-consuming and require elevated temperatures, but yield a polymeric material with integrated methylene bridges between the nuclei of the large aromatic groups. Because of the different molecular species present in pitch, chemical reactions for COPNA resins are difficult to visualize with certainty but can be conceptually idealized by a number of large aromatic subunits connected together as shown in Fig. 11.3.

COPNA Resins in Friction Materials

COPNA resins were found to be beneficial in a limited number of friction material applications because of their wear resistance and

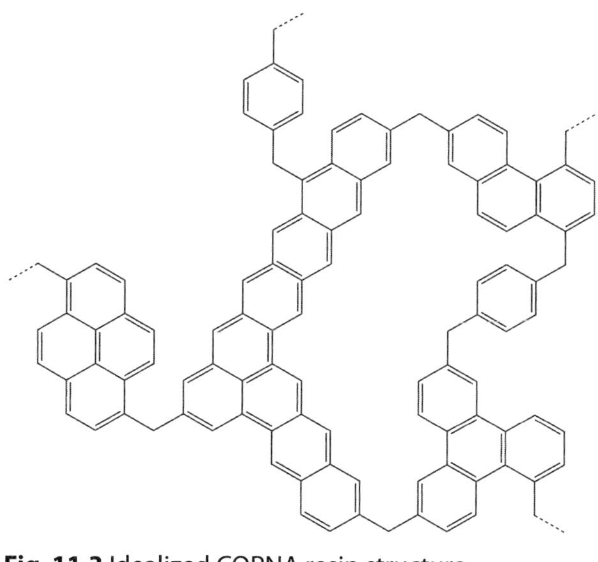

Fig. 11.3 Idealized COPNA resin structure.

ability to withstand elevated temperatures. These benefits are gained by the large carbonaceous moieties present in the structure, which not only limit the oxidation susceptibility of the methylene bridges by steric hindrance but also provide a molecular backbone rich in carbon that is intrinsically temperature resistant.

Komori and colleagues [11-12] used a COPNA resin to enhance a conventional friction material composite by improving the areas of heat and abrasion resistance. It was also found that COPNA resins were able to improve the friction performance in a semi-metallic friction material by enhancing porosity and thus enabling transport of the gaseous decomposition products formed at the interface during high temperature operation [11-13]. It was also interesting that the compositions listed by Komori's group were processed by conventional manufacturing techniques such as those listed in Chapter 5 of this book.

Washimiya-danchi [11-14] found COPNA resins to improve wear in an innovative carbon fiber friction material. COPNA resins improved bond strength between the matrix and the surface of the carbon fibers. The improved adhesion increased the strength and was considered to be the underlying cause for superior wear resistance.

The beneficial bonding characteristics between carbonaceous materials and COPNA resins were also reported by Teruyuki and co-workers at Tokico [11-15]. One root cause for poor wear performance had been linked to poor bond quality between graphite and phenolic resin in their compositions. This effectively limited the concentration of graphite. Replacement of phenolic resins with COPNA resins were shown to enhance the bond strength between graphite and the COPNA matrix, giving the composite with better wear resistance.

COPNA resins have exceptional potential for success. However, like many advanced polymer systems, they have not achieved extensive success outside of a few isolated attempts. It is thought that the cost of manufacture is simply too high for the incremental benefits gained from their use. Therefore, COPNA resins have seen little use in the last twenty years, and use is expected to remain dormant until a lower cost method of manufacture is developed.

Bismaleimide-Triazine (BT) Resins

Bismaleimide-triazine (BT) resins are complicated polymeric materials made by addition reactions of bismaleimide and triazines. The resin that develops is a three-dimensional polymer network

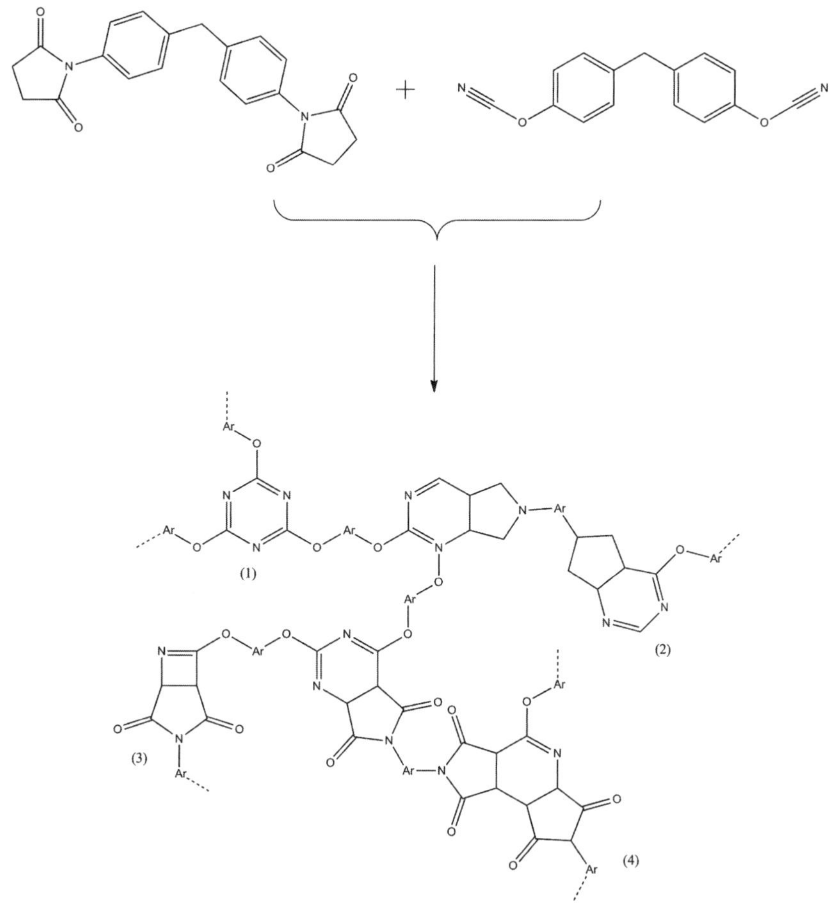

Fig. 11.4 Example of a BT polymerization reaction.

with excellent thermal and electrical resistance. An idealized reaction diagram is shown in Fig. 11.4 and illustrates the complexity inherent in the polymerization of these materials.

The number of potential outcomes for the reactions between bismalcimides and triazines can introduce at least four distinct types of molecular structures to the polymeric backbone. BT resins have not found widespread use as the primary binder of friction materials. In this example, the majority of recorded use of BT resins has been in the form of granulated organic filler, BT dust, intended to replace more conventional CNSL based granulates (i.e., CNSL dust or friction dust).

BT Resin Dust

Kamioka and colleagues [11-16] reported that the use of a granulated powder made from polymerized bismaleimide-triazine resin was superior for preventing brake noise, especially when compared to granulated dust manufactured from polymerized CNSL dust. The BT resin dust could be used in a wide range of concentrations and was not found to affect the friction performance adversely. It was further concluded that BT resin dust generally improved the wear performance of the brake lining composition.

Ohya and Shinohara [11-17] developed a semi-metallic friction material incorporating organic particles based on BT resin technology. The particles were produced by reacting BT resins with an epoxidized novolak phenolic resin. They cited significant improvements in the friction level as well as reduction in noise. It is also interesting to note that, when used in a clutch composition, the wear rates of the compositions with BT resin dust were dramatically lower than the comparable standard materials.

BT resins are another example of the use of the enhanced properties gained from improvements in polymer science. Like the other forms of polymer systems, the benefits of their use are not widely enjoyed because of cost factors. BT resins are not likely to experience growth in the friction materials industry in either the binder class or organic additive class of materials.

References

11-1. Edwards, W., and I. Robinson I. 1955. Polyimides of pyromellitic acid. United States Patent No. 2710853, filed October 19, 1953, and issued June 14, 1955.

11-2. Takekoshi, T. 1996. *Polyimides- Fundamentals and Applications*. New York: Marcel Dekker.

11-3. Kapton. Registered trademark of E. I. DuPont de Nemours and Company, Wilmington, Delaware; Registration No. 0798963, November 23, 1965.

11-4. Fusaro, R. 1982. "Fundamental Aspects of Polyimide Dry Film and Composite Lubrication - A Review." TM-82968. Washington, DC: National Aeronautics and Space Administration.

11-5. Nagahiro, T. et al. 1993. Polyimide based friction material and preparation process of the material. United States Patent No. 5258441; filed May 21, 1992, and issued November 2, 1993.

11-6. Boggs. R. 1991. "Composite Friction Material Thrives on Heat." *Design News* 12: 91-92.

11-7. Jacko, M. et al. 1989. "Wear Debris Compaction and Friction Film Formation of Polymer Composites." *Wear* 133: 23-38.

11-8. Sasaki, Y. et al. 2000. Friction material composition. United States Patent No. 6080230, filed November 13, 1998, and issued June 27, 2000.

11-9. Lamport, R. et al. 1998. Resin mixture for friction materials. United States Patent No. 5753018, filed April 14, 1997, and issued May 19, 1998.

11-10. Otani, S., and Y. Nagai. 1988. Thermosetting composition, method of producing fusible thermosetting resin using same and process for the production of condensed polynuclear aromatic hydrocarbon resin using same. United States Patent No. 4758653, filed February 14, 1986, and issued July 19, 1988.

11-11. Kurihara, S. et al. 2010. Thermosetting resin material. United States Patent No. 7687570, filed June 20, 2008, and issued March 30, 2010.

11-12. Komori, T. et al. 1990. Brake friction material. United States Patent No. 4954536, filed April 19, 1989, and issued September 4, 1990.

11-13. Herring, J. 1967. "Mechanism of Brake Fade in Organic Brake Linings." Technical Paper No. 670146. Warrendale, PA: Society of Automotive Engineers.

11-14. Washimiya-danchi, S. 1991. Heat resistant friction material. European Patent No. 443555,filed February 20, 1991, and issued August 28, 1991.

11-15. Teruyuki, K. et al. Brake friction material. Great Britain Patent No. 2217725, filed April 19, 1989, and issued November 1, 1989.

11-16. Kamioka, N. et al. 1995. Friction material containing BT resin dust. United States Patent No. 5384344, filed September 13, 1993, and issued January 24, 1995.

11-17. Ohya, K, and S. Shinohara. Friction material. United States Patent No. 5411773, filed December 2, 1993, and issued May 2, 1995.

Chapter 12
Inorganic Binder Systems

The fundamentals of ceramics are based on functionality.
—Elizabeth Moss

INORGANIC BINDER SYSTEMS ARE AN ATTEMPT to replace organic carbon based binder systems with more heat resistant materials. This can be typified by use of a ceramic material (i.e., a hard, brittle heat-resistant material produced by firing a mixture of selected chemicals at high temperatures), or it can include pozzolanic or other hydraulic types of cement. In addition, inorganic binder systems can also be metallic (such as the sintered metal class of products).

The use of an inorganic binder material is often motivated by the desire to improve functionality at elevated temperatures or to ensure structural integrity under high-energy conditions. Many early carbon-based binder systems were drying oils, rubber, phenolic resins, and asphalt. Therefore, the thermal resistance was not as high as desired and many attempts to replace organic binders began with friction materials bound together with hydraulic cement.

Hydraulic Cement

Portland cement, also known as Ordinary Portland Cement (OPC), is the most common type of hydraulic cement. It is a powdery mixture of metal silicates and oxides that, when mixed with water, begins a complex series of hydration reactions. The reaction products provide cement with strength and rigidity, especially under compressive loading. Because of its widespread availability and ease of processing, the idea of making a friction material composite bound with a hydraulic cement matrix is an old one dating back to the early days of the last century.

Cement Shoes for Railroad Brakes

Early attempts at manufacturing friction materials for railway use were based on casting mixtures into a thin metallic shell to form shoes that operated against the wheel of the brakes [4-17]. The process of casting a material allowed the use of castings made of metal (such as cast iron) and also provided a convenient manufacturing method for hydraulic cement mixtures.

Johnson and Dickinson [12-1] developed a composition of matter based on a mixture of coal cinders, asbestos pulp, salt, eucalyptus wood pulp, magnesium sulfate, creosote, and cement. The authors did not provide reasons for the use of many of these components. Additionally, they suggested that creosote, a hydrocarbon residue from incomplete combustion of wood, was able to react with magnesium sulfate and provide a means of reinforcing the binder system. The reaction product would have formed a carbonaceous binder filled with heat-resistant charred material and reinforced with cement.

Cement and Animal Blood

The Lima Brake Shoe Company [12-2] obtained a patent for a composition for filling brake shoes (e.g., casting) that were to be used on railway friction applications. The composition contained sawdust and relied on a mixture of Portland cement. Additionally, the composition was said to allow the addition of thermally stable fibers such as asbestos or mineral wool as a means to improve heat resistance. The authors also suggested that ground leather could be added to toughen the matrix, and that graphite could be added to provide lubrication. However, the most unusual aspect of the composition relied on the use of unconventional plasticizing agents (animal blood and sugar), which were added as additional proteinaceous binding agents to improve the integrity of the cement-based composite.

For approximately the next twenty years, the addition of chemically modified animal blood was a staple in friction material compositions based on Portland cement. For example, Norman [12-3] used asbestos, pulverized animal bones, and animal blood in his cement based composition of matter. Stinebaugh [12-4] developed a friction material, targeted for use on automotive applications, which was composed of Portland cement, charcoal, and plaster of Paris (a form of calcium sulfate). In addition, cow's blood that had been reacted with sulfuric acid was added to impart plasticity to the mixture and to aid in binding.

Laeufer [12-5] combined many of the previous findings including sugar, graphite, leather, saw dust, and asbestos in a cement mixture. To this mixture he added animal blood that had been treated with potassium carbonate and mixed with sodium benzoate, which was intended to reduce the flammability of the composite. Laeufer also used linseed oil for water resistance and juniper berries to offset the odor produced during high-temperature operation. In short, his attempt was one of the first methods to improve the fragrance of brakes during operation. After Laeufer's work, the use of animal-derived additives began to dwindle. However, as recently as 1985, animal-derived protein (collagen) [12-6] was still being considered for use in friction materials as a method for reducing the rate of lining wear.

Concerns with Cement

Although Portland cement is easily obtained and readily processed, and forms a rigid structure capable of surviving high degrees of compressive stresses, it is not an ideal binder for friction materials that are subject to tangential loads. The crystallization process that forms the structure is inherently brittle and requires reinforcement. Early attempts to reinforce tread shoes for railway applications used expanded metal structures [12-7] inside the castings to prevent cracking. Alternative methods involved the use of high concentrations of asbestos fiber and compression molding techniques to ensure that the compositions were thoroughly compacted [12-8], followed by treatment with a drying oil to impart water resistance.

Additionally, cementitious materials are chemically very basic and can attack materials that are not resistant to alkali. This tends to limit the selection of materials and requires knowledge of the impact of pH on all of the selected additives.

Cement as an Additive

In 1967, Eschen [12-9] attempted to incorporate cement as an additive to phenolic resin-bonded friction materials. He used high-temperature calcium aluminate cement powder as an additive to impart high levels of friction and enhance wear. Kaufman and co-workers [12-10] also listed the use of calcium aluminate cement as an additive in friction material compositions, but they did not expand on the benefits of using calcium aluminate as an additive.

In general, hydraulic cement is not suited as a binder for friction materials. It is inherently brittle and requires extensive reinforcement. Additionally, the material is caustic and can chemically degrade fillers and fibers that are not resistant to strong caustics. Finally, the nature of cement requires the formation of crystals from the hydration of oxides and hydroxides. As the operating temperature increases because of frictionally generated heat, the crystallites are subject to dehydration, and cement can revert to a powder with little intrinsic strength, resulting in rapid wear and poor friction performance.

Waterglass

Sodium metasilicate is readily soluble in water, forming an alkaline solution that is commonly known as waterglass or liquid glass. The benefit of using waterglass in friction materials is based on the conversion from liquid to solid at temperatures around the boiling point of water, which forms a heat-resistant amorphous glass matrix. Conceptually, the use of waterglass in friction materials is similar to that of cement mixtures where a liquid solution is combined with the necessary additives and then formed into a rigid heat-resistant tribological composite.

One of the first uses of waterglass was by workers [12-11] at the Royal Equipment Company in 1916. In this example, a cone clutch was produced from woven asbestos fabric and saturated with a mixture of waterglass and linseed oil. After saturation, the fabric was formed onto a cone-shaped mandrel and cured under pressure forming a rigid, heat-resistant friction material.

Stenberg [4-6] reported that by the mid 1930s, waterglass was still in use for automotive friction materials and was also being blended with various drying oils, which implies that this technique was in place for several years.

During the same time period, Eisenhardt [6-39] developed an asbestos-reinforced friction material using graphite, kaolin, and rubber in a sodium silicate solution that was used as a saturant to impregnate woven asbestos fabric with a compliant, lubricating, and heat-resistant binder mixture. Eisenhardt concluded that this material was acceptable for use in both drum brake systems and clutch systems.

Over the years, the use of waterglass declined, but there has been an increase in interest in recent years. Ho and Peterson [12-12] examined the use of silicate cements in the development of aircraft brake materials. They found the strength to be lower than desired, but also that wear could be improved by the addition of graphite to the mixture. Kapl and colleagues [12-13] developed a friction lining based on mixtures of soluble silicates and reactive oxides such as those used in high temperature cement. The product was a highly heat-resistant friction material, suggesting that this technology might have promise in selected applications.

More recently, Hörth [12-14] developed a coating for use on phenolic-bound friction materials that would give improved friction stability. The coating was comprised of a solution of sodium silicate with assorted abrasives and functional additives that were intended to reduce the variation observed in the friction levels during braking. It was further noted that friction materials coated with this solution gave improved friction performance immediately after installation (i.e., excellent green or pre-burnish friction levels) compared to those without the coating.

Today, waterglass-based friction materials are not widely used because of the caustic nature of the solution and the hard, inflexible nature of the amorphous glassy matrix formed after curing makes the compounds prone to brittle fracture. Waterglass-based materials seemed to suffer from many of the same ailments associated with the use of hydraulic cements and, therefore, further expansion in use is not expected until advances in flexibility can be achieved.

Ceramic Polymer Binder Systems

In the quest for a tribological composite with extreme thermal resistance, the next logical step is the development of polymeric matrices that can be converted into rigid ceramic materials. The technology behind ceramic precursors is an emerging field of polymer science based on the conversion of carbon and silicon containing molecules into a hard, wear-resistant, thermally stable matrix of silicon carbide.

One of the earlier methods of producing polymeric precursors was developed by researchers at Rensselaer Polytechnic Institute [12-15]. Their approach was to produce polycarbosilane compounds by reacting halomethylcarbosilanes in the presence of a Grignard coupling agent and then reducing with a metallic hydride. The resultant polycarbosilane compounds were typically non-cyclic stoichiometric mixtures of carbon and silicon, which could be converted to silicon carbide by heat treating to approximately 2900°F (1593°C) under nitrogen atmospheres.

Shen and MacDonald [5-39] developed a non-corrosive silicon carbide precursor known as 2,4,6-Trimethyl-2,4,6-trisilaheptane (TMTSH). TMTSH is unique in that it is a single molecule containing only carbon, hydrogen, and silicon. It is highly suited for chemical vapor deposition methods. This novel new precursor technology can be converted to substantially pure silicon carbide by using deposition temperatures between 1200 and 1800 °F (approximately 650 and 1000°C). When used with carbon fiber substrates, the product is a silicon carbide matrix formed around the carbon fibers that possesses exceptional strength and heat resistance, and that is low in weight compared to metallic structures of the same geometry.

The use of ceramic precursor compounds in friction materials has been targeted for use in high-energy aerospace brake systems. Prunier and Pyzki [12-16] discussed the energy levels associated with braking of an aircraft during landing. They rejected take-off scenarios where the temperature levels can reach between1100 and 1800°F (approximately 600 and 1000°C) at the brake interface. As a result, they disclosed the development of a reinforced ceramic matrix composite that has a melting point in excess of 2500°F (approximately 1400°C) and flexural strength of approximately 22,000 psi (150 MPa) at temperatures of 1652°F (900°C). This product provides the necessary strength at elevated temperatures to enable the product to survive aircraft braking during landing events.

Ceramic Technology Allows for Molding Operations

Strasser and Atmur [6-40] used high temperature fiber reinforcement to manufacture a ceramic brake pad and carrier plate in a single molding operation. Their technique relied on the ability to mix fiber with a ceramic precursor prior to forming it into shape. After the article was molded, the product was given a final thermal treatment to convert the polymer into a rigid matrix of ceramic material, which provided the necessary strength and frictional performance for elevated operating conditions.

Most recently, Bauer [12-17] developed a high-temperature friction material based on a combination of proven methods. The first portion of his method involved the use of a carbon fiber reinforced structure that was treated with a carbon precursor and formed a carbonaceous matrix upon heating. After the carbon structure was produced, a silicon carbide precursor was used to form a carbon-ceramic material capable of meeting the structural and thermal expectations of the most rigorous operating conditions.

Today, the use of ceramic precursor materials is very limited because of the extreme cost of these materials relative to the lower cost organic polymers. Additionally, the current methods of production involving vapor deposition and high-temperature conversion tend to restrict their use to those industries that positively demand this extreme level of high-temperature performance. Those arenas that can tolerate less performance tend to find lower cost solutions.

Metallic Binder Systems

There are a vast number of potential citations regarding friction materials that were developed based on properties of metallic binder systems, especially from the early patent literature of the United States. Specifically, the use of cast iron and sintered metal powders are two of the oldest forms of inorganic binder systems. In many cases, the methods of manufacture and the underlying causes for their use have changed little and have been described in prior sections of this book. This is a testament to their robust nature. Metallic binder systems remain a time-tested approach for the successful elimination of low-temperature organic binder materials.

Cast Iron

Cast iron is currently widely used in modern automotive tribological systems in the form of brake drums, discs (rotors), and assorted metal mating members. Its role as a mating member is based on a successful history of application that dates back to 1932 [12-18]. The continued use of cast iron mating members is a result of the well-known thermal characteristics, ease of fabrication, and relative low cost [12-19, 12-20]. Aside from being a mating member operating against a polymeric friction composite, cast iron is also a prominent friction material with a history of use in railroad applications that extends back to at least 1868.

Drake [12-21] used a cast iron brake shoe that was "more durable and less liable to injure the wheels." In this specific application, the cast iron shoe was formed with a groove that afforded an easy means of attachment and also provided the brake shoe with specific advantages over current brake shoes of this time. Namely, the groove admitted air that was intended to enhance cooling and thus prevented the wooden brake members from catching fire because of the radiant heat of the brake interface. Additionally, the groove provided a mechanical benefit and was capable of cleaning contaminants such as snow or mud from the wheel, allowing for better contact between the two bodies.

After 100 years, cast iron brake shoes are still in use as friction materials for railway applications. Wise and Lewis [12-22] examined the replacement of cast iron brake shows with composition shoes (i.e., those made with organic binder systems and comprised of various additives). It was reported that composition brake shoes were likely to cost as much as five times the price of cast iron. However, the higher cost could be offset by the longer life expectancy provided by composition shoes, which reduced maintenance costs.

A short time later, Mercer [12-23] reported on the use of medium friction non-metallic composition brake shoes as a direct replacement for railway applications operating in the mountainous terrain of Peru. After an evaluation period involving systematic testing and comparison of composition shoes to cast iron, it was reported that medium friction composition shoes provided the necessary static friction to ensure that the trains were not subject to roll away when parked. Furthermore, it was found that composition shoes gave higher coefficients of friction and provided longer life expectancies than cast iron. Ultimately, after six years of effort, composition brake shoes were found to be suitable for replacement of cast iron shoes.

In the last few years, Haddadi and colleagues [12-24] examined the wear and thermal effects of low modulus polymer-based friction material composites as a replacement for cast iron brakes shoes. It was generally concluded that cast iron brake shoes have poor friction performance when operating in dry environments. It was also noted that cast iron brake shoes generate high tread wear rates and unacceptable levels of thermal damage, and are noisy during braking.

Despite their use over a period of more than one hundred years, the use of cast iron as a friction material is beginning to see a gradual decline. It is expected that composition shoes, which are under a continual process of development and refinement, will eventually replace cast iron in virtually all railway applications. However, for the time being, cast iron brake shoes are still available and in use.

Sintered Metals

Sintered metallic friction materials were discussed previously in this book. They rely on the fusion of metallic particles to form a rigid and thermally resistant matrix. Their use dates back to the early part of the last century [6-44]. They generally use finely divided powders of copper and iron, which contain select tribological enhancing additives. Additionally, these compounds have been used in virtually all aspects of the friction material industry.

Uses of Sintered Metal Friction Materials

Sintered metal friction materials are used in a number of applications. Johnson and Stefl [12-25] discussed the use of full sintered metallic friction materials in automatic transmissions. Their targeted application involved high energy levels that were expected to occur during frequent use. The friction materials were thus subjected to very high operating temperatures. This high level of thermal abuse would degrade, if not completely destroy, conventional organically bound friction materials.

Ferrer and colleagues [12-26] examined the performance of novel sintered brake material comprised of an iron-copper-chromium-graphite alloy and compared the results to conventional cast iron brake shoes in railway applications. The sintered metallic lining provided higher coefficients of friction with better friction stability than the incumbent cast iron shoes. Additionally, the inherent porosity of the sintered metallic material prevented excessive transfer of thermal energy through the friction material.

In addition, Anderson [12-27] disclosed that sintered friction materials have found use on high-speed railway and racing applications. The success of these materials is based on their ability to survive the extreme temperatures generated during use. In general, the development of a metallic matrix allows the friction material to possess a higher heat capacity and a better ability to dissipate frictionally generated heat, which in turn affects the final temperature reached during use [12-28].

Sintered Copper

As mentioned previously, copper-based friction materials were one of the oldest forms, and dendritic copper was considered an ideal morphology for these processes of copper [6-44]. More recent work [12-29] regarding metal ceramic composites for use in high-energy brake systems suggests that material transfer is a common occurrence with copper-based materials and that a hard material such as mullite should be included to reduce or prevent material transfer.

Kryachek and Shvedkov [12-30, 6-49] evaluated a copper based friction material in the form of an aluminum bronze alloy. In general, their findings confirmed that sintered materials have a higher and more stable coefficient of friction, better wear and heat resistance, and a higher degree of resistance to seizing or jamming when compared to conventional friction materials. One interesting point is that the addition of aluminum affected the microstructure by forming solid inclusions in the eutectoid phase during cooling. This presence of inclusions was viewed as an important contributor to the final performance of the product.

Sintered Iron

Copper as an inorganic matrix is not ideal for all applications. Several studies on friction materials by Fedorchenko and colleagues [12-31, 12-32] suggested that copper is not suited for use in friction systems that operate in an unlubricated or dry state. The suggested alternative was an iron-based friction material that had been allowed with specific amounts of phosphorous. It was further stated that the addition of ferro-phosphorous to the melt was critical and should ideally be controlled to very precise levels to fully realize the benefits of using a sintered iron friction material.

It is quickly obvious from this information that the use of metallic materials requires an in-depth knowledge of the metallurgical effects of additives. For instance, graphite additions to iron-based friction materials were found to be critical to enhance friction stability. Levels of 3% were recommended [12-33]. Furthermore, additions of reactive metal sulfides such as copper or zinc sulfide were beneficial in the formation of a protective sulfidized iron graphite film that prevented the friction material from seizing at high rubbing speeds.

Finally, it should be noted that sintered metallic matrices have areas of concern. Ferrer's group [12-26] noted that sintered products were found to experience a decrease in shear strength when operating at high speeds. This resulted in poor wear performance under these conditions and can be attributed to thermally induced physical effects such as softening of the particles.

References

12-1. Johnson, L., and P. Dickinson. 1906. Composition of matter. United States Patent No. 827977, filed January 10, 1906, and issued August 7, 1906.

12-2. Laeufer, J. et al. 1909. Composition of matter. United States Patent No. 909617, filed February 14, 1908, and issued January 12, 1909.

12-3. Norman, J., 1909. Composition of matter. United States Patent No. 943157, filed February 8, 1909, and issued December 14, 1909.

12-4. Stinebaugh, C. 1913. Material for automobile brake-shoes. United States Patent No. 1076325, filed March 8, 1912, and issued October 21, 1913.

12-5. Laeufer, J. 1929. Brake-shoe filling and lining. United States Patent No. 1724718, filed June 11, 1927, and issued August 13, 1929.

12-6. Cartier, M., and G. Cros. 1985. Organic filler of animal origin to be used in the preparation of friction materials offering a very low rate of wear. United States Patent No. 4491542, filed September 30, 1982, and issued January 1, 1985.

12-7. Chipley, G. 1915. Brake-shoe. United States Patent No. 1136936, filed August 5, 1914, and issued April 27, 1915.

12-8. Andrews, P. 1925. Method of making artificially integrated shaped bodies and the product of said method. United States Patent No. 1559146, filed May 7, 1924, and issued October 27, 1925.

12-9. Eschen, F. 1967.Friction composition for high energy service. United States Patent No. 3297599, filed May 24, 1965, and issued January 10, 1967.

12-10. Kaufman, E. et al. 1983. Friction lining material. United States Patent No. 4388423, filed July 29, 1982, and issued June 14, 1983.

12-11. Kelley, E. 1916. Method of making cone clutch facings. United States Patent No. 1180503, filed February 3, 1915, and issued April 25, 1916.

12-12. Ho, T., and M. Peterson. 1974. "Development of aircraft brake materials." National Aeronautics and Space Administration, CR-134663, Washington, D.C., 1974.

12-13. Kapl, G. et al. 1995. Friction lining and process for the production thereof. United States Patent No. 5433774, filed November 15, 1993, and issued July 18, 1995.

12-14. Hörth, F. J. 1999. Stabilizing friction levels. United States Patent No. 5972090, filed July 9, 1998, and issued October 26, 1999.

12-15. Whitmarsh, C., and L. Interrante. 1992. Carbosilane polymer precursors to silicon carbide ceramics. United States Patent No. 5153295, filed July 20, 1990, and issued October 6, 1992.

12-16. Prunier, A., and A. Pyzik. 1999. Ceramic metal composite brake components and manufacture thereof. United States Patent No. 5878849, filed May 2, 1996, and issued March 9, 1999.

12-17. Bauer, D. 2008. Method of making carbon/ceramic matrix composites. United States Patent No. 7374709, filed January 11, 2005, and issued May 20, 2008.

12-18. Main, F. 1932. "Brake Drum Metallurgy." Technical Paper No. 320006. Warrendale, PA: Society of Automotive Engineers.

12-19. Zaidi, H. et al. 1999. "Behaviour of Metallic Additive in Composite Metal-Carbon/Steel Brake in Severe Friction Test." *In Lubrication at the Frontier* edited by D. Dowson et al. Amsterdam: Elsevier.

12-20. Cho, M. et al. 2003. "Tribological Study of Gray Cast Iron with Automotive Brake Linings: The Effect of Rotor Microstructure." *Tribology International* 36: 537-545.

12-21. Drake, G. 1868. Improvement in car brake shoes. United States Patent No. 75001, issued March 8, 1868.

12-22. Wise, S., and G. Lewis. 1970. "Composition Brake Blocks and Tyres." *Journal of Institute of Mechanical Engineers, Railway Division* 1: 386-443.

12-23. Mercer, C. 1987. "Medium Friction Non-metallic Composition Brake Shoes as Direct Replacement for Cast Iron Brake Shoes on Rolling Stock." Proceeding of Institute of Mechanical Engineers, 130/86, pp.11-20.

12-24. Haddadi, E. et al. 2005. "Wear and Thermal Effects in Low Modulus Polymer-based Composite Friction Materials." Journal of Applied Polymer Science 95: 1181-1188.

12-25. Johnson, J., and E. Stefl. 1956. "Friction Materials for Automatic Transmissions." Technical Paper No. 560308. Warrendale, PA: Society of Automotive Engineers.

12-26. Ferrer, C. et al. 2010. "Tribological Study of Fe-Cu-Cr-graphite Alloy and Cast Iron Railway Brake Shoes by Pin-on-disc Technique." Wear 268: 784-789.

12-27. Anderson, A. 1992. "Friction and Wear of Automotive Friction Materials." In *Friction, Lubrication and Wear Technology, ASM Handbook* ,vol.18, pp. 569-577.

12-28. Angus, H. et al. 1967. "Conditions Leading to the Failure of Cast Iron Brakes." *Metals and Materials* 1, pp.18-28.

12-29. Bill, R. 1974. "Some Metal Graphite and Metal Ceramic Composites for Use as High Energy Brake Lining Materials." TN-D-7756. Washington, DC: National Aeronautics and Space Administration.

12-30. Kryachek, V., and E. Shvedkov. 1969. "Sintered Friction Materials Based on Aluminum Bronze. Part II: Effects of Some Production Technology Factors on Properties of Materials." *Poroshkovaya Metallurgiya* 10: 66-70 (Russian).

12-31. Fedorchenko, I. et al. 1965. "Studies of Friction Materials - Part 1." *Poroshkovaya Metallurgiya* 5: 54-57 (Russian).

12-32. Fedorchenko, I. et al. 1965. "Investigation of Friction Materials - Communication 2." *Poroshkovaya Metallurgiya* 9: 65-68 (Russian).

12-33. Fedorchenko, I. et al. 1968. "Properties of Sulfidized Iron-Graphite Materials at High Rubbing Velocities." *Poroshkovaya Metallurgiya* 3: 28-31 (Russian).

Chapter 13
Organic Fiber

Leadership without support is like trying to make bricks without enough straw.

—Reed Markham

ORGANIC FIBER CAN BE SYNTHETICALLY produced or derived from natural sources. A large majority of naturally occurring organic fibers are derived from plants and are based on repeating cellulose units. The use of fibrous plant matter (such as straw in the manufacture of bricks) is one of the oldest methods for reinforcing a composite material and has been practiced since antiquity. Other naturally occurring organic fibers can be derived from animal origins such as hair or wool. Modern science tends to mimic nature, and synthetic organic fibers can include materials ranging from chemically modified cellulose (such as viscose rayon) to more advanced fibers (such as aromatic amides), which have exceptional thermal resistance and possess incredible strength.

Traditionally, the role of organic fiber in tribological composites was as an adjuvant to facilitate processing of asbestos. Examples include flax [4-6] and cotton [13-1, 6-24] typically in the form of thread used as the core of asbestos yarn to enhance production of woven friction materials. Today, organic fibers continue to be used as processing aids in the manufacture of friction materials.

Organic Fibers Are Processing Aids

Swinburn and Bartram [13-2] employed organic fiber to convey strength to the friction material mixture during production. The fiber (typically jute, sisal, or cotton linters) was thoroughly dispersed in the mixture and, when compacted, provided a means for retaining particles and preventing segregation. In addition, fiber entanglement provided green strength to a compacted mixture and thus improved handling of preforms. This was especially true when highly fibrillated fibers (i.e., pulp) were utilized.

Organic Fibers Affect Friction and Wear

Aside from acting as a processing aid, organic fibers can have a significant influence on the wear and friction performance of modern tribological composites. One of the more recent examples by Kim and Jang illustrated how modern synthetic organic fibers can provide a number of beneficial effects including enhancing particle retention, acting as an effective processing aid, and providing the composite with higher degrees of wear resistance and improved friction stability [3-42].

Organic fibers of all types can have pronounced effects on the performance characteristics of a product. Therefore, a deeper look into a few of these materials is in order. To limit the scope of the discussion surrounding organic fibers, will be limited to natural cellulose fiber and a few of the more common synthetic fibers such as polyacrylonitrile, polyethylene, aramid, and polybenzimidiazole.

Cellulose Fiber

Chemically, cellulose fiber is a natural polysaccharide based on a linear arrangement of repeating glucose units. It is a structural component of cellular walls in most plants and is exemplified by cotton fiber; commonly used in various parts of the world in textile and paper manufacturing. Cellulose is the most common organic fiber in the friction material industry and has seen continuous use for well over a century.

In some of earliest examples, cellulose fiber was not a specific compositional additive, but aided in improving a design for the brake industry. Shaw [13-3] used cotton wadding in a brake design as early as 1885. This design incorporated cotton wadding previously saturated with rosin and oil. The saturated cotton was embedded in the friction surface to act as a lubricant reservoir allowing the wheel brakes to operate with trace amounts of lubricating oil at the surface. This promoted longer life, less wear, and less dust generation. This design reappeared seven years later in a slightly different form [13-4] and still relied on the absorbent nature of a cellulose packing material to store linseed oil in recesses formed in cast iron brake shoes.

By 1908, cellulose fibers began to be incorporated in brake linings in more conventional ways. The team of Frood and Holmes [13-5] used bits of fabric made from cotton cloth to produce early brake linings comprised of metallic particles in a rigid matrix. A few years later, Knowles [13-6] incorporated cotton into an asbestos tape that was subsequently saturated with a drying oil and sulfur. The product was then vulcanized and used as a friction fabric for brake linings.

By the 1920s, the use of cotton in asbestos-based friction materials was exceedingly common. Brown and Higgins of the Bakelite Corporation

[13-7] carded together asbestos and cotton fiber to form a dry laid fibrous felt. The felt was coated with phenolic resin and folded to form a yarn, which was then woven to produce a highly durable friction fabric.

Workers at the World Bestos Corporation [13-8] developed a number of methods for the production of asbestos yarn for weaving friction materials. Many of their methods were described previously in this book and relied heavily on cellulose based fibers to provide strength. This allowed for twisting or wrapping of thin strips of moistened asbestos paper around a core to form yarn suitable for weaving.

Cellulose in Paper Friction Materials

Aside from yarn and thread, cellulose fiber has been a key constituent in the development of paper-based friction materials. The advent of cellulose-based paper as a friction material began as early as 1885 when Allen [13-9] developed a very unconventional friction composition by compressing strips of elemental lead with layers of cellulose paper. The combination of compressed paper with strips of soft metal provided a favorable balance of wet and dry friction performance that allowed it to be used across a range of operating conditions as a brake lining for railway applications.

More conventional methods for producing paper-based friction composites can be seen in the early patent literature of Raybestos-Manhattan, with Novak [13-10] and Judd [13-11] as two of the more pertinent examples. Novak described in great detail the process of papermaking and subsequent saturation processes for making a cellulose paper friction material. Judd took this process one step further by demonstrating the use of process aids such as starch or latex, which could be added during mixing (e.g., the pulping process). These additives provided wet pulp with a higher degree of strength and substantially improved the papermaking process, leading to a number of improvements in the manufacture of annular rings of friction material for use in clutch and brake applications.

Today, the manufacture of friction materials for use in wet friction applications remains dependent on the papermaking process and continues to be a major consumer of cellulose fiber. For instance, Mori and colleagues [6-58] described a simple friction paper with a mixture of fiber containing approximately 35% by weight cellulose pulp. Likewise, Dong [13-12] also used a blend of fibrous materials including a portion of cotton fibers at a concentration of about 10–15% by weight.

Cellulose in Dry Friction Materials

Cellulose fiber has also been used in many other types of tribological composites including brakc linings for use in automotive and modern railway applications. Aldrich [13-13] used cellulose in combination with steel fiber to effectively replace asbestos in automotive brake linings. He determined that cellulose was beneficial when used in conjunction with steel fibers and should be present at a concentration of at least 5% by volume.

Formolo [13-14] developed a self-lubricating brake shoe composition for use as a castable friction material for transit type railway vehicles. This composition used a mixture of inorganic filler materials, organic binders, crosslinking agents, and lubricating agents in combination with a mixture of inorganic and organic fibers. As part of the fibrous mixture employed in this design, cellulose fiber was reported to be included in the range of 4-9% by weight.

Despite widespread use, cellulose is limited by poor strength and low thermal resistance. In a series of studies designed to evaluate the performance of friction material composites based on comparisons between organic fibers, Satapathy and Bijwe [13-15 and 10-16] found cellulose to be inferior with regard to friction, specifically noting poor fade characteristics. Furthermore, composites based on cellulose fibers exhibited reduced wear resistance, and severe surface damage was observed after testing at elevated temperatures, thus implying a loss of structural integrity at the friction interface.

Sergienko [10-12] examined the impact of cellulose during the development of an asbestos-free material for fabricating brake shoes. He reported that excessive concentrations of cellulosic additives can result in higher wear rates (by as much as an order of magnitude), especially at higher operating pressures. However, it was noted that cellulose can impart better friction stability by providing an in situ means for developing internal porosity.

The mechanism proposed suggests that during braking, heat generated at the interface will decompose cellulose and increase the porosity level of the heat-affected friction layer. The resulting increase in porosity level enhances friction performance by providing a series of interconnected pathways for the transfer of gaseous decomposition products that, according to Herring [11-13], is the underlying mechanism for thermally induced reductions in the observed friction level (i.e., fade).

Polyethylene

Polyethylene fiber is a lightweight high-strength thermoplastic fiber with exceptional toughness. It has been used in a number of composite materials including ballistic materials. Although the benefits of polyethylene are numerous, the use of thermoplastic fibers in tribological composites has generally been avoided because of fear of melting at the friction interface, which could lead to a loss of friction. However, there is one notable exception.

Nakagawa [13-16] employed thermoplastic fibers as a component in a friction material underlayer. The underlayer was described as a second layer of material interposed between the steel backing plate and the friction material layer. The underlayer was incorporated with the intent of inhibiting thermal transport from the friction material interface deeper into the brake components. Nakagawa employed thermoplastic fibers with a prescribed melting point as a means of absorbing thermal energy during braking by endothermic absorption.

In other words, as the heat of braking migrates into the underlayer, the polyethylene fiber melts absorbing thermal energy. In addition, the process was stated to enhance the structure by acting as a secondary binder system to coat other materials in the underlayer. It was recommended that steel fiber be present in the underlayer to ensure adequate strength when the thermoplastic fiber softened, without which structural integrity was expected to suffer.

Polyacrylonitrile (PAN)

The use of fusible acrylic, or modacrylic, fibers as an additive in friction materials was reported by Le Lannou [13-17]. Le Lannou's method was intended to improve wear resistance of the composite. In this example, fusible fibers were present at a concentration that did not generate an excessively thick molten layer, but were of sufficient concentration to develop a wear resistant "crust" on the surface of the friction material. The "crust formation process" is thought to have been the result of the ability of the molten polymer to bind wear debris particles together and assist in retaining a relatively thick third body layer at the interface.

Highly Fibrillated Acrylic Fiber

Doxsee and colleagues [13-18] developed an acrylic fiber with small amounts of methyl methacrylate incorporated in molecular structure. The resultant fiber was easily fibrillated forming a pulp with a very high surface area. Additions of the high surface area pulp to a dry mix friction composition dramatically increased the strength of performs (i.e., pre-compacted powdery mixtures) and enabled more efficient manufacturing of tribological composites. A similar report by workers at BASF Corporation [13-19] described the manufacture of an acrylic fiber suitable for increasing green strength of compacts. They found that an essential attribute was a lack of resiliency upon compression (i.e., the fibrous preforms did not rebound or relax after compaction). In addition to improving the manufacturing process, acrylic fiber additions to friction materials can affect the tribological performance in several ways.

Tribological Impact of Acrylic Fiber

Horiguichi [13-20] replaced aramid fiber in a non-asbestos friction material for use in automotive drum brake applications. Compositions containing aramid fiber created friction instability resulting in an increase in noise levels generated during braking. Replacement of the aramid fiber with an acrylic fiber reduced the potential for noise and resolved the concern.

Satapathy and Bijwe [13-21] examined acrylic fibers as part of their study of organic fiber additions to asbestos-free friction materials. It was determined that inclusion of acrylic fiber would impart toughness and increase mechanical strength. However, the coefficient of friction was lower than composites made with aramid fiber. Additionally, the wear of the product was improved, but this might have been an effect of the reduced frictional output (i.e., less work). In this case, the acrylic fiber appears to have functioned as a lubricating fiber.

Acrylic Fiber as a Carbon Precursor

One of the largest industrial uses of acrylic fiber was as a precursor to the manufacture of carbon fiber. Evans, in his work with Trainor [13-22] and Crean [13-23], discussed the conversion of polymerized acrylonitrile fiber into carbon fiber and stated that the process begins by cyclization of the molecules at temperatures of 375–425°F (approximately 190–220°C) in the presence of oxygen. This generally involves shrinkage of the fiber, which can be as much as fifteen percent, and requires stretching to ensure that the final carbon fiber has the desired mechanical properties.

Subsequent heating results in carbonization at temperatures of 572°F (300°C). This conversion process is the basis of the carbon

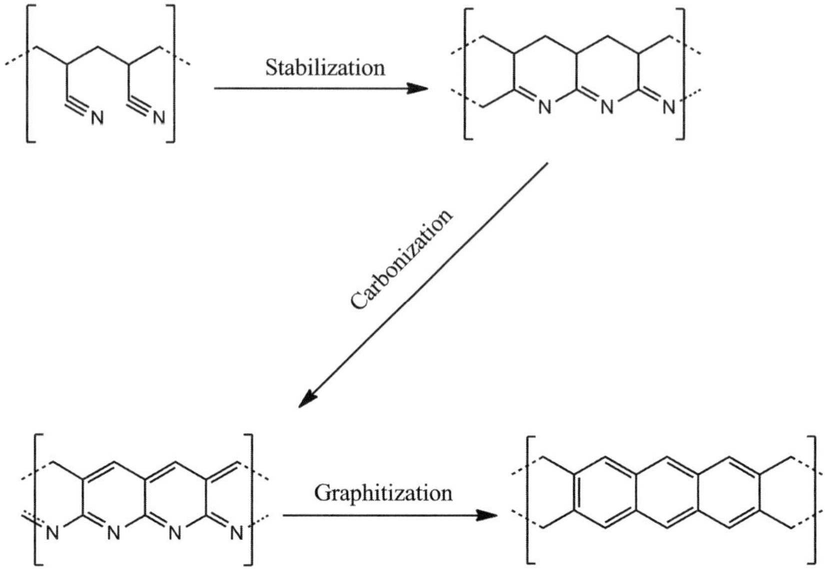

Fig. 13.1 Conversion of acrylic fiber into carbon (graphite) fiber.

fiber manufacturing process and can be visualized by the diagram shown in Fig. 13.1.

Byrne [13-24] discussed the use of carbon fibers derived from PAN fiber in the development of carbon-carbon composites for high-energy brake materials. He states that acrylic fibers are generally used to produce carbon fibers because they are lower in cost and less brittle. It was noted that the flexibility of the fibers allows them to bend during needle punching operations, which permits the production of a three-dimensional network of carbon fiber.

Because acrylic fiber can be pre-oxidized to form a carbon-like fiber, the use of acrylic blends with pre-oxidized fibers would be expected to provide improved handling and processing with an additional benefit of increased thermal resistance. This was confirmed by Patil and co-workers [13-25]. Their evaluation of a friction material containing a blend of acrylic pulp with pre-oxidized acrylic fibers was reported to provide a low-cost method for improving the performance of commercial friction materials without negatively impacting the manufacturing process.

Because pre-oxidized fibers would be expected to be thermally resistant, the work of Ogawa and colleagues [6-34] is also of interest.

Their method involved converting acrylic fibers into a fiber by heat treating fibers under tension and then carbonizing in a nitrogen atmosphere. The end result was a carbonized fiber with a bound nitrogen content of about 14 21% by weight. This pre-oxidized fiber was used in friction materials at about 20 parts by weight and was found to enhance the overall tribological performance.

Aramid Fiber

Aramid is the name of a class of chemical compounds known as aromatic polyamides. Aramid fibers are synthetic fibers in which at least 85% of the amide linkages are directly attached to aromatic moieties in the molecule. This class of engineered fiber possesses a combination of properties that has allowed it to occupy a place of distinction as one of the most important ingredients in the development of modern non-asbestos friction material composites.

The synthesis of aromatic polyamides suitable for spinning into fibrous forms was first reported by Kwolek and co-workers at E. I. du Pont de Nemours and Company in the early 1960s [13-26]. Chemically, the fiber was produced by combining selected aromatic diamines with aromatic acid dihalides in a suitable solvent, which ultimately led to the development of a novel new compound known as poly-paraphenylene terephtalamide.

Poly-paraphenylene terephtalamide is a linear polymer capable of extensive hydrogen bonding along the length of its structure. It typifies the unique chemistry behind the success of para-aramid fibers. The formation of poly-paraphenylene terephtalamide and a method of hydrogen bonding between the polymeric chains can be seen in Fig. 13.2.

After the reaction is complete, the polymer is spun from a solution of liquid crystals into a yellow colored fiber that has the highly desirable attributes of being low-density, high-strength, and resistant to thermo-oxidative decomposition at elevated temperature. Based on this unique combination of properties, the use of poly-paraphenylene terephtalamide fiber was readily adopted as a potential replacement for asbestos in a number of diverse applications by many friction material producers.

Aramid in Friction Materials

One of the first uses of aramid fibers in a tribological composite was intended as a means of replacing asbestos fibers in a railway brake lining. Littlefield [13-27] felt it critical to use aramid fiber in his

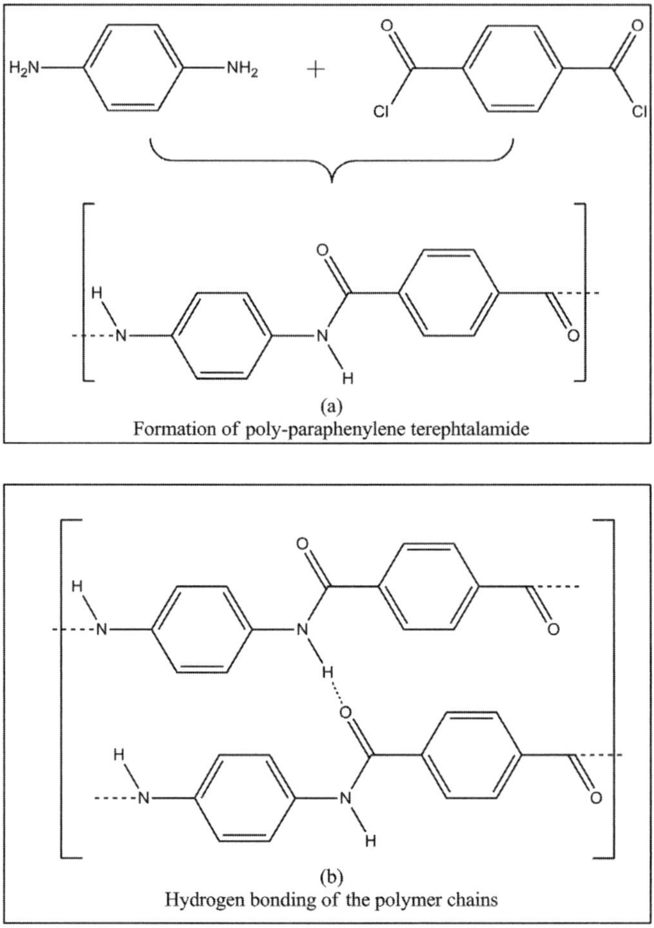

Fig. 13.2 (a) Formation of poly-paraphenylene terephtalamide; (b) Hydrogen bonding of the polymer chains.

composition because of the high energy levels involved in railway braking applications, which were capable of generating temperatures greater than 1832°F (1000°C) at the interface of the shoes and the steel wheel. It was reported that aramid fiber possessed the combination of high tensile strength and thermal resistance necessary for ensuring that the composite would retain functionality at elevated temperatures.

Gallagher and Dougherty [13-28] examined aramid fiber in the form of fibrillated pulp as an additive to both semi-metallic and non-asbestos organic friction materials for automotive brake linings. It was determined that aramid fiber additions enhanced flexural strength of the composite and aided in manufacturing by imparting green strength to the preforms. This level of benefit could be achieved by adding aramid pulp at concentrations up to 15% by weight, but an ideal balance of properties and cost was found at only 4% by weight, further supporting the use of aramid pulp at many different levels.

Parker [13-29] combined a number of manufacturing techniques including a slurry process, similar to the paper industry, followed by compression molding to manufacture a non-asbestos organic brake lining. The combination of manufacturing processes allowed him to use very high concentrations of aramid fiber at a level of 34% by volume. Loftipour [8-20] on the other hand, used a blend of aramid fiber in combination with polyacrylonitrile fiber to develop a friction material for modern railway use and used only 1% by volume of each of the fiber types.

Other manufacturing methods allowed the use of aramid fibers at even greater levels. Bortz [6-33] produced a non-woven air-laid mat of aramid fibers with a carding machine. The fibers were long enough to ensure entanglement, and needle punched to introduce three-dimensional arrangements of the fibers. The result was a product composed entirely of aramid fiber and bound together with a small amount of heat-resistant polymeric binder, which was found to be an ideal example of a nonwoven friction material for use in immersed brake applications.

The Many Benefits of Aramid in Friction Materials

The number of citations regarding the benefits of aramid fiber has grown steadily since its commercial introduction. Many of these sources consistently highlight specific areas where aramid fiber has played a critical role in resolving an issue associated with other forms of fibrous replacement. For example, Siegfried [13-30] developed a friction material suitable for use in brakes and clutches using aramid fiber as a replacement for asbestos. In this case, the aramid fiber demonstrated an ability to reinforce the composite structure effectively, enabling it to survive high-speed bursting tests. In addition, the fiber did not generate abrasion, which was a concern with harsh contaminants often present in asbestos.

The strength imparted to a composite by aramid fiber has been stated by a number of workers including Loken [13-31] and Kim and colleagues [13-32]. In all cases, the introduction of aramid fiber improved the mechanical strength of a composite and enhanced the ability of a composite to resist crack propagation. Additionally, pulp forms of aramid fiber are highly effective at retaining particles. They help to ensure the integrity of dry mixtures as well as provide an unmatched level of green strength to the preforms.

Aside from the remarkable strength of aramid fiber, a nearly universal observation finds that aramid fiber can reduce wear rates in virtually all types of friction materials. Anderson [12-27] reported that small amounts of aramid fiber (2-3% by weight) added to the improved wear rates of semi-metallic friction materials as a result of the increased strength provided by the fiber. Likewise, Satapathy and Bijwe [13-21] determined that aramid fiber was the best fiber choice for enhancing wear resistance of non-asbestos friction materials.

Gopal's team [13-33, 13-34] conducted a two-part study using aramid fiber and concluded that it also provided beneficial performance improvements by stabilizing the coefficient of friction, allowing for more consistent braking. In addition to the tribological effects, aramid fiber has the secondary effect of being able to dampen certain forms of high-frequency noise, resulting in quieter operation of the brakes.

Concerns with Aramid Fibers

Despite the benefits, the studies by Gopal's group also highlighted an area of concern with the incorporation of aramid fiber into a phenolic-bound friction material. Most notably, it was reported that aramid fibers can reduce the coefficient of friction. The reduced coefficient of friction can also be the source of the observed reduction in wear rates because of a reduction in work. The reduction in the coefficient of friction associated with using aramid fiber has also been confirmed by Kato and Magario [13-35] as well as Briscoe [13-36].

In fact, it was Briscoe who suggested that aramid fibers behave as a lubricant and, when used alone, do not have the fundamental tribological properties that are necessary in today's modern friction materials. Therefore, most modern friction materials have adopted a philosophy that was originally proposed by Loken and resorted to using various blends of fibrous materials.

Aramid in Fiber Blends

Workers at Honda [6-22] developed a friction material with a blend of fiber utilizing the particle retaining characteristics of aramid pulp with the strength attributes of chopped aramid fiber in combination with carbon fiber. Their non-asbestos organic friction material was reported to offer advantages in the area of high temperature stability and to possess favorable wear characteristics when compared to conventional friction materials. In addition the mixture of fibers provided the new brake lining composition with superior anti-fade properties when tested at elevated operating temperatures.

Satapathy's team [13-22] explored the effects of different fibers in a given friction material using Grey Relational Analysis and treated the problem as a simultaneous optimization of multiple criterion. Based on their analysis, a mixture of aramid fiber with basalt fiber was shown to be synergistic, giving their composition better performance with regard to friction (especially during fade and recovery).

More recently, Kim and co-workers [13-33] examined the synergistic effects of aramid pulp and potassium titanate whiskers in an automotive friction material. Improvements in green strength and particle retention were noted with the use of the aramid pulp, but the blend of potassium titanate and aramid was reported to have an optimal ratio of 1:3 that gave the lowest wear rates. It was also noted that the concentration of aramid fiber was inversely proportional to the measured porosity of the final product.

Aramid Summary

Aramid fibers can be either meta-aramid or para-aramid and have found use in virtually all types of friction material composites. They provide strength, heat resistance, and noise reduction. The effects on wear and friction are less clearly identifiable, and much of the confusion can be associated with other materials present in the composition. However, the effect of fiber additions such as aramid, steel, or mineral has been shown to affect development and regeneration of third body layers [13-37]. The effect of aramid fibers on transfer films is an area that is not clearly understood. Further research can hopefully eliminate some of the inconsistencies that have been reported during the past three decades.

Phenolic Fiber

Phenolic fibers are technically known as novoloid fibers produced by spinning molten novolac phenolic resin into filaments. The filaments are subsequently treated with curing agents to form a thermoset fiber that is resistant to chemical attack, possesses excellent thermal stability, and exhibits low thermal conductivity [13-38, 13-39, 10-28].

Phenolic fibers are another convenient method for replacing asbestos and have seen years of continual use. One of the first reports of successfully implementing phenolic fiber in place of asbestos was disclosed by Iwata [13-40]. Phenolic fibers were coated with a plastic material and combined with non-fibrous serpentine to replace asbestos in an automotive brake friction material.

Phenolic Fibers in Wet Friction Materials

The most widespread consumption of phenolic fiber in the friction material industry is for the manufacture of friction paper. Yamamoto [13-41] developed a simple paper comprised of aluminum powder, graphite, and phenolic fiber. In this example, the concentration of the phenolic fiber was 80% by weight. After the paper was formed, it was saturated with a phenolic resin, forming a phenolic reinforced phenolic composite. Santoso and DiPino [13-42] employed high-carbon yield fibers such as phenolic fiber in another wet friction material made with a matrix of fluoroelastomers.

Despite of these findings, the benefits of phenolic fiber were not fully disclosed until Chuluda [13-43] reported on their use in a novel wet friction material. Phenolic fiber is more than a reinforcing agent in wet friction systems. It is a functional fiber that is considered an essential component of the friction material composition. In atmospheric oxygen, the fibers degrade at temperatures of approximately 392°F (200°C), which limits their use in dry friction systems. However, the oil present in fully immersed tribosystems displaces air from the voids in the friction material and creates an atmosphere where the thermal limits of the fibers are increased. The efficient cooling and antioxidants present in the fluid further enable the fibers to provide beneficial friction and structural performance over a much longer period of operation because of the absence of available oxygen.

It is generally concluded that phenolic fibers offer some structural reinforcement characteristics but are typically brittle and tend to fracture easily. Additionally, the thermal resistance of novoloid fiber is on the same scale as phenolic resins and suffers from the same forms of oxidative degradation. Chemical modification to the fibers is possible. However, the costs tend to be prohibitive. Therefore, phenolic fibers are expected to be used only in wet friction applications or other non-oxidizing environments in the future.

Polybenzimidazole Fiber (PBI)

As early as 1982, heat-resistant organic polybenzimidazole (PBI) polymers were becoming commercially available and their use was likewise becoming more commonplace in the development of tribological composites. Tabe and co-workers [13-44] developed a friction material using a portion of pulp-like particles as a primary ingredient in their novel composition. They listed a range of products with varying chemistries including aromatic amides, polybenzimidazoles, and polyoxazoles. However, no specific merits were attributed to the use of polybenzimidazoles or polyoxazoles.

Brinzey [13-45] is perhaps the first to incorporate PBI fibers as a component in a mixture of non-asbestos fibers as part of the universal core of the example friction material. The unique blend employed a small amount of aramid pulp, carbon fiber, and a clean form of refractory ceramic fiber. The mixture also contained 10% by weight of the PBI fiber, which was stated to provide the final composite with excellent thermal resistance and low wear rates. Another of the benefits provided by the addition of PBI fibers was an increase in the friction level as the temperature increased (i.e., anti-fade).

Polyoxazole Fiber

Polyoxazole fibers are a class of thermoset liquid crystalline fibers possessing high strength and excellent thermal stability [13-46]. This type of product has seen use as a ballistic grade fiber and in the manufacture of flame retardant textiles. Because of the strength characteristics, it has also found application in the aerospace and composite industries. Despite these benefits, there is little definitive data regarding its performance in the design and development of friction materials. It is expected that polyoxazole fibers, would provide many of the same benefits of the aramid fibers, but this is an area where research has many questions left to answer.

Polypyridobisimidazole Fiber

Polypyridobisimidazole fibers are an emerging class of high modulus fibers that have appeared only recently. Despite their new introduction to the world, they have been employed by Merriman and Amma [13-47] in the development of a friction paper for use in immersed clutches and brakes. The fibers have a high affinity for water that allows them to be dispersed readily in aqueous slurries. High concentrations of the fiber were evaluated. It was found that when the paper formed on a traditional paper machine such as the Fourdrinier, the fibers interacted at the contact points of their surfaces, presumably by a form of hydrogen bonding, giving a notable increase in the strength of the paper. It is expected that polypyridobisimidazole fibers possess the necessary structural and chemical properties to provide a number of benefits to the final composite.

References

13-1. Soulis, R. 1929. "Development and Progress of Molded Brake Lining." Technical Paper No. 290045. Warrendale, PA: Society of Automotive Engineers.

13-2. Swinburn, R., and D. Bartram. 1981. Friction materials and their uses. European Patent No. 840, filed August 8, 1978, and issued July 1, 1981.

13-3. Shaw, T. 1885. Car brake. United States Patent No. 311262, filed October 23, 1884, and issued January 27, 1885.

13-4. Wohlfarth, C. 1892. Brake shoe. United States Patent No. 480982, filed March 14, 1892, and issued August 16, 1892.

13-5. Frood, H., and D. Holmes. 1908. Brake block. United States Patent No. 894946, filed June 9, 1906, and issued August 4, 1908.

13-6. Knowles, E. 1913. Brake liner fabric and process of making same. United States Patent No. 1053299, filed June 3, 1911, and issued February 18, 1913.

13-7. Brown, S., and S. Higgins. 1926. Impregnated fabric and process of producing same. United States Patent No. 1608165, filed May 26, 1925, and issued November 23, 1926.

13-8. Heany, J. 1928. Process and apparatus for making yarn. United States Patent No. 1681234, filed July 27, 1923, and issued August 21, 1928.

13-9. Allen, R. 1885. Brake shoe. United States Patent No. 315213, filed October 13, 1884, and issued April 7, 1885.

13-10. Novak, I. 1936. Manufacture of saturated fibrous products. United States Patent No. 2049469, filed August 16, 1933, and issued August 4, 1936.

13-11. Judd, M. 1939. Friction facing and method of making same. United States Patent No. 2175399, filed April 30, 1938, and issued October 10, 1939.

13-12. Dong, F. 2009. Wet friction material with pitch carbon fiber. United States Patent No. 7537824, filed October 24, 2002, and issued May 26, 2009.

13-13. Aldrich, F. 1978. Friction material reinforced with steel and cellulose fibers. United States Patent No. 4119591, filed July 5, 1977, and issued October 10, 1978.

13-14. Formolo, J. 1998. Self lubricating brake shoe material. United States Patent No. 5821282, filed October 26, 1995, and issued October 13, 1998.

13-15. Satapathy, B., and J. Bijwe. 2004. "Performance of Friction Materials Based on Variation in Nature of Organic Fibres: Part II. Optimisation by Balancing and Ranking Using Multiple Criteria Decision Model (MCDM)." *Wear* 257:585-589.

13-16. Nakagawa, M. 2001. Disk brake pad. United States Patent No. 6193025, filed November 17, 1998, and issued February 27, 2001.

13-17. Le Lannou, M. 1996. Friction lining. United States Patent No. 4418115, filed August 25, 1981, and issued December 17, 1996.

13-18. Doxsee, L. et al. 1989. Acrylic containing friction materials. United States Patent No. 4866107, filed July 26, 1988, and issued September 12, 1989.

13-19. Rush, H. et al. 1989. Fibrous polyacrylonitrile reinforcing mixture for friction product applications and method of making same. United States Patent No. 4886706, filed March 11, 1987, and issued December 12, 1989.

13-20. Horiguichi, K. 1992. Non-asbestos friction material. United States Patent No. 5106887, filed February 5, 1991, and issued April 21, 1992.

13-21. Satapathy, B., and J. Bijwe. 2005 "Fade and Recovery Behavior of Non-Asbestos Organic (NAO) Composite Friction Materials Based on Combinations of Rock Fibers and Organic Fibers." *Journal of Reinforced Plastic Composition* 24: 563-577.

13-22. Evans, R., and J. Trainor. 1990. "Acrylic Fibers in Non-asbestos Friction Materials." Presented at the 8th Annual Society of Automotive Engineers Brake Colloquium, Atlantic City, New Jersey.

13-23. Evans, R., and D. Crean. 1993. *Recent Developments with Acrylic Fibers: Nonwovens Applications*. Atlanta, GA: TAPPI Press.

13-24. Byrne, C. 2004. "Modern Carbon Composite Brake Materials. *Journal of Composite Materials* 38: 1837-1850.

13-25. Patil, A. et al. 1996. Fiber blend for low cost, asbestos free friction material. United States Patent No. 5508109, filed October 6, 1993, and issued April 16, 1996.

13-26. Kwolek, S. et al. 1962. Process for making wholly aromatic polyamides. United States Patent No. 3063966, filed February 5, 1958, and issued November 13, 1962.

13-27. Littlefield, J. 1980. Composite friction element. United States Patent No. 4219452, filed May 19, 1978, and issued August 26, 1980.

13-28. Gallagher, J., and P. Dougherty. 1982. Aramid containing friction materials. United States Patent No. 4374211, filed September 15, 1981, and issued February 15, 1983.

13-29. Parker, E. 1988. Friction materials and their manufacture. United States Patent No. 4656203, filed March 4, 1985, and issued October 4, 1988.

13-30. Siegfried, J.1979. A friction material for brakes, clutches and the like. Great Britain Patent No. 2000517, filed June 19, 1978, and issued January 10, 1979.

13-31. Loken, H. 1980. "Asbestos Free Brakes and Dry Clutches Reinforced with Kevlar® Aramid Fiber." Technical Paper No. 800667. Warrendale, PA: Society of Automotive Engineers.

13-32. Kim, S. et al. "Synergistic Effects of Aramid Pulp and Potassium Titanate Whiskers in the Automotive Friction Material." *Wear* 251: 1484–1491.

13-33. Gopal, P. et al. 1996. "Hybrid Phenolic Friction Composites Containing Kevlar® Pulp. Part I: Enhancement of Friction and Wear Performance." *Wear* 193: 199-206.

13-34. Gopal, P. et al. 1996. "Hybrid Phenolic Friction Composites Containing Kevlar® Pulp. Part II: Wear Surface Characteristics." *Wear* 193:.180-185.

13-35. Kato, T., and A. Magario. 1994. "The Wear of Aramid Fiber Reinforced Brake Pads: The Role of Aramid Fibers." *Tribology Transactions* 37: 559-565.

13-36. Briscoe B. et al. 1988. "Friction of Aramid Composites." Proceedings of Institute of Mechanical Engineering, C452/88, pp.15-30.

13-37. Cho, M. et al. 2005. "The Role of Transfer Layers on Friction Characteristics in the Sliding Interface between Friction Materials against Gray Iron Brake Disks." *Tribology Letters* 20: 101-108.

13-38. Hayes, J. 1981. "Kynol™ Novoloid Fibers in Friction and Sealing Materials." Product literature supplied by American Kynol, New York.

13-39. Hayes, J. 1993. "Novoloid and Related Fibers in Nonwoven Structures." Paper presented at Index 93 Congress, Session 2-C Fibers, Geneva.

13-40. Iwata, K. 1979. Frictional material for brakes. United States Patent No. 4145223, filed October 28, 1977, and issued March 20, 1979.

13-41. Yamamoto, Y. et al. 1981. Friction member of non-asbestos. United States Patent No. 4262788, filed September 29, 1978, and issued April 21, 1981.

13-42. Santoso, M., and M. DiPino. 1983. Fluoroelastomer composite friction material. United States Patent No. 4400434, filed November 18, 1981, and issued August 23, 1983.

13-43. Chuluda, P. 1981. Carbon fiber/flame resistant organic fiber as a friction material. United States Patent No. 4256801, filed December 14, 1979, and issued May 17, 1981.

13-44. Tabe, Y. et al. 1982. Friction material. United States Patent No. 4324706, filed January 14, 1981, and issued April 13, 1982

13-45. Brinzey, A. 1991. Friction materials with a universal core of non-asbestos fibers. United States Patent No. 5041471, filed October 29, 1990, and issued August 20, 1991.

13-46. Toyobo. "Toyobo Products." Accessed May 17, 2010. http://www.toyobo.co.jp/e/seihin/kc/pbo/menu/ fra_menu_en.htm.

13-47. Merriman, E., and A. Amma. 2006. Friction papers comprising PIPD fiber. United States Patent Application No. 2009/0151884, filed December 19, 2006.

Chapter 14
Carbon Fiber

I asked him what he was doing, and he told me he got into carbon fiber cellos.

—Paul Rosenthal

CARBON FIBER IS A MODERN ENGINEERING material that has found its way into a multitude of different composite applications ranging from aerospace structural components and sporting goods equipment to various forms of musical instruments. Aside from the structural benefits, carbon fiber is a thermally resistant fiber with exceptional tribological characteristics. It has been an integral part of the friction material industry for many decades.

A Short History of Carbon Fiber

Mankind's attempt to form fibers out of carbon-based materials dates back to 1881 during the development of the incandescent light bulb. Early forms of carbon fiber were first employed by several inventors including Thomas Edison [14-1]. During this period of industrious research, early fibers were pyrolyzed forms of cellulose fiber (often bamboo or cotton) and were brittle with very low strength. Despite these weaknesses, the fibers were sufficiently heat-resistant and therefore capable of providing the lamp with the incandescing source necessary for light generation. These early attempts were eventually discontinued and it was not until the 1950s that there was resurgence of interest in carbon fiber.

One of the more notable methods of producing fibrous carbon was recorded by Kauffman and colleagues at the Pittsburgh Coke and Chemical Company [14-2]. Their method involved the slow pyrolysis of hydrocarbon materials to deposit small acicular forms of carbon. This process yielded a mono-filament whisker form of high purity carbon (99.75%) with a diameter of approximately 0.00016 inches (approximately 4 microns) and a length that varied in size from a few microns to a few inches. This method was not widely successful as a commercial manufacturing method because of the poor strength of the fiber and low yield. Despite these challenges, the development of carbon fiber remained a priority for many groups.

Workers at Union Carbide [14-3] were the first to find that high-strength carbon fiber could be produced from synthetic fibers, namely rayon fiber intermediates. This discovery led to a number of developments and advances that allowed carbon fiber to emerge as a commercially available product in sizable quantities.

Classifications of Carbon Fiber

Modern forms of carbon fiber are classified in three distinct ways: thermal history, precursor material, and properties. The first form of classification is based on the thermal treatment of the fibers and divides carbon fibers into high heat treatment fibers (HTT), intermediate heat treatment fibers (IHT), and low heat treatment fibers (LHT).

The second form of classification is based on the precursor material. For instance, carbon fiber can be produced from rayon, polyacrylonitrile (PAN), or pitch, or can be grown from a gas phase such as described by Kauffman and co-workers [14-2]. In many cases, the selection of the precursor material will greatly affect the final properties of the fiber and the techniques used to select and convert precursor materials into high modulus fibers are often highly guarded trade secrets.

Another classification method is based on the strength and reinforcing properties of the fibers. Carbon fibers with a modulus in excess of 65 msi (450 GPa) are known as ultra-high modulus fibers. Fibers with modulus values of 50–65 msi (350–449 GPa) are high modulus, fibers with modulus values of 29–50 msi (200–349 GPa) are intermediate modulus fibers, and fibers with modulus values less than 14.5 msi (100 GPa) are low modulus fibers.

Additionally, fibers can also be classified by tensile strength. Fibers with a tensile strength of greater than 435 ksi (3.0 GPa) are high tensile (HT) fibers and fibers with strength values in excess of 653 ksi (4.5 GPa) are superhigh tensile (SHT) fibers [14-4]. As with all forms of classification, there is much debate about these categories, and the levels of acceptance typically varies within the respective domains of research.

To simplify matters, the discussion will focus on the type of fiber and will follow the classification based on the precursor material. Specifically, our discussion will focus on carbon fibers produced from rayon, pitch, polyacrylonitrile, and phenolic materials. For those who are interested in gaining a deeper understanding of the many different approaches to producing carbon based fibers, the text by Donnet and Bansal [14-3] is highly regarded as one of the best sources of knowledge.

Rayon-Based Carbon Fiber

The process of large-scale commercial manufacture of carbon fiber began with work at Union Carbide in 1967 [14-5]. Bacon and colleagues worked to convert rayon fabric into carbonaceous fabrics by treating the precursor fabric with a combination of acidic materials and heat. In some cases, heat alone was sufficient to achieve the desired results. The resulting graphitized fabric was found to be suitable for use in high-temperature applications.

This process involves three stages of thermal treatment: low-temperature decomposition, carbonization, and graphitization. The low-temperature decomposition phase allows for the elimination of water from the fibers. The carbonization phase is the destructive phase that begins to decompose the molecular backbone of the polysaccharide, leading to the formation of a number of small molecular species such as CO and CO2. Graphitization occurs when the temperature is elevated above 4532°F (2500°C) and the carbon residues begin to form graphene structures.

The process of carbon fiber from rayon or other suitable cellulosic fibers does not provide high yields—often less than 30%. Because of the nature of molecular disruption during pyrolysis, the process can include flame retardants and incorporate mechanical stress to aid in the formation of the graphite crystals during the graphitization phase. Other techniques of enhancing the yield involve lengthy pyrolysis methods that are also intended to reduce the degradation products. However, this incurs a penalty in the form of higher production costs, making it less than desirable in comparison to the other methods of production.

Pitch-Based Carbon Fiber

Pitch is a naturally occurring and readily available solid hydrocarbon that is lower in cost than other forms of carbon fiber precursor materials, which makes it ideal from an economic perspective. The process of converting pitch into carbon fibers is essentially that same as the rayon process described previously. However, the pitch process typically requires alteration of the pitch to prepare it for spinning into a fibrous form. The preparation phase and the spinning processes often offset the lower cost of the raw pitch feedstock.

Pitch Preparation

Pitch preparation involves chemical changes to the molecular species present and can be accomplished in a number of ways. Generally, the methods involve changes intended to enhance spinning and to convert the thermoplastic pitch into an infusible material that can be carbonized without melting. Common approaches include additives in the pitch that alter the rheological properties of the pitch during spinning, extraction of selected molecular species via solvent extraction, or distillation, which can be adjusted to selectively eliminate low molecular weight materials from pitch prior to the next phase of manufacture.

Otani [14-6] explored additives that increase the molecular weight of the pitch by removing insoluble matter followed with an addition of tetramethylthiuramdisulfide (TMTD) in the presence of heat. This attempt intended to increase condensation reactions between the molecular groups present in pitch and can be compared to the use of phenolic resins as an additive to COPNA resins reported by workers at Tokico [11-15].

Soehngen and Williams [14-7] employed a mixture of polystyrene and an inexpensive grade of pitch to form carbon fibers. The process began by hot drawing the polymeric pitch mixture into fibers and stabilizing them by sulfonation techniques. The stabilized fibers were then carbonized and graphitized. In a similar fashion, Cornuault and Giraudel [14-8] modified high molecular weight viscous forms of pitch by incorporating plasticizers to aid in flow during processing, followed by similar steps of stabilization and carbon formation.

Kimura and Habata [14-9] employed several methods of oxidative conversion that included comparisons between chemical treatment with peroxides and subjection of the spun fiber to ozone-rich atmospheres at elevated temperatures. These oxidation reactions increased the viscosity and assisting in rendering the fibers infusible.

The list of potential methods involves an array of approaches where the previous techniques are combined to progress the pitch to a more stable material capable of fiber production. Other approaches might provide further insight for those who are interested in these different methods [14-10, 14-11, 14-12, 14-13].

Pitch-Based Fibers in Friction Applications

As can be imagined, the use of pitch-based carbon fibers in the friction material industry has historically been closely linked with the use of pitch as a carbon matrix precursor in carbon-carbon friction materials. In many sources, reports of the incorporation of carbonaceous fiber do not exclude or specify a single type of carbon fiber. However, there are a few limited examples that describe the benefits of using a pitch-based carbon fiber.

Huang and colleagues [14-14] used carbon fiber derived from mesophase pitch in combination with pulverized pitch to manufacture a molded carbon composite. The manufacturing method was essentially a form of compression molding conducted at temperatures in the vicinity of 1472°F (800°C). The molded bodies were subsequently impregnated with additional pitch and carbonized. This process can be repeated as necessary to increase the density of the final product and is completed by a final graphitization step. This method shares many similarities to those previously described by Kosuda and Niijima [5-35].

Dong [13-12] described the use of pitch-based carbon fibers in wet friction material composites and examined their use at concentrations up to approximately 40% by weight. It was found that the wet friction material using pitch fiber had a favorable friction response attributed to a low propensity for generating unacceptable vibrations (i.e., shudder) during use. Furthermore, pitch fibers were found to contribute improvements by elevating the friction level and reducing the wear rates.

Polyacrylonitrile-Based (PAN-Based) Fiber

The importance of polyacrylonitrile-based (PAN-based) carbon fiber has already been touched on in previous chapters of this book. The well-defined manufacturing process begins by cyclization of the pendant nitrile groups as described by Evans [13-22]. The conversion process is also given an extensive treatment in the text of Donnet and Bansal [14-3]. This treatment has been adequately described with a comprehensive listing of excellent, highly recommended references for those interested in further expanding on the subject of manufacturing PAN-based carbon fibers.

The benefits of PAN-based carbon fibers are similar to the advantages of the other forms of carbon fiber, but PAN-based fibers are often superior in many ways. The benefits can also offset the higher cost of the precursor materials because of improved carbon yields and a fiber that is generally more flexible [13-24].

The increased flexibility of PAN-based fibers allows them to be woven into complicated structures and assorted forms of structural fabric. Like all aspects of manufacturing carbon fiber, the carbonization phase can be controlled to different levels (such as the so-called pre-oxidized fiber state), which has also been suitably described in previous chapters [6-34, 13-25].

Phenolic-Based Carbon Fiber

As discussed in previous chapters, phenolic resins can be formed into filaments as a convenient means of replacing asbestos fibers in friction materials [13-38, 13-39, 13-40]. The char yield, or carbon yield, of phenolic resins tends to be superior to other carbonaceous precursors such as hydrocarbon pitch [14-15], and fibrous forms lend themselves to conversion into carbon fibers.

Kawamura and Jenkins [14-16] developed a glassy carbon fiber by thermal treatment of an extruded phenolic resin fiber. The fibers were reported to exhibit a tensile strength greater than glass fibers, but less than PAN-based carbon fibers. In addition, the fibers were considered to posses the advantageous characteristics of abrasion resistance and inertness to chemical attack. The method described provides a lower cost option for the production of lower strength carbon fiber.

Economy and Lin [14-17] examined the effects of stretching during the carbonization phase of the process and determined that significant enhancements could be made to glassy phenolic-based carbon fibers. By applying a tensile force to the fiber during the carbonization and graphitization phases, the final carbon fiber was found to possess enhanced mechanical strength; an effect that has been attributed to preferential orientation of the carbon crystallites.

Carbon Fiber Use in Friction Materials

The largest area of carbon fiber use in friction material composites is based on structural reinforcement associated with the use of chopped fibers or with laminated fabric structures [4-53]. Other uses involve the addition of carbon fibers to non-woven and paper-type friction materials that have been described by Fei and co-workers [14-18] as well as Dong [13-12]. In virtually all of the cited cases, the desired outcome is an enhancement in the specific strength and stiffness [14-19] as well as a desire for improved wear resistance of the composite as a result of the favorable thermal characteristics of carbon fiber reinforced composite [14-20, 14-21].

Carbon Fiber in Automotive Brake Linings

Because of the expense and resilient nature of carbon fibers, the most commonly observed method of incorporating carbon fiber into a brake lining composition is based on blending of various fibrous materials like those described by a number of previously cited research groups such as Ho [6-10].

In addition, there have been a number of comparative studies between various fiber types. For example, Lhymn and Light [14-22] described the effects of carbon fiber and glass fiber on wear rates of fibrous composites at various sliding velocities. Likewise, Birch and colleagues [14-23] devised a composition containing carbon fiber and aramid fiber with additions of ceramic and metallic fibers. The final product was stated to give improved friction and wear performance. It was further noted that fibrous clusters (i.e., poorly dispersed fiber bundles) allowed gas generated at the surface to penetrate into the body of the composite and enhanced fade performance.

Fujimaki and his team [14-24] developed an organic friction material combining various approaches in a novel low metallic composition. Their compound used a blend of steel and carbon fibers with a significant portion of activated carbon. The role of the activated carbon was to inhibit the development of vapor lock (a form of thermally initiated gas evolution in brake fluid) by reducing the conductivity of the overall composite.

Issues with Carbon Fiber

The many benefits of carbon fiber are highly sought after. However, in specific applications the addition of carbon fiber can bring about unwanted performance concerns. The most notable of these concerns is a lower-than-desired level of friction generation because of the interaction of the carbon surface with adsorbed low-molecular weight materials such as water vapor.

Buckley [3-34], Byrne [13-24], and Bill [14-25] described the impact of environmental influences and the rapid frictional transitions that can accompany the use of carbon composite brake materials in an open atmosphere. Additionally, the environment can affect the development of the third body layer because of alignment and orientation of graphite crystallites. The orientation of the basal planes of graphitic constituents was shown to affect both friction and wear behavior in these types of composites and was intrinsically linked to the use of carbon fiber.

Additionally, the single largest factor that limits carbon fiber usage is the high cost of the product. Advances in manufacturing that can bring down the cost will undoubtedly open new avenues for the use of carbon fibers in the friction materials of the future.

References

14-1. Edison, T. 1881. Design for an incandescent electric lamp. United States Patent No. D12631, filed November 30, 1880, and issued December 27, 1881.

14-2. Kauffman, H. et al. 1957. Process for making fibrous carbon. United States Patent No. 2796331, filed June 9, 1954, and issued June 18, 1957.

14-3. Donnet, J. B., and R. Bansal. 1990. *Carbon Fibers*, 2nd ed. New York: Marcel Dekker.

14-4. Donnet, J. B. et al. 1987. "Editorial Section International Committee for Characterization and Terminology of Carbon—First Publication of Five Further Tentative Definitions." *Carbon* 25: 449-450.

14-5. Bacon, R. et al. 1967. Process for manufacturing flexible carbonaceous textile material. United States Patent No. 3305315, filed September 20, 1962, and issued February 21, 1967.

14-6 Otani, S., and A. Yokoyam. 1969. "Characteristic Chemical Constitution of Pitch Materials Suitable for the MP Carbon Fiber." *Bulletin of Chemical Society of Japan* 42: 1417-1424.

14-7. Soehngen, J., and A. Williams. Carbon fiber production. United States Patent No. 4020145, filed March 12, 1975, and issued April 26, 1977.

14-8. Cornuault, P., and B. Giraudel. 1971. Procédé de fabrication de fibres carbonées. French Patent No. 2067619, filed November 12, 1969, and issued July 26, 1971.

14-9. Kimura, I., and H. Habata. 1972. Method of producing carbon fibers. United States Patent No. 3639953, filed August 3, 1970, and issued February 8, 1972.

14-10. Shea, F. et al. 1972. Pitch treatment means. United States Patent No. 3668110, filed October 28, 1970, and issued June 6, 1972.

14-11. Toyoguchi, M. et al. 1973. Making of carbon fibers from solvent extracted and airblown vacuum distillation residues of petroleum. United States Patent No. 3767741, filed December 10, 1970, and issued October 23, 1973.

14-12. McHenry, E. 1977. Process for producing mesophase pitch. United States Patent No. 4026788, filed December 11, 1973, and issued May 31, 1977.

14-13. Powell, J. et al. 1974. Manufacture of carbon fibres. United States Patent No. 3852428, filed August 24, 1971, and issued December 3, 1974.

14-14. Huang, D. et al. 2005. Manufacture of carbon composites by hot pressing. United States Patent No. 6878331, filed February 24, 2003, and issued April 12, 2005.

14-15. Thomas, C. 1993. "Essentials of Carbon-Carbon Composites." *Royal Society of Chemistry*, Cambridge.

14-16. Kawamura, K., and G. Jenkins. 1970. "A New Glassy Carbon Fibre." *Journal of Materials Science* 5: 262-267.

14-17. Economy, J., and R. Lin. 1971 "Carbonisation and Hot Stretching of Phenolic Fiber." *Journal of Materials Science* 6: 1151-1156.

14-18. Fei, J. et al. 2008. "Carbon-fiber Reinforced Paper-based Friction Material: Study on Friction Stability as a Function of Operating Variables." *Journal of Tribology* 130: 041605_1-7.

14-19. Devi, G., and K. Rao. 1993. "Carbon-Carbon Composites - An Overview." *Defence Science* 43: 369-383.

14-20. Wang, M. et al. 2009. "Thermal Conductivity Enhancement of Carbon Fiber Composites." *Applied Thermal Engineering* 29: 418-421.

14-21. Decuzzi, P. et al. 2002. "The Effect of Material Properties on the Thermoelastic Stability of Sliding Systems." *Wear* 255: 311-321.

14-22. Lhymn, C., and R. Light. 1987. "Effect of Sliding Velocity on Wear Rate of Fibrous Polymer Composites." *Wear* 116: 343-359.

14-23. Birch, S. et al. 1991. "Honda's Non-asbestos, Non-metallic, Non-glass Brake Pad Composite." *Automotive Engineer* 99: 45-46.

14-24. Fujimaki, H. et al. 1983. Organic friction material. United States Patent No. 4386168, filed May 23, 1982, and issued May 31, 1983.

14-25. Bill, R. 1975. "Friction and Wear of Carbon-graphite Materials for High-Energy Brakes." TN-D-8006. Washington, DC: National Aeronautics and Space Administration.

Chapter 15
Inorganic Fiber

Of steely colour and of wondrous might Arcadia's hill produce th' asbeston bright. . .

—The Lapidarium of Marbodus (XXXIII)

THERE ARE MANY BENEFITS GAINED BY the addition of an inorganic fiber to a given tribological composite. This can readily be seen in the long history of successful utilization of asbestos. In fact, the historical significance of asbestos fiber and the level of success associated with its use in friction materials have made it virtually impossible to separate the two. Because inorganic fibers continue to play a role of pivotal importance in the various phases of friction material development, the use of asbestos fiber will be examined to establish an historical foundation upon which to investigate alternative materials.

Asbestos

Asbestos is a term derived from the Greek word meaning inextinguishable. It is not a single material, but is actually a group of fibrous materials that has typically been limited to six naturally occurring mineral species. These six species can also be subdivided into two parent mineral groups: amphibole and serpentine. The six forms were also of commercial importance [15-1] and were widely used in the development of many industrial applications for several decades. Members of the asbestos group were prized for their combination of properties including high tensile strength and flexibility.

Amphibole Asbestos

Amphibole asbestos contains five fibrous varieties that are used in commercial applications: anthophyllite, tremolite, actinolite, grunerite, and riebeckite. Anthophyllite, tremolite, and actinolite are typically very short whisker forms that were employed in industrial applications such as paint, fireproof cement, and chemical filter media [15-2]. Grunerite and riebeckite were longer forms that enjoyed large scale commercial use in the manufacture of tribological materials. Grunerite and reibeckite are therefore considered more relevant to discussion of the friction material industry than the other forms of the amphibole family.

Grunerite

Grunerite can be found in a number of locations worldwide, but a majority of the commercial mines were located in the Transvaal region of South Africa. Commercial grades were sold under the name of Amosite, which was a brand name chosen to represent the Asbestos Mines of South Africa (AMOSA).

According to Postel [15-3], amosite fibers could be as long as 11 in. (almost 28 cm) and were widely sought after for insulation of naval vehicles and other thermal barrier applications. The longer fiber length also made them ideal for weaving into heat-resistant fabric and, therefore, perfect for early brake band and drum brake friction linings.

Riebeckite

Riebeckite, which is also known as crocidolite asbestos, was mined in Australia and Africa, specifically in the regions of Transvaal and Cape Providence. The bluish color of this amphibole form gave it the common name Cape Blue and, in some cases, Transvaal Blue. The fiber length was generally longer, like those of Grunerite, and could be easily spun into fabric suited for fire-retardant clothing, rope, and brake band linings [15-4]. Crocidolite was also available in shorter forms and found a host of uses in many other industrial applications.

Serpentine Asbestos

The only member of the serpentine group that was sold widely as a commercial product is chrysotile. Chrysotile, also known as white asbestos, is perhaps the most common form of asbestos fiber used in the production of friction material composites with a history spanning more than one hundred years. The extensive use of chrysotile in friction materials can be attributed to its availability in longer forms (making it highly suitable for weaving) and to its fibrillar nature (making it perfect for dry mix processes). In addition, chrysotile was easily accessible because of widespread geographic availability.

One of the largest sources of chrysotile asbestos was from mines dispersed throughout Quebec. Other secondary sources included a host of sites from Africa, various parts of the former Soviet Union, and a number of smaller mines in the United States, including a now abandoned mine and ghost town known as Chrysotile, Arizona.

The majority of commercial chrysotile asbestos was mined in Canada. One example describing the location and ready availability of Canadian chrysotile asbestos was recorded by Erastus Smith [15-5] as early as 1885. Smith described the occurrence of chrysotile asbestos from Shipton as being generally found in veins that ranged from 1 to 4 in. (2.5 to 10.1 cm) in width and varied in color from a deep green to pale yellow. The fibers can be easily separated and formed into a "voluminous wool-like mass."

The Characteristics of Asbestos

The unique nature of the different forms of asbestos becomes readily apparent under microscopic examination. In Fig. 15.1, chrysotile is compared to crocidolite. It demonstrates the inherently flexible nature of the longer chrysotile form in contrast to the shorter whisker form crocidolite. In addition, when viewed under higher magnification, crocidolite can be seen as a bundle of microscopic fibrils that are easily separated into individual whiskers.

Aside from the morphological distinctions between the types of fiber, the thermal resistance of asbestos made it vastly superior to naturally occurring organic fibers such as cotton that were commonly used during the early days of the friction materials industry. A comparison between chrysotile asbestos and cotton is shown in the thermogravimetric analysis in Fig. 15.2.

Based on this simple comparison of weight loss versus temperature, it is easy to understand why asbestos fibers were prized by friction material producers of this era.

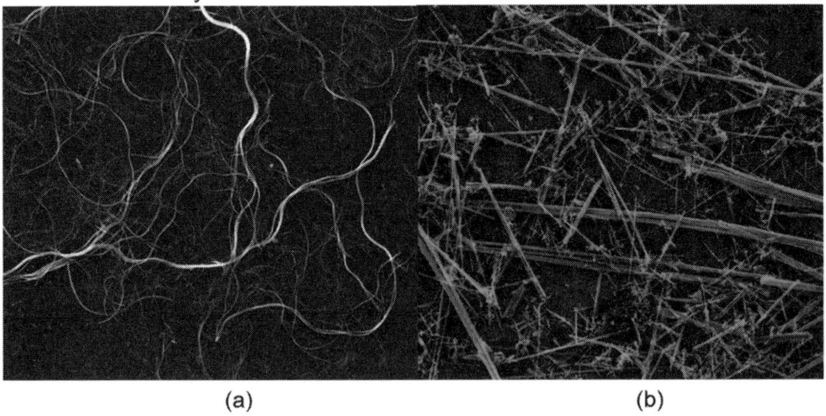

Fig. 15.1 (a) Chrysotile asbestos; (b) Crocidolite asbestos.

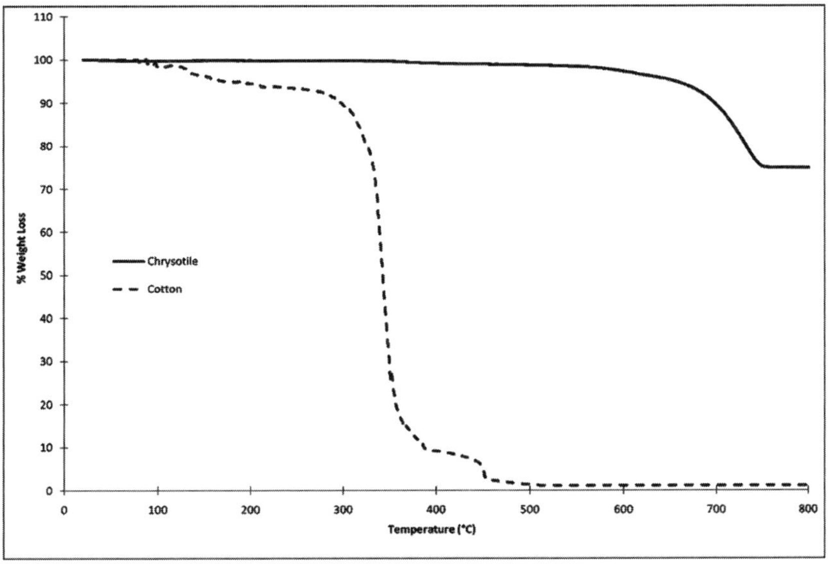

Fig. 15.2 Thermogravimetric comparison of chrysotile and cotton in air.

Asbestos in Friction Materials

The various manufacturing processes used in the production of a brake lining or clutch facing were capable of using asbestos in its various forms. In addition to the ease of manufacturing, the most frequently cited benefit of asbestos brake linings was undoubtedly the high heat resistance of the fiber. Asbestos fiber also had a relatively high coefficient of friction when used against an iron or steel mating member. This combined ease of processing, heat resistance, and high friction output made asbestos the perfect material for early mechanically actuated brake systems.

In a more recent survey, Longley and Gardner [6-35] examined the beneficial nature of incorporating asbestos into friction materials. Their conclusion was that asbestos provided friction materials with an ideal set of desirable attributes. The fibers are heat-resistant, flexible, and easily processed by various means. They impart high strength and damping characteristics to the final composite, and they are generally wear resistant. In addition, the fibers were typically considered non-abrasive to the mating metals and were considered a "God given material for inclusion in friction materials" [15-6].

Issues with Asbestos in Friction Materials

As ideal as asbestos may have seemed, it was not a perfect material. One of the most notable concerns was dehydration at elevated temperatures and the associated loss of strength. The loss of the "water of crystallization" was reported as early as 1916 by Bobaricov and Mramornov [15-7] during a study of asbestos rope. It was noted that heating asbestos rope for a sustained period at elevated temperatures resulted in the conversion of asbestos fibers from a supple, flexible fiber into a brittle fiber with a loss of nearly 90% of the original tensile strength.

Rohrer [15-8] described the dehydration of asbestos fibers during his report on friction materials and stated that the dehydration of asbestos begins near 650°F (343°C).

It was further determined that chrysotile asbestos, which was stated to be the most commonly used form of asbestos fibers in the friction material industry, contained approximately 12–14% by weight water. This water would be driven off during high-temperature operation and observed as a significant loss in weight as well as an associated drop in mechanical strength.

Several years later, the work of Tanaka and colleagues [7-31] demonstrated that dehydration reactions of asbestos were also capable of impacting the friction performance of the brake linings. Asbestos was shown to play an adverse role that reduced friction output at elevated operating temperatures. The loss in friction because of dehydration of asbestos fibers has also been confirmed by Mansfield and co-workers at the National Aeronautics and Space Administration [6-28].

Degradation of Asbestos during Use.

In an effort to understand the degradation of asbestos because of dehydration, there have been a number of studies focused on examining wear debris from brake linings made with asbestos fibers. Spurr [7-6] examined brake linings and determined that asbestos was converted to forsterite, a non-asbestiform magnesium silicate. The conversion to forsterite was also confirmed by studies focused on understanding the toxicological implications of wear debris from asbestos-based brake linings [15-9].

Mizutani and co-workers [7-35] examined the conversion process by employing a comprehensive x-ray analysis method to examine the wear surfaces of resin-bonded asbestos composites based on antigorite asbestos, a polymorph of serpentine (i.e., chrysotile). During this study, the effects of filler materials and friction induced heat were determined to be significant factors in the development of thermally induced structural changes to the antigorite, leading to degradation in the composites performance.

Similar findings have also been reported by Bark and colleagues [15-10] during their studies on the chemical changes that occur during the operation of asbestos-based friction materials. In addition, Jacko [3-41] reported the conversion to olivine during the operation of organic disc brake pads in service. In all cases, the degradation of asbestos fibers is based on dehydration reactions where water present in the crystalline structure of asbestos is vaporized. The loss of the water alters the crystal structure, resulting in a severe loss of tensile strength, and is accompanied by a pronounced shift in the observed friction and wear performance of the brake linings.

Other Issues with Asbestos

Aside from the thermal degradation of the fibers during service, the natural variation in the varieties of asbestos obtained from different sites was another source of concern. Variation in the fiber because of inconsistencies in the geologic processes resulted in fibers that differed not only in chemical composition and size but also in purity. Halstead [6-23] discussed the fundamentals of chemical differences between the various forms of asbestos in friction materials in relation to the observations of severe scoring of brake drums.

It was determined that amosite and crocidolite forms would more readily generate severe damage to the surface of brake drums during sliding than the same friction material composition using chrysotile. The observed mechanism was attributed to variations in the hydration level where chrysotile contained a higher percentage of water than the other forms and was thought to be the cause of the reduced propensity for drum scoring.

Contamination was also a concern when using asbestos. Workers at Raybestos-Manhattan [15-11, 15-12] employed a number of chemical, mechanical, and magnetic methods to purify asbestos. In virtually all cases, the objective was to remove ferromagnetic iron compounds. For instance, Jensen developed a process specifically targeted toward purification of chrysotile asbestos fibers mined in Canada. The specific grades were found to contain approximately 3% magnetite (a magnetic form of iron oxide).

The magnetite impurities were also relatively large in size, with a nominal particle size of 60 mesh (approximately 250 microns). The combination of concentration, hardness, and size of these particles was shown to adversely impact the wear rate of friction materials. It was demonstrated that purification by removing the magnetite particles would improve lining life.

Asbestos contamination continued to plague the industry with the most commonly cited complaint being excessive abrasion of the mating member. During a quest to replace asbestos, Siegfried [13-30] found that one of the more successful methods also provided key reductions in abrasion to the mating member, which was obtained by substituting aramid fiber for asbestos.

Asbestos Bans and Continued Use

Chan and Stachowiak [6-11] reported that the United States banned the use of asbestos in 1989. However, the ban was repealed in 1991 because of difficulties in finding suitable replacement materials. Even though asbestos has been banned from new product development, it may still be used in existing applications such as friction materials for brakes and clutches. Because of the potential for hazards associated with inhalation and an elevated increased potential for litigation concerns, many companies have elected to restrict the development process by using only non-asbestos fibers.

Methods of Replacing Asbestos

Non-asbestos inorganic fibers provide one of the most logical approaches to the replacement of asbestos in friction materials. However, much of the testing that has been conducted to date reiterates a common theme: no single fiber can entirely substitute asbestos, and the replacement of asbestos requires the simultaneous incorporation of fibers of different sizes, shapes, and types to achieve the desired balance of strength, thermal resistance, and tribological performance [13-33].

One of the earliest attempts at replacing asbestos using an inorganic fiber was the incorporation of mineral wool, a vitreous fiber spun from a molten mixture of slag and limestone. This was not successful because the short, low-strength mineral wool fibers could not be spun into fabric [15-13] and thus could not be directly employed in the manufacture of woven linings and clutch facings. Along a similar line of reasoning, Grossman [15-14] developed a "silicose" fiber by spinning sodium silicate. This process was complicated and did not gain wide acceptance as a potential replacement product.

Today, there are virtually hundreds of methods of replacing asbestos using combinations of various fibers including blends of glass and mineral fibers [15-15], direct replacement with sisal [15-16], substitution with basalt [10-12], and many unique methods such as the use of thermally degraded asbestos fiber (i.e., forsterite) in place of chrysotile [15-17]. All of these methods require extensive study and a thorough analysis, but the fundamental conclusion is that asbestos can be successfully replaced in a number of ways using a blend of materials.

Wollastonite

Wollastonite is a pearly white calcium inosilicate named in honor of the English chemist William Hyde Wollaston (1766–1828). Geologically, wollastonite is formed by thermal metamorphosis of limestone in the presence of silica. It can be chemically described by reactions between calcium carbonate and silicon dioxide. Wollastonite is triclinic and, because of the repeating tetrahedral arrangements of the crystalline unit cells, is generally found as masses of crystals that cleave into shorter acicular segments [15-18]. The acicular nature provides wollastonite with a fibrous attribute, making it suitable for structural reinforcement in ceramics and polymeric systems [15-19].

Wollastonite in Friction Materials

The role played by wollastonite in friction materials began as early as 1958 [15-20] when Batchelor and colleagues employed a mixture of silica and wollastonite in a sintered copper friction element. In this example, the addition of silica and wollastonite were large-particle mineral fillers between 20 and 60 mesh (approximately 850 and 250 microns) intended to enhance performance. These were found to prevent heat checking of the mating metallic member; a benefit attributed to the hardness of the minerals. It was believed these compounds were capable of generating small amounts of abrasion that wore away the cracked surface region before it could propagate deeper into the bulk of the material.

The hardness of wollastonite is approximately 5 on the Mohs scale, and it is capable of functioning as a friction enhancing agent. Falcettoni [15-21] blended acicular wollastonite and powdered boron carbide and reported that the mixture would efficiently stabilize friction levels of carbon matrix friction materials. Santoso and Anderson [15-22] incorporated wollastonite into an elastomer for use in a wet friction system. The benefit of wollastonite was reported as a reduction in the static friction level relative to the dynamic friction level. It was considered desirable to have a ratio where the two friction levels approach unity. Wollastonite was found to be a favorable component for achieving this objective.

Structural Aspects of Wollastonite

The acicular character of wollastonite makes it ideal as a short fiber polymeric reinforcing material. This was described by Kaufman [15-23], Yasushi and Kunio [15-24], Takeshi and Takao [15-25], and Nguyen [15-26].

Kaufman incorporated wollastonite into his compositions with the intent of preventing surface cracks from forming at the friction interface during service. Yasushi and Kunio employed high aspect ratio wollastonite to improve the flexural resistance and impact strength of a non-asbestos organic friction material. This was also a practice employed by Takeshi and Takao during their development of a novel non-asbestos composition targeted for use on commercial automotive brake applications.

Alternative approaches rely on the use of wollastonite to ensure dimensional stability of the molded polymeric material. This technique has been applied to polyimide compositions for reducing the coefficient of linear thermal expansion [15-27] and to polyoxymethylene resin compounds for reducing the warpage of the final product [15-28]. It has also been used in polyphenylene sulfide compositions [15-29] and in more conventional friction materials [15-30, 15-31].

Additional Uses of Wollastonite

Separate from the structural and frictional properties of wollastonite, there are a host of possible applications and uses in the world of friction materials. Emmet [15-32] incorporated wollastonite into an underlayer material for brake pads. This specific underlayer composition was designed to ensure structural integrity of the friction assembly by inhibiting thermal energy generated at the friction material interface from being transferred into the adhesive boundary between the composite and the carrier plate. The use of wollastonite, as well as other insulating fibrous materials, allowed the underlayer to possess high strength and excellent thermal resistance, thus providing an effective solution to this problem.

Aside from underlayer materials and sintered metallic friction materials, wollastonite has also found use in railway applications [15-33]. In addition, there is a growing majority of compositions showing the benefits of using wollastonite in non-asbestos friction material compositions [15-34, 15-35].

Glass Fiber

Glass fiber, commonly known as fiberglass, is produced by two methods: through extrusion and by melt spinning from a molten mixture of silica and assorted chemicals. There are a number of methods for the production of commercial grades of glass fiber and many trace their origin back to the work of Games Slayter [15-36, 15-37]. During the 1930s, Slayter described several methods for producing fibers from molten solutions of glass where the final product could have a number of physicals forms, including crimped versions.

The most common form is E-glass, an alumino-borosilicate glass, given the E designation because of the potential for use in electrical applications. S-glass is another form with a unique composition that has higher tensile strength. T-glass is a chemically resistant variety. T-glass might also be known as C-glass in other countries. E-glass is more commonly used and has a rich history in all forms of plastic composites including tribological composites.

Glass Fiber in Tribology

Glass fiber is one of the most commonly used fibrous reinforcing materials in today's world of non-asbestos organic friction materials [7-26]. However, glass fiber has also been involved in the scientific study of friction for nearly a century. Tomlinson [15-38] used single filaments of glass to explore friction generation mechanisms. He observed that the level of friction between two filaments was larger in magnitude immediately after heating (i.e., the adsorbed chemical species had been driven off allowing for direct contact between the surfaces). From these observations, Tomlinson concluded that cohesion must be present at an atomic level and that cohesion between atoms might be capable of explaining the observed friction responses between bodies in contact.

Glass Fiber in Friction Materials

Glass fiber has been used in the manufacture of woven friction materials since 1939 [6-19] and it was re-investigated during the 1970s in other brake lining composites [15-39]. In addition, glass fiber has been used in rubber-bound cone clutch materials [6-20] and in a number of annular ring clutch facings [15-40, 15-41, 15-42].

Glass fiber provides excellent impact and heat resistance [15-43]. It is capable of being woven into fabric and annular facings from continual filaments of glass. In molded composition, glass fiber can be found in the form of chopped and milled fiber varieties. The benefits of using glass fiber as an asbestos replacement is based on many of the same characteristics found in asbestos fiber such as high strength and excellent thermal resistance. However, glass fiber has several areas of concern that have limited its use.

Areas of Concern with Glass Fiber

The most dominant area of concern with glass fiber is based on the brittle failure mode of the fibrils during use. The significance of brittleness resulted in Chan and Stachowiak [6-11] stating that glass fibers cannot be the sole fibrous reinforcing additive and must be blended with other fibers in friction materials.

The brittle nature of the glass fiber has also been linked to the generation of abrasive wear debris that can adversely impact the performance of the final composite. Bohrer [15-44] reported the use of glass fiber in brake lining compositions as being overly aggressive, which manifested itself as objectionable levels of brake noise.

Littlefield [13-27] attributed the breakdown of glass fibers at the friction interface as a primary cause of high wear in composite friction elements for use in railway applications. In addition, Bijwe [15-45], Xin [15-16], and Viswanath [15-46] described similar concerns of elevated friction and wear rates in automotive friction materials, specifically non-asbestos organic friction materials reinforced with glass fibers and fabrics. Again, the dominant factor was attributed to fracture of the brittle glass fibers present at the friction interface resulting in poor wear performance.

As an alternate perspective, the abrasive nature of fractured particles of glass fiber was harnessed in the design of Marzocchi and Jannarelli [15-47]. In this work, glass fiber was an active friction-generating additive bound with a phenolic resin, which had been modified with calcium cations provided by the addition of small amounts of Portland cement. In this example, glass fiber was used at a concentration of 40% by weight in combination with copper chips to provide a friction material with unique and novel performance characteristics.

Fiberglass and Blends of Other Fibers

Aside from the inherent brittleness and generation of abrasive wear debris, fiberglass is typically blended with other fibrous materials for a number of reasons. One primary reason for the use of fiber blends is to enable effective manufacturing.

Rush [13-19] described a sample preforming operation used in the production of molded friction materials and stated that one of the key characteristics of a suitable fibrous material was the ability to be compacted easily. Because of the resilient nature of glass fiber bundles, soft organic pulp fibers such as polyacrylonitrile or aramid are often used to overcome this hurdle and should be used in conjunction with glass fiber.

Other reasons for the use of blends depend on the structural needs of the final product. Harunobu [15-48] developed a molded resin-bound composite friction material for use in clutch systems. This product used a blend of short chopped glass fiber and aramid pulp at a total concentration of 50% by volume of the composition. The benefit of a blend of aramid and fiberglass was the development of a strong material capable of surviving the high rotational speeds present in modern clutch systems.

Blends of glass fiber and other fibers are also common outside of the automotive friction materials market. One example by workers at Daehwa Brake Company [15-49] utilized a fibrous blend of materials at a concentration up to 40% by weight. The blend employed a mixture of aramid fibers, steel wool, glass fiber, and wollastonite in a phenolic resin and rubber-bound composite. Interestingly, this example also incorporates a fragrance to overcome the odor produced during braking.

Generally, the use of fiberglass is more prevalent in non-asbestos organic friction materials and typically includes mixtures of glass fiber with aramid fiber [15-50, 15-51], wollastonite [15-52], potassium titanate [15-53], or vitreous inorganic fibers such as rock wool [15-54]. In some cases, all of these are combined to achieve the necessary balance of strength, wear, and frictional output.

The list of fiberglass composites and compositional variations is limited only by the designer's imagination. However, one critical aspect regarding the use of fiberglass is the poor compatibility of the fibers for phenolic resins. This dictates that glass fibers first be coated with a suitable finish (typically a suitable silane coupling agent) to ensure proper adhesion and structural reinforcement [15-55]. An alternative approach would be to employ a silicone-based matrix system (i.e., a silicone modified phenolic resin) to provide enhanced adhesion to the surface of the glass fiber [15-56].

Basalt Fiber

Basalt fiber is produced from basalt rock, a dark colored, finely grained igneous rock composed of a mixture of plagioclase and pyroxene. Basalt fiber production began as early as 1922 when Paul Dhé [15-57], a Parisian engineer, formed a filament by drawing molten basalt rock into a fine thread, which was subsequently annealed to develop a crystalline structure. The fibers were very flexible, capable of being woven or braided, and were chemically inert to reaction with a majority of chemical agents.

A more modern example by Rittler [15-58] described a method for producing an alkali-resistant basalt-ceramic fiber that is ideally suited for use in highly alkaline environments such as cement reinforcement. This process begins by heating a finely divided mixture of basalt and selected metal oxide additives until a homogenous melt is obtained. The molten mixture is then formed into fibers and heat treated to convert the fibers from a vitreous state into a polycrystalline state. A simple diagram illustrating the process is shown in Fig. 15.3.

Like glass fiber, basalt fibers can also be chopped to form short discontinuous fiber bundles. They can also be produced by melt spinning to give a wooly form. Both types of fibers have found use in friction materials for a number of years.

Basalt Fiber in Friction Materials

One of the earliest attempts to use basalt fiber in place of asbestos was recorded by Longley [15-59]. Longley reports that basalt fiber was suitable for replacing asbestos in friction materials by providing heat resistance. Work by Izyumova and colleagues [15-60] examined the use of basalt fiber as a replacement for asbestos in automotive disc brake pads and concluded that the performance of

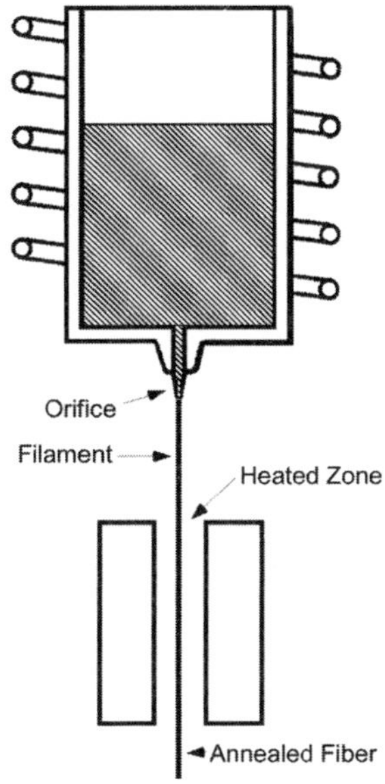

Fig. 15.3 Extrusion of basalt-glass fibers

basalt fiber mixtures was similar to results produced with asbestos. This reiterates the findings of Sergienko [10-12]. Basalt is generally considered suitable for replacing asbestos. However, basalt fiber compositions are often reported to be structurally weaker than the asbestos fiber reinforced materials.

Satapathy and colleagues [15-61, 13-21] assessed the contribution of various fibers to the performance of friction materials using Grey Relational Analysis and combinations of rock fibers with organic fibers. From these studies, it was demonstrated that basalt fibers gave higher friction levels than other synthetic spun rock fibers, and that blends of aramid fiber and basalt were synergistic regarding friction performance (most notably fade and recovery).

In general, basalt fiber can be compared to many other forms of mineral fiber. The method of manufacture will dictate specific characteristics of interest. For example, one of the reasons provided by Satapathy's group for the higher friction level observed with spun basalt fiber was based on the high concentration of non-fibrous material (i.e., shot). The shot present in the fibers acted as a large abrasive particle that created two-body abrasion or plowing during sliding. The content of non-fibrous material is therefore of critical importance during the selection of the proper grades of spun basalt fiber.

Pele's Hair and Pele's Tears

The origins of manufacture for spun mineral fiber can be found in natural processes associated with volcanoes. The ejection of lava from the volcano's caldera in a vertical plume can combine with local wind currents to attenuate ejected globules of molten liquid into fiber. These fibers can vary in color and length, and have become known as Pele's hair in honor of the Hawaiian goddess of fire and volcanoes. The factors that affect the production of the fibers can also produce non-fibrous forms, which are called Pele's tears because of the spherical and teardrop shapes formed upon cooling.

The critical characteristics of these processes were studied by Shimozuru [15-62] and Moune [15-63]. Shimozuru used an ink jet and varied experimental conditions to produce droplets of fluid instead of a continuous flow whereas Moune and colleagues conducted a detail petrographic analysis of debris ejected from the Masaya volcano in Nicaragua.

Both Shimozuru and Moune's group shared the view that fibrous forms are produced by deformation of low-viscosity magma and can be affected by the surface tension of the molten fluid, the ejection velocity, the presence of acidic gases in the vapor of the volcanic plume, and resistance to motion from the surrounding atmosphere. Unlike the fiber form, Pele's tears are high-viscosity forms that do not attenuate readily and will form larger globules with a non-fibrous form upon cooling.

Pele's hair and tears have not been used in friction materials per se. However, the techniques and role of molten fluid viscosity have been applied to the development of synthetic vitreous fiber forms, which have seen great use in the development of tribological composites.

Synthetic Vitreous Fiber (Spun Fiber)

One of the earliest forms of synthetically produced mineral fibers was patented in 1927 by Koberle [15-64]. The Koberle process involved blowing a molten mixture into fiber by using a series of well-placed high-pressure streams of compressed air. The streams of air were positioned to enhance the dispersion of the liquid and increase the yield of fibrous material with concomitant reductions in non-fibrous forms (i.e., shot). Aside from blowing the molten material into fiber, the majority of manufacturers seem to have abandoned this method in favor of adopting a process of making a spun fiber.

Spun Synthetic Vitreous Fiber

Buss [15-65] developed an apparatus and method for forming mineral wool from a mixture of fusible solid materials. This process used a conical turntable with a corrugated surface. During operation the molten mixture was pumped onto the rotating surface. Centrifugal force ejected liquid globules away from the cone and, as a result of the air resistance, formed fibers of the material.

This method was widely used by workers at Johns-Manville [15-66, 15-67, 15-68, 15-69, 15-70] and was the source of patents as recently as 1970 [15-71]. Many of the benefits associated with the process are based on the flexibility of potential raw materials and the use of metal oxides that can be added to the melt to impart unique characteristics to the fiber. Many of these methods could use slag, a waste product from foundry operations, to form heat resistant fibers. Mechel and Kummermehr [15-72] developed a mineral wool from silicate glass with a precise mixture of iron oxides and limited the concentration to very specific levels for the various iron oxides.

The use of virtually any molten substance can be spun into a fiber using this process. There are a number of producers worldwide and a host of various fiber chemistries that are commercially available with or without surface coatings to promote adhesion. In addition, many of the modern producers have invested heavily in the post-processing aspects of manufacture and can provide fibers in bulk and cleaned varieties (i.e., with very low levels of non-fibrous material present).

Synthetic Vitreous Fibers in Friction Materials

The importance of spun synthetic vitreous fibers in a conventional friction material has been partially described in the work by Satapathy [15-61], but a number of other sources help to illustrate the diversity in which this product can be used.

Wagner [15-73] described a process where vitreous fibers are used in a non-asbestos friction material composition and allows for both slag and basalt fibers to be used. It is interesting that Wagner also used a method of bag molding where the mixtures are pre-weighed into thin, thermally degradable polymeric bags and molded without the aid of preforming. This process suggested a novel method of manufacturing, but might also have been necessary if the fiber length was not sufficient to ensure adequate green strength.

Morita and Matsukawa [15-74] developed a dry friction material based on the benefits of a specific alloy of tin and a mixture of fibers. This example illustrated the use of aramid pulp fibers in addition to a blend of equal parts glass fiber and spun slag fiber. In addition, the composition also used a significant amount of potassium titanate whisker fiber. This practice of blending synthetic vitreous fibers with other fibrous materials was also used in bismaleimide-triazine, resin-based, non-asbestos friction materials [11-17] and recently in work by researchers at Hitachi Chemical [10-26].

Concerns with Spun Synthetic Vitreous Fibers in Friction Materials

The largest concern associated with synthetic vitreous fibers is the level of non-fibrous materials (i.e., shot) and subsequent scoring of the mating members that can occur if the levels or particle size is uncontrolled. Additional concerns with abrasion were noted by Leroy and Pilache [15-75] during a comparison between slag fiber reinforced friction materials and basalt fiber reinforced friction materials.

It was determined that basalt fibers were more aggressive toward the metal mating member. Slag fiber reinforced friction materials were found to be more brittle than friction materials reinforced with basalt; an attribute determined by chemical analysis to be affected by the concentration of lime (CaO) used in the melt phase of fiber production. Adjustments to the concentration of calcium oxide allowed for the development of a fiber with superior products.

Bio-Soluble Inorganic Fiber

Concerns with inhalation of fibrous materials resulted in the development of a glass fiber with improved bio-solubility [15-76]. This fiber uses a blend of silicon dioxide and calcium oxide, and other alkali metal oxides, to produce a fiber with a high rate of dissolution in a simulated extra cellular fluid. Likewise, the calcium oxide concentration (CaO) and the magnesia (MgO) content of synthetic inorganic fibers have been linked to their solubility in physiological saline solutions.

Workers at Thermal Ceramics, a division of Morgan Crucible [15-77], developed a thermally resistant inorganic fiber known as Superwool [15-78]. Superwool has a maximum use temperature of 2912°F (1200°C) and is based on specific combination ratios of calcium and magnesium oxides. In addition, the patents specifically expressed an upper limit on the content of other materials present in the mixture. This Superwool possesses excellent thermal characteristics that make it ideal for use in friction materials. It has the additional benefit of having been exonerated from classification and labeling regulations in Europe.

There is a limited amount of information that suggests bio-soluble fibers can have a significant impact on the tribological composites market. In a recent patent, workers at Hitachi Chemical [15-79] used bio-soluble fibers in a non-asbestos friction material in a range of 1.5–7.5% by mass, but little definitive information regarding performance was given. More work is definitely needed in this area of product development.

References

15-1. Virta. R. 2002. "Asbestos: Geology, Mineralogy, Mining and Uses." U. S. Geological Survey, Open File Report 02-149. Washington, DC: Department of the Interior.

15-2. Loughlin, G. 1918. "Mineral Resources of the United States. Part II: Nonmetals." U. S. Geological Survey, Washington, DC.

15-3. Postel, A. 1943. *The Mineral Resources of Africa*. Philadelphia, PA: University of Pennsylvania Press.

15-4. Roush, G., ed. 1922. *The Mineral Industry: Its Statistics, Technology, and Trade during 1921*. Vol. 30. New York: McGraw-Hill.

15-5. Smith, E., 1885. "On the Chrysotile from Shipton, Canada." *The American Journal of Science* 7: 169-174.

15-6. Smales, H. 1995. "Friction Materials - Black Art of Science?" Proceedings of Institution of Mechanical Engineers 209: 151-157.

15-7. Bobaricov, J., and W. Mramornov. 1916. "The Tensile Strength of Asbestos Rope When Exposed to Fire." Tomsk Institute Technical Engineering 102: 451-452.

15-8. Rohrer, D. 1954. "Friction Materials. 1954." Technical Paper # 540069. Warrendale, PA: Society of Automotive Engineers.

15-9. Lemen, R. 2004. "Asbestos in Brakes: Exposure and Risk of Disease." *American Journal of Industrial Medicine* 45: 229-237.

15-10. Bark, L. et al. 1975. "Chemical Changes in Asbestos-Based Friction Materials during Performance - A Review." *Wear* 34, 131-139.

15-11. Novak, I. 1953. Removal of magnetite from asbestos. United States Patent No. 2662639, filed May 28, 1949, and issued December 15, 1953.

15-12. Jensen, W. et al. 1969. Friction material. United States Patent No. 3462387, filed October 17, 1966, and issued August 19, 1969.

15-13. Davol, R., 1922. *Raw Products of the World: Volume 1 – Africa*. Taunton, MA: Davol Publishing Company.

15-14. Grossman, M. 1929. Process of manufacturing synthetic mineral fibers. United States Patent No. 1730609, filed October 17, 1927, and issued October 8, 1929.

15-15. Chand, N. et al. 2004. "Development of Asbestos Free Brake Pad." *IE(I)-ME* 85: 13-16.

15-16. Xin, X. et al. 2007 "Friction Properties of Sisal Fibre Reinforced Resin Brake Composites." *Wear* 262: 736-741.

15-17. Delvaux, P. et al. 1993. Organic friction material composition for use to produce friction linings. United States Patent No. 5250588, filed January 16, 1990, and issued October 5, 1993.

15-18. Amethyst Galleries, Inc. The Mineral Wollastonite. Accessed June 5, 2010. http://www.galleries.com/minerals/silicate/wollasto/wollasto.htm.

15-19. Ciullo, P. 1996. *Industrial Minerals and Their Uses: A Handbook and Formulary*. Westwood, NJ: Noyes Publications.

15-20. Batchelor, C. et al. 1958. Sintered copper friction elements containing a mineral filler. United States Patent No. 2818634, filed March 17, 1954, and issued January 7, 1958.

15-21. Falcettoni, J. 1969. "Method of stabilizing friction coefficient of carbonaceous base materials and the products thereof. United States Patent No. 3421913, filed November 12, 1964, and issued January 14, 1969.

15-22. Santoso, M., and J. Anderson. 1985. Fluoroelastomer-based friction material having improved frictional properties. United States Patent No. 4530881, filed February 24, 1983, and issued July 23, 1985.

15-23. Kaufman, E. et al. 1984. Process for molding a brake lining friction product. United States Patent No. 4432922, filed March 15, 1983, and issued February 21, 1984.

15-24. Yasushi, T., and S. Kunio. 1981.Friction material. Japanese Patent No. 56-016578, filed July 23, 1979, and issued February 17, 1981.

15-25. Takeshi, O., and S. Takao. 1983. Friction material. Japanese Patent No. 58-013684, filed July 17, 1981, and issued January 26, 1983.

15-26. Nguyen, C. et al. 1987. Method of manufacture of a molded friction pad. United States Patent No. 4652415, filed February 11, 1985, and issued March 24, 1987.

15-27. George, D. 1994. Low thermal expansion polyimide compositions. United States Patent No. 5284904, filed May 27, 1993, and issued February 8, 1994.

15-28. Horio, M. et al. 2002. Polyoxymethylene resin composition. United States Patent No. 6391956, filed March 14, 2000, and issued May 21, 2002.

15-29. Morikazu, K. et al. 1998. PPS resin composition for molding material. Japanese Patent No. 10-017770, filed June 28, 1996, and issued January 20, 1998.

15-30. Zhang, J. 009. Method for producing brake flat with acicular wollastonite. Peoples Republic of China Patent No. 20081160243, filed November 11, 2008, and issued April 9, 2009.

15-31. Shingo, M. et al. 2008. Brake friction material. Japanese Patent No. 2008-057693, filed August 31, 2006, and issued March 13, 2008.

15-32. Emmet, R. et al. 2003. Brakc pad. United States Patent No. 6634470, filed December 19, 2001, and issued October 21, 2003.

15-33. Hong, S. H. 1995. Non-asbestos brake shoe producing method for railcar. Republic of Korea Patent No. 950002563, filed March 3, 1990, and issued March 23, 1995.

15-34. Yang, S. 1995. High temp resistance non-asbestos friction material. Peoples Republic of China Patent No. 1110286, filed April 4, 1994, and issued October 18, 1995.

15-35. Gladun, V. et al. 1998. Polymer friction composition and method manufacturing friction material. Russian Federation Patent No. 2119511, filed February 21, 1996, and issued September 27, 1998.

15-36. Slayter, G. 1938. Method and apparatus for making glass wool. United States Patent No. 2133235, filed November 11, 1933, and issued October 11, 1938.

15-37. Slayter, G. 1941. Method of producing glass fibers. United States Patent No. 2230272, filed August 4, 1938, and issued February 4, 1941.

15-38. Tomlinson, G. 1928. "Atom to Atom Cohesion." *Philosophical Magazine* 6: 697-712.

15-39. Bohrer, G. 1989. Friction material for brake linings and the like. United States Patent No. 4130537, filed February 2, 1977, and issued December 19, 1978.

15-40. Bentz, L., and H. Hilton. 1970. Clutch facing. United States Patent No. 3526306, filed February 8, 1968, and issued September 1, 1970.

15-41. Lowry, B. 1978. Glass fiber friction facing. United States Patent No. 4118528, filed July 28, 1977, and issued October 3, 1978.

15-42. Barnett, I., and W. Peters. 1973. Glass fiber friction element. United States Patent No. 3743069, filed June 24, 1971, and issued July 3, 1973.

15-43. Simonds, H. 1959. *Source Book of the New Plastics*. New York: Reinhold Publishing.

15-44. Bohrer, G. 1978. Friction material for brake linings and the like. United States Patent No. 4130537, filed February 2, 1977, and issued December 19, 1978.

15-45. Bijwe, J. 1997. "Composites as Friction Materials: Recent Developments in Non-Asbestos Fiber Reinforced Friction Materials - A Review." Polymer Composites 18: 378-396.

15-46. Viswanath, B. et al. 1993. "Effect of Reinforcement on Friction and Wear of Fabric Reinforced Polymer Composites." *Wear* 167:.93-99.

15-47. Marzocchi, A., and A. Jannarelli. 1976. Friction material for brake linings and the like. United States Patent No. 3967037, filed April 12, 1974, and issued June 29, 1976.

15-48. Harunobu, K. 1988. Molded resin composition of frictional material for use in clutches. Great Britain Patent No. 2196977, filed August 26, 1987, and issued May 11, 1988.

15-49. Hong, G. et al. 2002. Brake lining for electric train. Republic of Korea Patent No. 20020080807, filed April 17, 2001, and issued October 26, 2002.

15-50. Chinda, N. et al. 2008. Friction material. United States Patent Application No. 2009/0011962, filed June 19, 2008.

15-51. Shigeru, K. et al. 1993. Friction material composition. Japanese Patent No. 05-194936, filed January 23, 1992, and issued August 3, 1993.

15-52. Tamotsu, H., and I. Tatsuya. 1987. Friction material for brake. Japanese Patent No. 62-215678, filed March 17, 1986, and issued September 22, 1987.

15-53. Morita, K. et al. 1994. Dry friction material. United States Patent No. 5308392, filed October 23, 1992, and issued May 3, 1994.

15-54. Shingo, M. et al. 2008. Brake friction material. Japanese Patent No. 2008-057693, filed August 31, 2006, and issued March 13, 2008.

15-55. Abdel-Magid, B. et al. 2003. "Flexure Creep Properties of E-glass Reinforced Polymers." Composite Structures 62: 247–253.

15-56. Ebnesajjad, S. 2008. *Adhesives Technology Handbook*. Norwich, New York: William Andrew Incorporated.

15-57. Dhé, P. 1922. Filament composed of basalt. United States Patent No. 1438428, filed April 15, 1921, and issued December 12, 1922.

15-58. Rittler, H. 1980. Method for making basalt glass ceramic fibers. United States Patent No. 4199336, filed September 25, 1978, and issued April 22, 1980.

15-59. Longley, J. 1975. Friction materials. United States Patent No. 3896075, filed November 3, 1972, and issued July 22, 1975.

15-60. Izyumova, V. et al. 1991. "Asbestos-free Frictional Compositions for Automotive Disk Brake Pads." *Trenie i Iznos* 12: 114-117. (Russian).

15-61. Satapathy, B. et al. 2006. "Assessment of Fiber Contribution to Friction Material Performance Using Grey Relational Analysis (GRA)." *Journal of Composite Materials* 40: 483-501.

15-62. Shimozuru, D. 1994. "Physical Parameters Governing the Formation of Pele's Hair and Tears." *Bulletin of Volcanology* 56: 217-219.

15-63. Moune, S. et al. 2007."Pele's Hairs and Tears: Natural Probe of Volcanic Plume." *Journal of Volcanology, Geotherm. Res.* 164: 244-253.

15-64. Koberle, A. 1927. Means for blowing mineral fiber. United States Patent No. 1650136, filed June 10, 1923, and issued November 22, 1927.

15-65. Buss, C. 1939. Apparatus for and method of forming mineral wool from fusible solids. United States Patent No. 2153739, filed December 10, 1936, and issued April 11, 1939.

15-66. Powell, E. 1950. Method and apparatus for fiberizing molten material. United States Patent No. 2520168, filed September 22, 1944, and issued August 29, 1950.

15-67. Powell, E.1950. Method and apparatus for fiberizing molten material. United States Patent No. 2520169, filed September 29, 1948, and issued August 29, 1950.

15-68. Powell, E. 1952. Method and apparatus for the manufacture of mineral wool. United States Patent No. 2594799, filed July 18, 1946, and issued April 29, 1952.

15-69. Powell, E. 1959. Method and apparatus for producing fibers. United States Patent No. 2884659, filed November 9, 1956, and issued May 5, 1959.

15-70. Rudin, G. 1966. Apparatus for producing fibers. United States Patent No. 3246971, filed June 14, 1962, and issued April 19, 1966.

15-71. Baier, J. and J. Demaison. 1970. Method for spinning wool from igneous rock. United States Patent No. 3533769, filed September 1, 1967, and issued October 13, 1970.

15-72. Mechel, F., and H. Kummermehr. 1972. Mineral wool made from silicate glass comprising iron oxides. United States Patent No. 3679443, filed December 17, 1969, and issued July 25, 1972.

15-73. Wagner , H. 1981. Process for making friction products. United States Patent No. 4268467, filed October 12, 1979, and issued May 19, 1981.

15-74. Morita, K. and T. Matsukawa. 1994. Dry friction material. United States Patent No. 5308392, filed October 23, 1992, and issued May 3, 1994.

15-75. Leroy, P. and M. Pilache. 1984. Friction lining suitable for use in brakes, clutches, and other applications. United States Patent No. 4465796, filed February 1, 1983, and issued August 14, 1984.

15-76. Bauer, J. and K. Schall. 1994. Glass fiber composition with improved biosolubility. Japanese Patent No. 06-321578, filed September 17, 1993, and issued November 22, 1994.

15-77. Jubb, G., and J. L. Martin. 2005. High temperature resistant saline soluble fibers. United States Patent No. 6861381, filed August 24, 2000, and issued March 1, 2005.

15-78. Superwool. Registered trademark of Morgan Crucible Company Plc, Madeira Walk, Windsor Berkshire, England, Registration No. 1896941, February 6, 1992.

15-79. Kota, N. et al. 2009. Friction material composition and friction material using the friction material composition. Japanese Patent No. 2009-091422, filed October 5, 2007, and issued April 20, 2009.

Chapter 16
Metallic Fiber

The hardest metal yields to sufficient heat.
—Mohandas Gandhi

METALLIC FIBERS PLAY A PIVOTAL ROLE in the development process of modern tribological composites and have enjoyed a long history of commercial success based on their beneficial qualities. Incorporation of a metallic fiber has generally been found to impart strength and improve thermal transport, which moves heat into the bulk of the composite, away from the friction interface. This reduces thermal gradients and helps inhibit "hot spot" formation at the surface.

A metallic fiber can be functionally defined as chopped wire (e.g., continuous wire drawn from a die and chopped into smaller segments). A metallic fiber can also have the form of shaved metallic wool. The use of metallic fiber in a friction material can include a number of metals and their alloys. To make this subject manageable, the discussion will include the more commonly used materials, namely steel, lead, aluminum, copper, brass, and zinc.

Shaved Metal Fiber Production Methods

The production of metal wool in the United States was patented as early as 1908 by Harry Gamage [16-1]. Gamage's method drew a hard knife blade across the perimeter of stacked metallic sheets to shave a thin fiber from the edge of the sheets. The thin cross section of the sheet metal made this an ideal method to use various metals and produce wool blends. By the 1920s, the development of steel wool producing machines and large volume production methods had increased dramatically.

Stern [16-2] described steel wool as "steel wool, which, as is well known, is an abrasive material consisting of long, curly, filamentary steel shavings." His steel wool machine formed wool into a compacted mass suitable for use in scouring and polishing applications, such as those used in typical household kitchens.

Galvin [16-3] developed an alternate means of shaving metal that is similar to modern methods and used serrated knives held against the surface of a moving metal rod. The action of the moving rod against the edge of the blade drew thin shavings from the surface.

This process allowed for the use of various serration shapes and multiple knives, which could be arranged in sequence to generate higher yield.

In recent years, the number of patents describing the production of metallic wool has declined, but one of the more interesting recent additions to the science of steel wool production is from Boeckenhoff and co-workers in Germany [16-4]. This group described a method using a plurality of specially formed steel wires with a unique profile. The primary benefit of using the special wire was that it gave higher yields with reduced waste. Much like the other processes, this process could also be applied to various other metallic materials such as copper or brass.

Metal Fiber in Friction Materials

Oswald's use of steel and aluminum fiber has already been discussed and was one attempt to use metallic fiber as a potential replacement for asbestos [6-6]. Aside from Oswald, the use of metallic fiber and wire dates back to a much earlier time when Painter [16-5] described his design for a rubber friction material.

Painter's concept incorporated a ductile metallic fabric as reinforcement in a sheet of rubber. The reinforced rubber sheet formed the tractive surface of a pulley system relying on the strength and bi-directional aspects of woven gauze to generate high strength across the surface. The compliance of the rubber in combination with the enhanced strength enabled this product to perform well under high loads.

Bonner [16-6] described a method for producing a friction material based on a structure reinforced by wire gauze. However, Bonner's method began by folding several layers of wire gauze together. To prevent unfolding, the layers were welded together and impregnated with rubber that was subsequently vulcanized to produce the final friction element. Bonner's method was analogous to the folded and compressed method used with rubberized fabric and overcame the delamination concerns by using welds to bond the plies.

Acres [16-7] reported that woven asbestos friction materials began to use wire reinforcement as early as 1908. However, these materials did not begin to see widespread acceptance and commercial success until "well into the 1920s." Many woven friction material production methods described the use of wire as a core for asbestos yarn or as a winding in multiple-strand asbestos yarns [5-5]. Wire was a convenient method for imparting strength and thermal conductivity to the highly insulative asbestos friction materials of the time.

Short Metal Fibers in Friction Materials

Discontinuous or chopped metal fibers in friction materials began with Eisenhardt [5-11]. The composition used short segments of wire derived from the waste stream of a wire-reinforced woven friction material. The waste material was pulverized and added to a laminated sheeter material. The end result was a friction material with small chips of wire dispersed throughout the matrix.

Metal fiber has also been used in sintered metal friction materials. DuBois and Roth [16-8] described a number of sintered metal friction material examples that exploited the thermal capacity and high strength of metal fibers. This technique allowed fibers of similar elemental composition (i.e., ones that were highly compatible with the metallic matrix) to be incorporated into the composition, which aided in the development of the matrix during sintering. DuBois and Roth did not limit themselves to a single fiber type and described examples using low carbon steel fiber (such as SAE 1010 grades) and stainless steel fiber. In most cases, the fibers were up to 0.5 in. (approximately 1.25 cm) in length.

Haddadi [16-9] examined wear and thermal effects associated with the operation of low modulus polymeric tribological composites for railway applications. The composition contained a blend of steel fibers and iron particles at a combined concentration of less than 10% by weight of the total. This blend was deemed ideal for use in composite railway shoes.

Many modern friction materials rely on the benefits of short metallic fibers that have been distributed throughout the matrix. Steel fibers are perhaps the most common form and are extensively employed in semi-metallic and low steel materials, but copper, brass, and other non-ferrous metallic fibers are also used in the other material types to a great extent.

Steel Fiber

Steel fiber is perhaps the most widely used metal fiber in today's friction material industry. Its prevalence is due to a combination of desirable characteristics including high strength, excellent thermal capacity, and moderately low cost. Many friction material producers use small to moderate concentrations of shaved steel fiber in low steel compositions, whereas manufacturers of semi-metallic friction materials use a significantly larger amount in their compositions, many with concentrations often exceeding 50% by weight.

Matejka and co-workers [16-10] examined aspects of the tribological performance of a semi-metallic brake lining and focused on the formation of a third body layer with this type of composite. It was determined that steel fibers are one of the best reinforcing materials used in modern semi-metallic friction compositions, and can affect both friction and wear performance.

Kumar and Bijwe [16-11] examined different metallic filler materials in a non-asbestos friction material with interest in controlling the sensitivity of the coefficient of friction during operation, specifically with regard to changes in the load and sliding speed. Steel fiber was determined to be beneficial in friction materials, especially when used in concentrations as high as 20% of the total.

Steel fiber also has secondary benefits that can improve processing. Gopal [16-12], who stated that the fibrous portion of a friction material can have a pronounced effect on the performance attributes of the composite, evaluated a blend of glass and steel fiber. Steel fiber enhanced the mix quality by assisting in the dispersion of glass fiber. The steel fiber also improved mixing action and aided in breaking apart the bundles of glass filaments. Additionally, steel fiber was found to enhance the thermal conductivity of the mixture and could be a means of accelerating the manufacturing process by enabling a reduction in the molding times.

Concerns with Use of Steel Fiber

According to Stenberg [4-6] steel wire was rarely used to reinforce woven friction materials. This restriction was due to concerns about excessive abrasion and scoring of brake drums. Because brake drums of this era were often made from soft steel and were manufactured by conventional stamping operations, they were prone to wear readily when operated against composite structures containing harder steel fibers. Modern woven friction materials continue to use non-ferrous metal wire reinforcement and, therefore, provide support for this area of concern.

Steel fiber additions to tribological composites can lead to excessive wear of the mating member. Bijwe [16-13] reported increased aggression associated with steel fibers because of an increase in the plowing component of sliding friction. The increase was attributed to a physical transformation where the fiber becomes separated from the matrix and a portion rises above the plane of sliding, acting as an abrasive asperity. This concept has also been supported by Xin and colleagues [16-14] who ascribe the friction generation mechanism as fibers standing upright and plowing into the other surface. It was recorded that this also produced higher wear rates and excessive noise generation.

Generally, the most cited deleterious effects of steel fiber additions include elevated abrasion of the mating surfaces [6-10] and the introduction of friction variation associated with degradation of the third body layer at the friction interface [6-11], which was also described by Cho [13-8] and by Jang [16-15].

Steel fibers are beneficial for imparting strength. They provide wear resistance while maintaining the friction level during operation at elevated temperatures, but steel fibers can elevate wear rates, especially of the mating member. Jang [16-15] also reported that steel fibers increase disc thickness variation. He considered thickness variation as one of the primary causes of judder (i.e., an unacceptable level of vibration during use).

In spite of the concerns, the strength and thermal properties of steel fiber are well-suited for use in friction material composites. Additionally, the cost of steel fiber is generally much lower than other non-ferrous metals, which will enable steel fiber to experience continued use.

Lead Fiber

Metallic lead has been used in friction materials since 1883 [16-16]. Shimer cast a brake shoe with recesses formed into the body of an iron friction member. The recesses were filled with molten lead and allowed to cool to form a dual-metallic brake lining. This material utilized the softness of lead to give an element of compliance to the friction surface, which ensured that the brake shoe made adequate contact with the wheel. It was further noted that lead acts as a solid lubricant to improve the longevity of this design.

Lead has also been used in a paper-based friction material comprised of layers of compressed paper with intermingled layers or strips of lead [13-9]. In this low cost concept, the layers of compressed paper provided excellent friction during dry operation, but the friction level tended to drop rapidly when wet. Thin layers of elemental lead enabled better friction when surfaces were wet, and allowed for effective operation over a wider range of conditions.

A major use of lead was as a wire reinforcing agent in woven friction materials [4-47]. Lead also found use in other non-woven friction material composites. For example, Leonard [16-17] used ribbons of lead foil braided together with ribbons of other soft metals such as aluminum or copper to form a friction material. The soft lead ribbons were formed and compressed into the desired shape prior to being fused together. When the fusion process was complete, the final product was metallic friction material based on the beneficial combination of softness and lubrication.

Other Benefits of Lead in Friction Materials

Bluhm [16-18] used elemental lead in a woven friction as early as 1929 and listed the following benefits of lead in friction materials:

> *The lead augments and stabilizes the coefficient of friction at all temperatures, pressures and speeds. It interposes between the metal of the drum or like friction piece, and the facing portion, a very thin film which prevents the adhesion of the metals in use thus eliminating the cause of the erosions.*

The prevention of "adhesion" and "eliminating the cause of erosions" refer to a severe scoring associated with metallic transfer at the friction interface. Elemental lead prevents this from occurring and improves wear rates of both tribosurfaces.

Along a similar line of thought, Wild [16-19] used combinations of lead and brass wire in woven friction material with differing weave patterns. The role of lead wire was listed as a friction stabilizing agent. Wild also described elemental lead as an inhibitor of slag formation processes that occurred when using asbestos friction fabric. According to Wild, the degradation of asbestos led to the formation of a strongly abrasive slag at the interface and resulted in severe damage to the drums. The only solution was to incorporate lead into the friction material.

Aluminum Fiber

Aluminum fiber was one of the potential replacement fibers proposed by Oswald in 1939 [6-6]. It was said to provide reductions in the rate of wear of friction materials. This was attributed to an increased ability to transport thermal energy away from the interface that prevented degradation of the organic binder material.

Other than this source, there is very little in the literature regarding the role of aluminum fiber in friction materials with the exception of a recent study of the effect of various metal fibers on the friction performance of automotive brake friction materials by Jang [16-15]. Jang simply states that compositions containing aluminum fiber gave reduced wear rates relative to compositions containing copper and steel fiber.

Concerns with Aluminum Fiber in Friction Materials

Bowden and Thomas [16-20] were perhaps the first to point out a singularly important concern with using metallic aluminum. From their study of the surface temperatures reached by sliding solids, it was determined that certain metals such as aluminum will undergo an exothermic reaction. The exothermic reaction elevates the interface temperature and can lead to hot spots with temperatures in excess of 3600°F (approximately 2000°C).

The concern with exothermic reactions with aluminum was also noted by Oehl and Paul [3-12] with an additional concern related to the generation of oxides. In this case, the aluminum oxide served as a hard abrasive and led to issues relative to abrasion and wear. This mechanism also suggested that aluminum fiber can improve the recovery of a friction material after operating at elevated temperatures (i.e., under fade conditions).

With a slightly different perspective, Blau [2-7] examined the role of metals during sliding and reported certain metallic materials such as copper or aluminum were commonly found to transfer to the mating member. Transfer of aluminum during braking could lead to uneven deposits around the periphery of the brake disc, which could lead to unstable vibration during operation.

Copper Fiber

Copper fiber (or chopped copper wire) is a commonplace metallic component in many dry friction materials and has seen continued use since 1913 [13-6]. The importance of copper wire in woven friction materials for use in band and drum brakes is seen in many early patents and was clearly described by Reuter [16-21] as well as many of the previously cited patents that described the formation of asbestos yarn [5-4, 5-5]. However, DeLaski [16-22] developed a friction facing for use in clutch applications that merged asbestos and copper wire by coating wire with short fiber asbestos. The shorter fiber length and the inherent flexibility of the copper allowed the coated wire to be formed into a circular woven friction material. Aside from the flexibility and ease of use in the weaving process, copper played a significant role in enhancing the frictional performance of many early band brake materials.

For instance, Greenwood [16-23] used copper fiber in asbestos yarn to improve the wet friction performance of band brake materials. Copper wire was found to provide the benefit of imparting strength to the yarn, which enabled weaving and also stabilized friction levels.

Norton [16-24] developed an asbestos friction material that could be produced by weaving or felting operations, and then subsequently saturated with a synthetic resin binder. During operation, the synthetic binders were very effective thermal insulators and the friction materials suffered extensive weight loss at higher operating temperatures because of carbonization. In order to offset this concern, non-abrasive metallic wires (i.e., copper) were positioned normal to the friction surface and dispersed throughout the material. This provided an effective means of transporting heat throughout the material, which enabled effective curing and also reduced the interface temperatures during braking.

The Role of Copper in Modern Friction Materials

The use of copper in modern tribological composites mirrors use from prior art. The majority of findings illustrate the beneficial improvements in thermal conductivity of a non-asbestos friction material because of incorporation of copper fiber [16-25]. The inclusion of copper reduces the rates of wear [6-10 and 12-12] and provides an overall improvement in frictional performance, which is most notable in the area of fade resistance [6-16 and 10-16]. For these reasons, copper is a very common additive in modern friction materials and is often blended with other fibrous materials, which also serves to illustrate another commonality in the design of modern brake lining materials.

Birch [14-23] described a blend of copper fiber with carbon, aramid, and ceramic fibers. Hikichi and colleagues at Akebono [16-26] described a non-asbestos organic friction material based on mixtures of potassium hexatitanate and potassium octatitanate with a blend of copper and aramid fiber. Their composition was stated to improve friction stability and generate less noise during use. Aside from tribological considerations, another critical point regarding copper fiber that must be considered is taken from the work of Shojaei [16-27].

Copper Fiber as a Processing Aid

Shojaei described the importance of thermally conductive elements as an expedient to the processing of composite friction materials. It was determined that higher molding temperatures rapidly develop a cured skin on the exterior of the part and inhibited heat from propagating throughout the composite. This prolonged the cure rate of the mixture, especially at the interior of the composite. Due to the low thermal diffusivity of the composite mixture, this induced a thermal gradient that sealed in trapped gases and led to damage to the composite upon release of the trapped gases (i.e., cracking during molding).

It was admitted that this problem could be solved by implementing new equipment to preheat the mixture prior to molding; however, the cost of new equipment is prohibitive. The use of thermally conductive elements (i.e., copper) was found to reduce thermal gradients between the surface and the interior, which allowed for a more uniform cure of the final product and eliminated the need for purchasing additional equipment.

Concerns with Copper

In spite of the many benefits gained by incorporating copper fiber into a friction material, there are also a few very critical concerns. One of the more interesting, and technical, concerns was reported by Kristkova [9-24]. Kristkova's work suggests that copper, and the oxides of copper formed during sliding, can have an influence on the degradation of the phenolic resin binder system. This chemical effect can be seen as a catalytic degradation of the polymeric network because of the presence of copper and its oxides.

Today, the primary concern with copper is the potential environmental impact of brake lining wear debris. This concern has led to the promulgation of a recent restriction [16-28] imposed as part of a

California Senate Bill (No. 346 introduced by Senators Kehoe and Simitian) which expressly states:

> *The bill, commencing on January 1, 2025, would prohibit motor vehicle brake friction materials exceeding 0.5% copper by weight from being sold in California.*

In a similar fashion, a recent bill [16-29] passed by the legislative branch in the state of Washington (HB 3018- 2009-10) restricts the use of copper in brake friction materials much earlier but does not restrict the content of the linings as stringently:

> *Beginning January 1, 2015, after-market brake friction material exceeding five percent copper and its compounds by weight may not be sold in Washington State.*
>
> *Beginning January 1, 2020, new vehicles with brake friction material exceeding five percent copper and its compounds by weight may not be sold in Washington State.*

Therefore, the role of copper in friction materials is expected to see a sharp decline in use over the next few years and will most likely be eliminated because of toxicological concerns.

Brass Fiber

Brass fiber is an alloy of copper and zinc that was originally produced with a small amount of lead to aid in machining and drawing. Brass wire, much like copper wire, was used in the manufacture of woven friction materials [4-37] in early rubber-bound sheeter materials [16-30] and continues to be used in modern non-asbestos friction materials [16-31, 16-32].

The benefits of this copper alloy are similar to copper and, in many cases, the metals are described interchangeably. The role of brass is similar to that of copper, and brass provides improved thermal dissipation because of the favorable conductivity of the metal [16-33]. However, Kumar and Bijwe [16-25] determined that additions of brass fiber to a non-asbestos organic friction material would improve frictional output, but at the cost of incurring poor wear performance.

Another interesting role for brass is that of a scavenger [6-24]. According to Batchelor, small segments of brass at the friction interface scrape the surface of the mating member and remove deposits. This ensures effective contact and less likelihood of unacceptable vibration occurring during high temperature operation. This report is consistent with Jacko and Rhee [6-25] but is in direct opposition to the findings of Kerridge and Lancaster [16-34].

Although Kerridge and Lancaster did not study brass as an additive in friction materials, their study of a pure sliding couple between brass and a harder metal member was shown to exhibit transfer of brass to the harder mating member. The process of material transfer and back transfer would continue until equilibrium was reached and was critical to an understanding of severe wear mechanisms between sliding metal pairs.

In a recent study focused on polymer matrix friction materials reinforced with brass fibers, Xian and co-workers [16-35] found brass fiber to have an effect on both the friction and wear performance of the composite. It is interesting to note that both friction and wear increased as the concentration of brass increased. When concentrations reached approximately 20% by weight, the concentration of copper began to increase in the third body layer. This increased concentration of copper was also coincident with an increase in the friction level and a marked improvement in wear resistance.

Zinc Fiber

Stenberg and Heldt [4-6, 4-7] tell us that zinc wire was used in the production of wrapped asbestos yarn for use in early woven friction materials. Like many soft metals, the importance of zinc was based on the need for enhanced thermal conductivity and ease of manufacturing. Additionally, zinc has a number of ancillary benefits in friction materials.

Rohrer [15-8] reported that early brake drum materials were prone to score during use. One solution to the scoring concern was the implementation of zinc wire as a reinforcing agent in woven asbestos friction facings. Zinc wire did not generate the same degree of severe surface damage as other metallic reinforcing agents.

Zinc can impact wear rates [6-23] but stabilizes the coefficient of friction in many tribological composites. Perhaps the most commonly cited improvement echoes the sentiment by Halstead more than sixty years ago that zinc improves friction after fade.

Consistently, the addition of zinc to a brake lining composition has been shown to improve recovery of friction materials after fade [6-25, 6-29 and 10-16]. This is most likely because of a combination of effects based on oxidation and transfer from the friction composite that allows the tribosystem to quickly develop a third body layer upon cooling.

Zinc as an Anode

Zinc has also been incorporated into friction materials as a sacrificial anode [6-11]. The incorporation of zinc was intended to inhibit corrosion of other more reactive metals present in the composite. This is most logically associated with semi-metallic friction materials, which possess a large concentration of easily corroded ferrous metals, and would benefit from the introduction of a small amount of a sacrificial anode material. It is not known how effective this approach is in comparisons to other methods (such as by increasing the pH of the composite), and more work is needed in this area.

Other Alloys

Any metallic material or alloy can potentially be used as a metallic fiber in friction materials. The only limitation is based on the ability of the metal to be drawn into wire or shaved to form wool. The importance of metallic fiber additions to friction materials continues to be a means for imparting structure while simultaneously improving thermal transport. It is expected that new allows could become an integral part of future friction products, especially in light of the upcoming restrictions on copper and copper-based alloys.

References

16-1. Gamage, H., 1908. Apparatus for manufacturing steel or other metal wool. United States Patent No. 881852, filed March 20, 1903, and issued March 10, 1908.

16-2. Stern. L. 1924. Preparation of steel wool. United States Patent No. 1490544, filed February 1, 1922, and issued April 15, 1924.

16-3. Galvin, J. 1926. Metal-shaving machine. United States Patent No. 1598814, filed February 21, 1924, and issued September 7, 1926.

16-4. Boeckenhoff, H. et al. 1976. Wire and method for making steel wool. United States Patent No. 3969085, filed June 9, 1975, and issued July 13, 1976.

16-5. Painter, W. 1883. Pulley covering. United States Patent No. 283356, filed April 2, 1883, and issued August 14, 1883.

16-6. Bonner, W. 1909. Friction member. United States Patent No. 931988, filed January 20, 1909, and issued August 24, 1909.

16-7. Acres, F. 1946. "Some Problems in the Design of Braking Systems." *Journal of Institute of Automotive Engineers* 15: 19-49.

16-8. DuBois, W. and D. Roth. 1963. Brake element having metal fiber reinforcing. United States Patent No. 3114197, filed June 17, 1960, and issued December 17, 1963.

16-9. Haddadi, E. et al. 2005. "Wear and Thermal Effects in Low Modulus Polymer-Based Composite Friction Materials." *Journal of Applied Polymer Science* 95: 1181-1188.

16-10. Matejka, V. et al. 2009. "Semimetallic Brake Friction Materials Containing ZrSiO4: Friction Performance and Friction Layers Evaluation." *Journal of Composite Material* 43: 1421-1434.

16-11. Kumar, M., and J. Bijwe. 2010. "Role of Different Metallic Fillers in Non-asbestos Organic (NAO) Friction Composites for Controlling Sensitivity of Coefficient of Friction to Load and Speed." *Tribology International*. Article in press doi:10.1016/j.triboint.2009.12.062.

16-12. Gopal, P. et al. 1994. "Fade and Wear Characteristics of a Glass Fiber Reinforced Phenolic Friction Material." *Wear* 174: 119-127.

16-13. Bijwe, J. 1997. "Composites as Friction Materials: Recent Developments in Non-Asbestos Fiber Reinforced Friction Materials - A Review." *Polymer Composites* 18: 378-396.

16-14. Xin, X. et al. 2007. "Friction Properties of Sisal Fibre Reinforced Resin Brake Composites." *Wear* 262 .736-741.

16-15. Jang, H. et al. 2004. "The Effect of Metal Fibers on the Friction Performance of Automotive Brake Friction Materials." *Wear* 256: 406-414.

16-16. Shimer, G. 1883. Brake shoe. United States Patent No. 272911, filed November 28, 1882, and issued February 27, 1883.

16-17. Leonard .J. 1926. Friction metal. United States Patent No. 1588757, filed January 15, 1926, and issued June 15, 1926.

16-18. Bluhm, A. 1929. Friction material. United States Patent No. 1722890, filed January 3, 1927, and issued July 30, 1929.

16-19. Wild, R. 1931. Friction material. United States Patent No. 1803448, filed November 29, 1929, and issued May 5, 1931.

16-20. Bowden, F., and P. Thomas. 1954. "The Surface Temperature of Sliding Solids." Proceedings of the Royal Society of London, A 223, pp.29-40.

16-21. Reuter, F. 1935. Brake band. United States Patent No. 1990880, filed March 13, 1931, and issued February 12, 1935.

16-22. DeLaski, A. 1914. Friction facing for clutches and the like." United States Patent No. 1122081, filed January 29, 1914, and issued December 22, 1914.

16-23. Greenwood, T. 1920. Brake. United States Patent No. 1758253, filed September 30, 1927, and issued May 13, 1930.

16-24. Norton, R. 1932. Friction facing. United States Patent No. 1879433, filed December 20, 1930, and issued September 27, 1932.

16-25. Kumar, M., and J. Bijwe. 2010. "Role of Different Metallic Fillers in Non-asbestos Organic (NAO) Friction Composites for Controlling Sensitivity of Coefficient of Friction to Load and Speed." *Tribology International* 43: 965-974.

16-26. Hikichi, A. et al. 2003. Friction material. United States Patent No. 6670408, filed December 14, 2000, and issued December 30, 2003.

16-27. Shojaei, A. 2006. "A Numerical Investigation on the Cure Characteristics of Railroad Composite Brake Blocks during the Compression Molding Process." International Polymer Processing 21: 421-431.

16-28. State of California Legislative Council. "Official California Legal Information." Accessed June 29, 2010. http://leginfo.ca.gov/pub/09-10/bill/sen/sb_0301-0350 / sb_346_bill_20100621_amended_asm_v92.pdf

16-29. Washington State Legislature. "HB3018-2009-10." Accessed June 29, 2010. http://apps.leg.wa.gov/billinfo/summary.aspx?bill=3018&year=2010

16-30. Schmidt, C. 1929. Lining for brakes, clutches, and the like. United States Patent No. 1716090, filed May 14, 1924, and issued June 4, 1929.

16-31. Fujimaki, H. et al. 1982. Organic friction material. Japanese Patent No. 57-205474, filed June 10, 1981, and issued December 16, 1982.

16-32. Hayashi, T. 1996. Brake lining for rotor made of stainless steel. Japanese Patent No. 08-135703, filed November 15, 1994, and issued May 31, 1996.

16-33. Crosa, G., and I. Baumvol. 1993. "Tribology of Polymer Composites Used as Frictional Material." In *Advances in Composite Tribology* edited by K. Friedrich: Amsterdam: Elsevier.

16-34. Kerridge, M., and J. Lancaster. 1956, "The Stages in a Process of Severe Metallic Wear." Proceeding of the Royal Society of London, A 236, pp.250-264.

16-35. Xian, J. et al. 2004. "Friction and Wear Characteristics of Polymer-matrix Friction Materials Reinforced by Brass Fibers." *Journal of Materials Engineering and Performance* 13: 642-646.

Chapter 17
Abrasives

Only in the frictionless vacuum of a nonexistent abstract world can movement or change occur without that abrasive friction of conflict .
—Saul Alinsky

ABRASION CAN BE VISUALIZED AS A FORM of conflict that arises when asperities collide. The addition of abrasive materials to tribological composites is an effective technique for generating the very high, stable level of friction required by the end user. An abrasive material will generate friction by the mechanism of abrasion, which can take the form of either plowing or cutting and includes the associated wear mechanisms of chip liberation and ribbon formation.

Therefore, the selection process and correct use of abrasive additives in tribological composites is an area with exceptional promise. However, it also brings a list of concerns that must be fully appreciated to suppress the negative attributes while simultaneously amplifying the benefits. To examine abrasive additions in detail, we must look at the historical use of these materials.

Historical Use of Abrasive Additives

It is difficult to define precisely the first time that an abrasive material was deliberately added to a friction material. However, it appears that early attempts were associated with the development of a composition brake lining for railway applications. For example, Morrison and Allen [17-1] experimented with a cast composition brake shoe containing a mixture of comminuted cast iron particles and pulverized rock bound in a matrix of asphalt and sulfur. Sulfur was needed to increase the thermal resistance of asphalt. This enhancement gave the lining with longer life. The addition of pulverized rock was found to be beneficial for improving friction. However, it was considered an abrasive material that was prone to generate unacceptable levels of wear at the wheel surfaces. The solution to this problem was careful control of the concentration to ensure a reduced rate of wheel wear with an optimum friction output.

Another historical example is seen in the work by Soulis [13-1]. Soulis stated that abrasives were added to friction materials to overcome difficulties with operation in wet weather conditions.

Recent investigations by Blau [17-2] concerning the frictional behavior of an automotive brake material operating under wet conditions found friction material compositions that contained high abrasive concentrations in combination with a low lubricant concentration had the highest friction output under wet conditions. This confirmed the findings of Soulis.

A different approach was taken by Borchers [17-3, 17-4] who used abrasive additives called roughening agents during the manufacture of a woven friction material. Borchers utilized a technique of coating strands of fiber, or reinforcing wire, with adhesive-bonded abrasive particles prior to weaving. The woven product was subsequently heat treated to seal any open pores in the structure—a vital step that aided in retention of the abrasive grains throughout the bulk of the friction lining. Borchers further stated that this type of friction material gave a "highly superior braking action" even when used against "discs which have been very heavily soiled with oil."

Similarly, Sanders [17-5] incorporated hard refractory materials in an early molded friction material. This application illustrated many of the salient advantages of an abrasive additive. In his compound, the hard refractory materials (exemplified by zircon, alumina, rutile, and silica) were deemed necessary for generating the desired friction level and resistance to fade. In addition to the abrasive material, a metal powder was added to improve thermal conductivity. The metal powder served to stabilize the temperatures at elevated conditions, and additions of small amounts of carbon lubricated the surfaces. All were combined with asbestos in a phenolic resin matrix. Sanders also noted that the size of the abrasive particles should be very small to reduce excessive abrasion.

Many modern friction materials share a similar philosophy and employ an abrasive material (often a hard refractory ceramic or other suitable metal oxide) in the form of discrete grains.

Matthews [17-6] used discrete grains of an abrasive material in an asbestos-based drum lining with the intention of cleaning glazed drum surfaces. Matthews noted that the composition must be carefully balanced to ensure sufficient cleaning of the surfaces while maintaining low wear rates. The balance between abrasion and cleaning remains a critical aspect of tribological composite design.

Abrasion, Cleaning, and Wear

Rhee [17-7] evaluated automotive brake drum materials using an inertial dynamometer. The investigation focused on brake drums formed with three distinct compositions. As part of the general conclusions reported, mildly abrasive linings gave less drum wear than more abrasive linings. This may seem overly simplistic. However, abrasion is not a simple matter, and friction transitions are constantly at work during the interactions of the surfaces in motion.

For example, Georges [17-8] stated that when sliding friction begins, wear is mainly an abrasive process. After a period of time, however, wear processes generate debris that transforms the topography of the sliding surfaces and leads to wear by a milder "adhesional" process. The transition from abrasive to adhesive friction is often another fundamental challenge that must be overcome during the selection of abrasive materials. Unfortunately, there are no simple rules that allow for an effective balance of abrasion and cleaning.

Schallamach [17-9] reported that abrasion of rubber compounds was not strictly dependent on the particle size of the abrasive material. It was experimentally determined that a fivefold increase in particle size did not produce a fivefold increase in abrasion. This is one example of a situation when a lack of strict particle size dependence can make the use of abrasive models difficult to apply in real world situations.

Likewise, Sin [7-15] examined abrasive wear mechanisms and the effects of various grit sizes. Wear rates were shown to rapidly increase with particle size until an upper limit was reached, which was found to be approximately 80 microns. Furthermore, the number of contacting abrasive particles was shown to increase in a nearly linear fashion with applied load, which was responsible for the linear dependence of the wear rate on the applied load. It should also be noted that variation in the total number of contacting particles for a given grit diameter under an applied load varied inversely as the square of the grit diameter.

Additionally, Sin pointed out that the effect of grit size on wear and friction can be explained in terms of the "relative bluntness" of the abrasive particles. An idealized model of an abrasive particle as a cone with a spherical tip explains, at least partially, the variation of friction and wear with grit diameters, including the transition of wear mechanism from a cutting mode to sliding wear. A similar finding by Wirth and colleagues [3-46] also served to illustrate the importance of particle geometry and its significance on wear rates.

In addition to bluntness, other factors are also important. The size of the grains can be important to removal rates of third body layers [17-10], but sliding speed and the thickness of the lubricant film are areas that have an influence and can affect criteria for abrasive selection [3-6]. Generally, slower speeds are worse for abrasion. Likewise a very thin lubricant layer does not require large particles to experience direct contact, thus leading to rapid wear and higher material removal rates.

Common Abrasive Materials

The most common abrasive materials are those that are readily available and sufficiently hard enough to generate a high level of friction and effective surface cleaning. Perhaps the most common abrasive in friction materials is sand (i.e., quartz). Other hard minerals include aluminum oxide, zirconium silicate (zircon), silicon carbide, and the numerous oxides of common metals. Because the list can be extensive, a few of the more relevant examples are discussed here beginning with silicon dioxide.

Sand (Silicon Dioxide—Quartz)

Sand is a commonly occurring granular mineral mixture of varying composition. Sand, which is predominantly silicon dioxide (also known as silica), is typically quartz with other particles of rock. Natural sand particles range in size from silt (very fine particles) to individual sand grains that are fractions of an inch in diameter. Larger particles can also be called sand gravel and are generally unsuitable for use in friction material composites.

Quartz has high melting point— greater than 3000°F (1648°C)—and occupies the 7th position of Mohs scale of relative hardness. This combination of hardness and thermal resistance makes quartz an ideal friction augmenting additive in a range of compositions. Sand has been used in friction materials for more than 100 years.

Silica in Early Friction Materials

One early patent for the use of sand was granted in 1892 to Buell [17-11]. Buell did not give specific attributes supporting its use in his cement-based brake shoe, but used sand as an example of one of many "similar hard cutting materials as consolidate in a bond of cement."

Several early patents by DeFrance [17-12, 17-13, 17-14] also described the use of sand as a raw material in a cast brake shoe for railroad applications. Like Buell, DeFrance did not provide specific benefits or concerns about the use of sand in a metallic friction material.

In the mid-1940s, Halstead [6-23] discussed the fundamental aspects of asbestos-based friction materials. During his extensive discourse on these organically bound friction materials, the author disclosed that compositions containing silica were considered unorthodox and that the use of silica was not commonplace. Yet, despite this lack of information regarding an unorthodox abrasive, the role of silica in friction materials continued until modern times.

Silica, Sand, and Quartz in Modern Friction Materials

Nakagawa and Nitto [17-15] developed a resin-bonded semi-metallic friction material using a blend of iron powder and steel fiber to effectively replace asbestos. Iron powder was deemed necessary for generating the desired friction level while steel fiber provided structural reinforcement and mechanical strength. Rubber powder was added to this blend to increase the real area of contact. A very small amount of silica (only 1% by weight) was also added.

Wood and Clark [17-16] incorporated silica at 10% by weight during the development of a friction material bound with a matrix of copper sulfide. A similar patent by Clark and Summerling [17-17] used a comparable level (10.4 parts by weight) in a product based on the development of a matrix produced by reactions of iron and sulfur. Unlike organically bound friction materials, the use of a metallic sulfide binder system is an intermediate step in full metallic matrix materials where silica has experienced a long history of successful use.

Silica in Sintered Metallic Friction Materials

Silica was commonly used in sintered metallic friction materials. Wellman [5-33] developed a composite machine element formed by the fusion of copper, lead, and tin with small additions of graphite and silica. Lowey and Charles, both of the S. K. Wellman Company [17-18], developed a copper-tin-bronze compound containing iron (for strength) and graphite (for lubrication). Additionally, elemental lead was used as a seizure preventing agent and silica as a polishing agent. It is interesting to note that molybdenum disulfide was added to improve fade and the coefficient of friction. This seemingly contradictory application of a well-known lubricant was attributed to an impurity level of 10% silica in the molybdenum disulfide.

Genkin, Dmitrovich, and Fishbein [6-50] used silicon oxide in a sintered iron-based friction material. The hard particles (present at only 3%) were added to improve the coefficient of friction. Likewise, Fedorchenko [17-19] described an iron-based "cermet" sintered friction material with silica at a concentration of 3%, sharing many of the same design features as Genkin and colleagues.

Concerns with Silica in Friction Materials

Silica additions have a number of benefits including an enhanced friction level as previously noted. However, they also impart wear resistance to the final product [7-30]. In elastomeric compounds, silica additions can improve a number of the physical properties such as abrasion resistance and creep [17-20]. Generally, finer particles provide better wear resistance with a reduced friction effect as compared to additions of larger particles.

Despite the benefits, silica (especially crystalline silica, also known as cristobalite or tridymite) is becoming an area of significant concern regarding respiratory ailments such as silicosis. Inhalation of silica can lead to acute and chronic forms of silicosis, with long-term exposure resulting in an elevated risk of lung cancer as noted in the following quotation from Rice [17-21]:

> *The results of our analyses suggest a positive, linear, and significant increase in risk of lung cancer with increasing estimated cumulative exposure to respirable crystalline silica dust (mainly cristobalite).*

As a result of these concerns, several organizations are taking steps to eliminate silica, in all of its forms, from use in friction materials.

One example by Loftipour [17-22] reported that silica concentrations in railway friction materials can be very high. During operation, the wear of these particles liberates dust of finely divided silica particles. The use of larger particles decreases this concern, but introduces excessive wear of the mating components, thereby forcing the replacement of silica with other, softer mineral species such as magnesium oxide.

Impure Forms of Silica

Silica is present in a number of various geologic families and, because of natural variation, can be found in varying levels of purity. Many of the impure forms of silica have been used in the friction material industry for a number of applications and are generally regarded as mild abrasives. In addition, specific forms of impure silica can also provide secondary performance benefits when used in wet friction material applications, with the most notable candidate being diatomaceous earth.

Tripoli

Tripoli is a very soft and readily friable form of silica that occurs as deposits of metamorphic crystalline rock. It is used in a number of abrasive and filler applications. Tripoli particles are typically fine grains that can occur as aggregates of fine particles, some of which can be present at the nanometer range of sizes [17-23]. Tripoli is commercially available in purity levels of 98–99% with minor amounts of aluminum oxide present. Iron oxides can also be present, which accounts for variation in color of the various forms of tripoli.

Tripoli is an extremely fine form of crystalline silica and therefore subject to inhalation concerns. Additionally, fine particle sizes of commercial tripoli can also be of concern regarding PM-2.5 regulations regarding air quality. Tripoli in friction materials is expected to be limited and can ultimately be eliminated from consideration.

Rottenstone

Rottenstone is a pulverized form of weathered limestone. Rottenstone has been used as an abrasive material in the same fashion as Tripoli (i.e., for polishing of metallic surfaces and glass). The history of tripoli and rottenstone are commingled. An early description of rottenstone (circa 1862) described it as "a decomposed stony substance resembling tripoli, and used for polishing" [17-24]. The composition of this early form was analyzed and found to contain 86% alumina, 10% carbon, and 4% silica.

The use of the terms rottenstone and tripoli continue to be somewhat interchangeable, and the role of rottenstone as an abrasive mineral is typically compared with tripoli [17-25]. Rottenstone has found limited use in friction materials [6-24] as a mild abrasive that gives slight increases in the friction output without incurring many of the negative issues associated with harder abrasive materials.

Generally, tripoli and rottenstone are considered similar materials and their roles in friction materials can be reasonably anticipated to give similar characteristics. In addition, rottenstone also faces similar concerns associated with respirable fractions and particle size effects, which can limit further applications and future growth potential.

Diatomaceous Earth

Diatomaceous earth, also known as diatomite or kieselguhr, is a naturally occurring soft rock formed by sedimentation of skeletal remains of unicellular aquatic plants. These micro-fossils are composed predominantly of silica and possess a wonderful array of beautiful shapes and forms that have generated interest for centuries. An example of the skeletal structure of diatoms is provided in Fig. 17.1.

The benefit of using diatomaceous earth in friction materials dates back to 1935 [17-26]. Longley's composition incorporated 5–7 parts by volume to generate friction during operation when wet. The porosity of diatomaceous earth allowed the boundary layer of water to have an avenue of transport away from the interface, allowing for more effective contact between the two friction surfaces.

Nanfeldt [5-28] reported that an effective brake lining should contain density elements to resist the normal loads experienced during braking. This addition of high-density, shear-resistant particles was exemplified by square particles of clay or diatomaceous earth. The

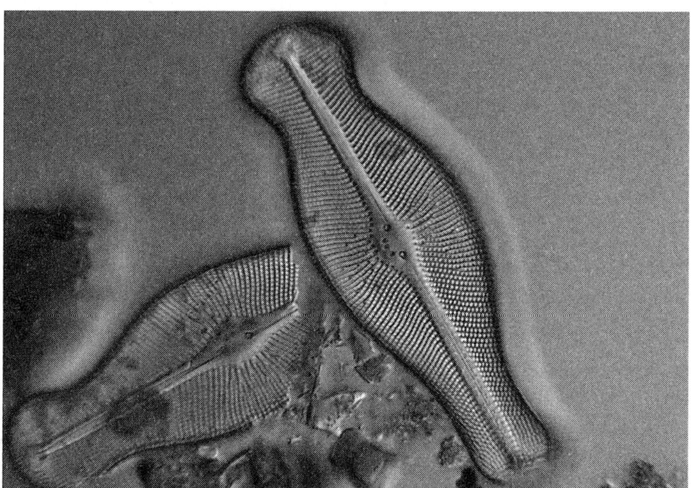

Fig. 17.1 Image of diatoms: Didymosphenia geminate cell wall.

use of these materials helped maintain the ability of the brake lining to generate the necessary friction level during operation.

Arledter [6-57] developed a wet friction material for operation in fluid that was based on a mixture of cellulose and asbestos fiber. This composition was augmented with additions of diatomaceous earth and subsequently saturated with phenolic resin. Diatomaceous earth continues to find widespread use in the wet friction material market [6-58] today, but does not seem to be as widely prevalent in dry friction material development.

Final Words on Silica

Silica in all forms is heat-resistant, inorganic material that is capable of producing a significant increase in the friction level and that can aid in reducing wear. However, the particle size and degree of contamination must be carefully controlled to achieve the highest levels of success without incurring severe penalties.

Additionally, long-term exposure data has indicated that silica can pose a serious threat to the respiratory health of workers who use the material daily. This will require workers to be equipped with effective personal protective equipment, and can tend to limit silica use during the next decades.

Aluminum Oxide

Aluminum oxide, also commonly known as alumina and Alundum [17-27], is a hard refractory material that can be found naturally as the mineral corundum or, in its more pleasant forms, as rubies and sapphires. A large portion of aluminum oxide is produced synthetically by treatment of bauxite ore, which provides the major supply of commercially available aluminum oxide used in the friction industry.

Production of Aluminum Oxide

The first commercial method of aluminum oxide production began in 1887 when Bayer developed a three-step process for converting bauxite into aluminum oxide. The method outlined in Bayer's patents [17-28, 17-29] began by digesting the bauxite ore in a strong solution of base under elevated temperature and pressure. Upon digestion, the impurities were allowed to settle out, leaving a highly alkaline solution of aluminum hydroxide.

The solution was decanted and allowed to cool until alumina trihydrate began to precipitate out. Seeding could also be used to accelerate the precipitation process [17-30]. The trihydrate was filtered, washed, and calcined.

It is during the calcination phase of the manufacturing process that the particles undergo phase transitions and take on various forms. There are a number of forms that aluminum oxyhydroxide and aluminum hydroxide can take, but the most thermodynamically stable is alpha-aluminum oxide, which forms when the calcining temperature exceeds 1832°F (1000°C). It is also important to note that the process can be tailored to form tabular aluminum oxide, where the particles are sintered together, and can be taken to full molten conditions yielding fused alumina. Fused alumina is generally higher in density, and particles have a lower permeability.

The Norton Process

According to Ingalls [17-31], synthetic versions of aluminum oxide were first produced in North America by the Norton Emery Wheel Company in 1904. The Norton process consisted of heating bauxite until the material was molten. After the melt became homogenous, it was allowed to cool. It was then pulverized and classified into various sizes and sold as abrasive grains for polishing or for the production of bonded abrasive wheels, paper, or similar products. This early form of fused aluminum oxide was ideal for material removal.

Aluminum Oxide in Friction Materials

The history of aluminum oxide use in friction material dates back to Buell [17-11] who suggested that aluminum oxide, in the form of corundum, might be a suitable replacement for silica sand in his brake shoe composite. There are several patents after Buell's work that demonstrate the use of aluminum oxide in virtually all types of friction materials.

For instance, Biggs [6-7] developed a resin-bonded friction material based on the benefits of iron powder, which allowed for the incorporation of small amounts (less than 1%) aluminum oxide. Komori and colleagues [11-12] developed a steel fiber reinforced composite based on COPNA resins and included a trace amount of aluminum oxide (only 0.1%). Asano and Iwata [17-32] reported a non-asbestos organic friction material using organic fiber and cork with a blend of aluminum oxide and magnesium oxide (both at

1%). On the other hand, Myers [6-13] developed a fused metallic friction material with a much higher abrasive content (8% by weight aluminum oxide).

Batchelor [6-24] described the importance of aluminum oxide as a friction augmenting agent and cited the importance of selecting grades with particles finer than 600 mesh (approximately 22 microns). Even at this fine size, 1% aluminum oxide was reported to be able to generate a 15% increase in the friction output.

Crosa and Baumvol [16-34] examined the role of fused alumina in a polymeric friction material composite. They reiterated the critical concern of selecting the correct size and they outlined the pronounced effects of aluminum oxide on friction and rate of wear. At concentrations of only 2%, notable increases in the friction level were observed during high temperature fade and under cold operating conditions.

Satapathy and Bijwe [17-33] analyzed the simultaneous influence of several operating variables on the abrasion resistance of phenolic resin-bound composites. Phenolic resins filled with aluminum oxide were found to provide enhanced wear resistance where the minimum wear level was obtained by careful optimization of the concentration of hard particles.

More recently, Fan and colleagues [7-13] examined the role of aluminum oxide in semi-metallic friction materials. Their study focused on the effect of the abrasive concentration on the friction and wear performance. As the concentration of aluminum oxide increased, the abrasion of the mating surface was also found to increase. However, this was generally considered to have a more stable friction level than formulations that did not contain aluminum oxide additions.

Aluminum Silicates

The aluminum silicate class of abrasive additives has typically been limited to the sillimanite class of minerals, most often in the form of kyanite (also known as cyanite) and its polymorphs (andalusite, and sillimanite). Kyanite is a unique refractory material that can vary in hardness across the same crystal face ranging from 4.5 to 6.5 on the Mohs scale [17-34]. It possesses outstanding thermal resistance, thus making it ideal for high energy friction applications. Generally, the role of kyanite and its polymorphs has been directed toward the synthetic production of mullite. In friction materials, there is little that supports their use that does not incorporate mullite as a potential replacement.

Mullite

Mullite is a rare, naturally occurring mineral form of alumino-silicate. The name is derived from its locality, the Island of Mull [17-35] off the west coast of Scotland. The science and synthesis of mullite has been comprehensively examined and documented by Schneider and Komarneni [17-36], who generally described the production by either chemical means (i.e., sol-gel) or by solid state/liquid state methods (whereby fused and sintered mullite are formed). The distinctive characteristics of mullite have made it a favorite in friction materials for more than sixty years.

Mullite in Friction Materials

Stedman and Pocock [17-37] were perhaps the first to develop a commercial friction material using an aluminum silicate as the abrasive material. Their sintered friction lining composition was enclosed in a metallic shell and used in aircraft braking applications. The composition relied on a metallic matrix containing particles of frangible mullite that would splinter if the compressive stress became excessive. This enabled the composite to generate fresh cutting surfaces and facilitated the retention of friction level during sustained operation at elevated temperatures.

A similar example by Batchelor used mullite, but allowed for the use of kyanite in the form of large particles within the matrix of a sintered copper friction material [15-20]. Batchelor sited the material removal properties of these hard abrasive minerals as a beneficial agent for preventing excessive structural cracking (known as heat checking) in the mating member of the braking system. Furthermore, mullite is an effective means for preventing material transfer in copper-based sintered metallic friction materials [12-29]. Aside from friction generation and material removal, mullite additions have other benefits.

For example, Luther and Koehring [17-38] developed an iron-based sintered friction material with a small amount of mullite (0.8 parts by weight). In this case, the small amount of mullite provided an improvement in wear resistance over the base composition that did not contain mullite.

Spokes and Littlefield [17-39] incorporated mullite in a composition friction element for use in railroad applications with the objective of improving overall performance. Aldrich and Deane [6-8] described a semi-metallic mixture for modifying an organic base friction material that was comprised of iron powder and graphite with a large concentration (approximately 15%) of ceramic powder, where mullite was listed as the preferred ceramic powder.

Mullite has become a widely used abrasive and is effective across a wide range of operating conditions including high-speed railway and racing applications [12-27]. For these numerous reasons, mullite is expected to see continued use for many years to come. The only limitation is that designers are often forced to keep the concentration low (on the order of only a few percent).

Silicon Carbide

Silicon carbide is one of the hardest materials currently in use in the friction materials industry. It has a hardness of approximately 9–9.5 on the Mohs scale. Like many ceramic particles, the combination of thermal properties and hardness makes silicon carbide ideally suited for high-energy friction materials, especially in sintered metallic compositions.

For example, Cox developed an iron-based friction material [17-40] containing silicon carbide and graphite at 10% by weight in a matrix of iron, where iron was roughly 80% of the composition. The beneficial aspect of this patent was based on the large grain size of graphite, which was found to improve wear dramatically over finer particles. This simple composite was intended for use in extreme braking applications such as with aircraft.

Matejka [17-41] examined the effects of silicon carbide additions to a semi-metallic brake material and concluded that silicon carbide increased the friction level (even at small concentrations) and improved friction stability. This finding was confirmed by Satapathy and Bijwe [17-42], who treated the process of abrasive additions as a multi-criteria optimization problem, but also reported that silicon carbide gave increased wear rates regardless of speed.

Anderson [12-27] cites that small amounts of silicon carbide (only a few tenths of a percent) in graphite produced by the Atchison method, was linked to excessive disc scoring and severe wear. This distinct difference in material hardness is an extreme example of a hard contaminant present in a soft material. Despite this concern, silicon carbide is considered a very effective abrasive addition when used in the correct concentration and when sized properly.

Beneficial Applications of Silicon Carbide

Horth [17-43] developed a coating comprised of abrasive powder in a sodium silicate solution. This unique application of an abrasive coating was found to stabilize the friction levels of a conventional phenolic resin-based composite, especially during use immediately after installation, which is commonly referred to as green friction or as the pre-burnish friction level. The list of preferred abrasives included aluminum oxide, iron oxide, and silicon carbide.

Yamamoto [17-44] recently developed a tribological composite that incorporated silicon carbide in conjunction with aluminum oxide. The silicon carbide particles were much smaller than the aluminum oxide particles but were used in a slightly higher concentration. This approach of blending harder abrasive materials with larger particles of a softer abrasive material suggests great promise and may be an avenue for further exploration in the near future.

Zirconium Silicate

Zirconium silicate, commonly known as zircon, is another hard abrasive material with a hardness of approximately 7–7.5 on the Mohs scale. Zirconium silicate has been used in both clutch and brake friction materials and can be stated to provide many of the same benefits, and pose the same concerns, as other hard mineral additives.

For instance, Harunobu [15-48] developed a molded friction material for clutch applications based on the enhanced fade resistance of unmodified phenolic resins. The preferred friction additive was zirconium silicate, which could be used at a concentration of 0.1– 5 % by volume of the composite. Harunobu reported the ability to tune the friction output of his composite by adjusting the concentration of zirconium silicate in conjunction with other mineral fillers such as barytes.

Like many of the other abrasive materials, zirconium silicate is typically used in lower concentrations. Jang and colleagues determined that 1.5% was suitable for improving the creep groan behavior of a brake friction material [17-45]. On the other hand, Kim and Jang [17-46] reported that zirconium silicate additions will affect the formation and regeneration of third body layers, which can also lead to variation in noise levels and altered performance at high temperatures. It was noted that despite enhanced stabilization of the coefficient of friction because of increased abrasive content, the wear rates also increased with zircon additions.

Garnet

Garnet is used in many abrasive applications, but little is found regarding its use in friction materials. A few sources mention garnet but include it in a list of hard mineral species that are suitable for use in railway friction materials [17-47, 17-48]. These citations do not provide additional information regarding the benefits of garnet over other abrasive materials.

Metal Oxides

Virtually any metal oxide that possesses sufficient hardness can be added to a tribological composite to influence friction and wear characteristics. Generally the more common metal oxides, other than those previously described, include oxides of chromium, iron, magnesium, titanium, zinc, and zirconium.

Chromium Oxide

Chromium oxide, also known as green chome oxide, has a hardness of approximately 8 on the Mohs scale and is typically available in small particle sizes for use as a colorant. Chrome oxide has excellent thermal resistance and possesses a stark green color that has been widely recognized in artistic circles for well over one hundred years [17-49]. However, the role of green chrome in friction materials is not well defined.

Typically, the inclusion of chromium oxide is reportedly similar to that of any suitable metal oxide abrasive [17-50, 17-51]. It has been used in asbestos-containing friction materials [17-52], but often with little or no record of its benefits. In a side-by-side comparison with aluminum oxide [7-12], rounded particles of green chrome oxide were shown to provide less wear than the more granular particles of aluminum oxide. However, there is little additional information available about its use.

Chromite

Chromite is an iron chromium spinel with a moderate hardness of approximately 5.5 on the Mohs scale. Chromite is very heat stable and is used in refractory ceramic materials. The use of chromite in friction materials was most likely associated with impurities in asbestos [4-6], but it has become a functional abrasive in friction materials for rail and automotive applications.

Pollard [17-53] described a cast brake shoe for rail applications with a formulation containing chromite sand at 12.5% by weight. Chromite was evaluated as one of many abrasive materials that possessed sufficient hardness in combination with a level of thermal resistance necessary to remain in the matrix as a discrete particle that did not melt during the casting process.

In organically bound friction materials, the concentration of chromite has generally been significantly lower, but is often one of many abrasives materials present in the same compound. This can be seen in the work of Harold [17-54] where blends of chromite with silicon carbide were listed. This is similar to the composition outlined by Eschen [17-48], which used silicon carbide, chromite, and iron oxides in a dual matrix of rubber and cashew nut shell liquid (CNSL).

Kaufman [15-23] patented a friction material composition comprised largely of inorganic minerals, namely vermiculite and wollastonite. The composition was modified with graphite to improve processing. However, graphite introduced an excessive drop in the friction level, which was offset by incorporation of chromite that increased friction during operation.

Iron Oxides

Generally, iron oxides in friction materials can be used for a number of purposes including fillers, pigments, or abrasive agents. Common forms include hematite, ferric oxide, and magnetite (a ferrous-ferric oxide spinel). The use of iron oxides in tribological composites began with Mayall and Mayall [3-29] and has been a part of the industry for many years in a number of applications.

Headson [17-55] coated woven asbestos fabric with a rubber compound filled with a mixture of finely divided iron powder and iron oxide. Brahs [17-56] coated braided asbestos ropes with a resinous mixture of iron oxide, rubber, and barytes. The ropes were formed into an annular friction facing and compressed after saturation to ensure complete consolidation.

Iron oxides are common soft abrasive additives in asbestos friction materials. In many cases; the role of iron oxide was more for space filling [17-57] than as the principal abrasive. Iron oxide is commonly used today and is found in many modern non-asbestos organic friction materials [17-58, 11-8].

Hematite

Hematite is red iron oxide with a Mohs hardness of roughly 5.5. It is available in both natural and synthetic varieties and in a wide range of sizes and purity levels. Hematite additions to friction materials must be carefully evaluated with regard to surface area of the finer particle sizes commonly found in commercially available pigment grades.

Magnetite

Magnetite is the magnetic iron oxide spinel with a hardness of approximately 5.5 on the Mohs scale. It was commonly found as an impurity in asbestos [15-11]. From the early work performed on purified asbestos products, we find that magnetite was capable of adversely affecting wear rates because of abrasion associated with larger particulate grain sizes. Today, the performance enhancements of magnetite are not fully understood, but magnetite in fine particle sizes, such as those used in colorants, might provide an array of interesting benefits.

Spinel

Spinel is an oxide of magnesium and aluminum with the chemical formula $MgAl_2O_4$. However, the family of minerals known as the spinel group shares a similar structure with a generalized empirical formula $A^{2+}B_2^{3+}O_4^{2-}$ where A and B can be ions of magnesium, zinc, iron, manganese, aluminum, chromium, and titanium. The different ions occupy the octahedral and tetrahedral sites in the lattice of a cubic crystal system and give a wide range of hardness values, where Chromite ($FeCr_2O_4$), Jacobsite ($MnFe_2O_4$), and Magnetite (Fe_3O_4) are at the lower end of the hardness scale (approximately 5.5) and Gahnite ($ZnAl_2O_4$) and Hercynite ($FeAl_2O_4$) at the higher end of the hardness scale (above 7).

As shown above with magnetite and chromite, members of the spinel class of minerals have been commonly used in friction materials for a number of years. Their role is typically that of a friction augmenting agent. However, spinel additions have been shown to generate enhanced wear resistance.

One example of the enhanced wear resistance provided by spinel additions to a sintered copper friction material was found to have

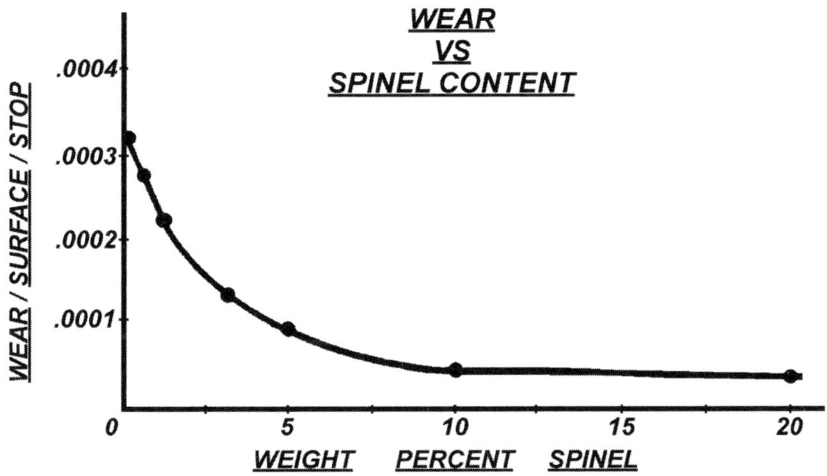

Fig. 17.2 Wear reductions by spinel additions to sintered friction materials from Roth [17-59].

an optimum minimum wear rate at addition levels of 10% by weight [17-59]. This marked reduction in wear rates is shown in Fig. 17.2.

Baskin [17-60] also employed spinel compounds in resin-bonded asbestos-based friction materials. The spinel compounds were sized to have particles of 0.5 to 10 microns, with the majority in the 2 micron range, and provided beneficial enhancements in friction (especially fade) and reduced wear rates.

Magnesium Oxide

Magnesium oxide is a hygroscopic mineral that has been widely used in friction materials of all types for nearly a century. Fisher [17-61] described an "unwoven" friction facing comprised of asbestos, cork, magnesium oxide, and casein (a phosphoprotein found in bovine milk). Magnesium oxide was stated to be an example of a filler material suitable for imparting the necessary friction requirements. This composition was produced as millboard and saturated with a phenolic resin solution. The concentration of magnesium oxide is not described in Fisher's work; however, magnesium oxide has been used in an incredibly wide range of concentrations.

Nanfeldt [17-62] developed a flexible molded rubber friction material with a small amount of magnesium oxide present (3.6%) and prepared the linings by calendaring and molding operations. Asano and Iwata [17-32] described a non-asbestos cellulose-based friction material that contained only 1% magnesium oxide, which is in stark contrast to the 45% level used by Loftipour [17-22] in his attempts to replace silica in railway friction materials.

Magnesium Oxide—More Than an Abrasive

Magnesium oxide is more than a simple abrasive additive and has many interesting points for consideration. For instance, Miyoshi [17-63] observed plastic flow in the surface layers of magnesium oxide ceramic materials in solid-state contact during sliding. Bowden and Hanwell [17-64] examined the friction of clean crystal surfaces and found that magnesium oxide crystals wear more rapidly than other materials (e.g., sapphire). Aside from the interesting deformation properties outlined by Miyoshi, and the novel tribological characteristics described by Bowden and Hanwell, magnesium oxide has interesting chemical effects that must be considered during its use, especially in organically bonded friction material composites.

Chemical Effects of Magnesium Oxide

Magnesium oxide is a well-known additive to elastomeric compounds. Nanfeldt's previously mentioned may have been associated with effecting a change in the final properties of the elastomer. Brady [17-65] disclosed that magnesium oxide can soften rubber compounds and speed up vulcanization.

The role of magnesium oxide as an accelerator in the compounding [17-66] and processing of rubber is essential in modern manufacturing of filled elastomers. However, Knop and Pilato [17-67] reported that magnesium oxide can affect cure rates in rubber-modified phenolic resin systems, most notably those containing acrylonitrile-butadiene copolymers. In addition, divalent salts of magnesium can also interact with phenolic resins, forming chelate systems and assisting in the development of high ortho (i.e., softer) phenolic resins.

Fused Magnesium Oxide

Fused magnesium oxide was originally developed for use in the manufacture of resistive heating electrodes such as those described by Kercher [17-68]. The process of fusion resulted in a glassy material that was readily crushed and sized. Fused magnesium oxide has been incorporated in some forms of friction material, but at a moderately low concentration [17-69]. The amount of information is limited and, therefore, this appears to be an area where more work is needed.

Titanium Dioxide

Titanium dioxide is a stark white mineral with a hardness of approximately 6.5 on the Mohs scale. It has widespread use in the pigment industry. The most common form is rutile, which was listed by Sanders [17-5] as a potential hard refractory material suitable for improving friction. Marzocchi and Jannarelli [15-47] utilized 15% by weight rutile in a glass fiber friction material, but considered it to be an inorganic filler material.

Researchers at Allied Signal [17-70, 17-71] developed a surface coating for enhancing the friction output of a disc brake pad. The coating consisted of abrasive materials in varying sizes and hardness values, exemplified by a mixture of rutile and aluminum oxide. This blend of abrasive powders was suspended in a phenolic adhesive and applied to the surface of the friction material prior to a flash cure, which permanently bonded the particles to the surface of the brake pads.

Titanium oxide and titanium compounds have been used more in friction materials in recent years [17-72]. It is expected that smaller size fractions of pigment grade titanium oxide will provide thermal resistance and impart a mild polishing action to the friction lining. Titanium dioxide can be reasonably anticipated to become a commonplace additive for the advancement of non-asbestos organic friction materials in the near future.

Zinc Oxide

Zinc oxide is a white powder that has excellent heat resistance and is relatively soft (only 4.5 on the Mohs scale). Zincite is the naturally occurring mineral form, but the majority of zinc oxide is synthetically produced for use in the rubber industry where zinc oxide, in combination with stearic acid, is an excellent activator for rubber-based compounds. As an additive to friction materials, zinc oxide plays the role of a rubber curative. It been a part of the friction material industry since 1924 [17-73].

Zinc oxide continues to be used in friction materials today [17-74], but the softness of elastomeric materials relegates it to a secondary abrasive that stabilizes the coefficient of friction but does not generate excessive wear and abrasion. Additionally, zinc oxide can alter the physical characteristics of elastomeric species present in a dry mix compound. However, this aspect of its performance has not been adequately described. Therefore, more research is needed.

Zirconium Oxide

Zirconium oxide, commonly called zirconia, is a white crystalline mineral known as Baddeleyite. Baddeleyite has a hardness of 6.5 on the Mohs scale and is monoclinic at room temperature. Zirconia can be stabilized into tetragonal or cubic forms by addition of metal oxides such as calcium, magnesium, or yttrium. Upon cooling, the presence of these ions in solution helps to stabilize the structure into the cubic and tetragonal phases, which tends to increase the hardness dramatically (often reaching 8 on the Mohs scale).

Zirconia does not appear to be well-studied or documented in the various literature sources. Cho [13-37] listed zirconium oxide as a material that will decrease the thickness of the third body layer during sliding. This effect was attributed to the hardness of the material and shown in direct comparison to other metal oxides. In many examples, zirconium oxide typically follows the same category as many of the other materials and is often classified as "any suitable refractory inorganic, non-metallic" filler material. Batchelor [17-75] listed zirconium oxide as one such filler in a very long list of other materials, but does not specifically point out any of its attributes.

Zirconia-based fibers have also been produced and were used by Birchall [17-76] in a simple resin-bonded friction material at concentrations of 65% by weight. The fibers were less than 10 microns in diameter and moderately free from non-fibrous material (i.e., shot).

More recent examples from patent literature have emerged showing zirconium oxide as the primary abrasive in non-asbestos organic friction materials. Nakajima and Hikichi [17-77] described the use of abrasive granules with identical chemical compositions to the ceramic fibers in brake lining compositions. Their examples incorporate zirconium oxide at levels of 2–6% as a primary friction generating additive.

Honma and colleagues [17-78] illustrated several aspects of friction material development in their work with blends of polyamide fiber and fiberglass. Their discussion described the importance of zirconium oxide as a friction enhancing raw material and limited it to a concentration of 10% by volume. Levels in excess of this resulted in an increased rate of wear. It is interesting to note that Phipps and Vyletel [6-30] also limited the content of zirconium oxide in their potassium titanate-based friction compound to a maximum of 10%.

References

17-1. Morrison, J., and J. Allen. 1900. Brake shoe. United States Patent No. 646301, filed July 26, 1899, and issued March 27, 1900.

17-2. Blau, P. et al. 1996. "Frictional Behavior of Automotive Brake Materials under Wet and Dry Conditions." ORNL/M-5824. Oak Ridge, TN: Oak Ridge National Laboratories.

17-3. Borchers, H. 1929. Brake band. United States Patent No. 1700759, filed May 25, 1925, and issued February 5, 1929.

17-4. Borchers, H. 1930. Brake band. United States Patent No. 1771204, filed May 22, 1925, and issued July 22, 1930.

17-5. Sanders, V. 1939. Brake lining. United States Patent No. 2159935, filed August 25, 1936, and issued May 23, 1939.

17-6. Matthews, J 1940. Friction material. United States Patent No. 2201191, filed February 24, 1939, and issued May 21, 1940.

17-7. Rhee, S. et al. 1970. "An Inertial Dynamometer Evaluation of Three Alloys for Automotive Brake Drums. SAE Paper No. 700138. Warrendale, PA: Society of Automotive Engineers.

17-8. Georges, J. 1985. "Nature of the Surface and its Effect on Solid-state Interactions." In *New Directions In Lubrication, Materials, Wear and Surface Interactions: Tribology in the '80s* edited by W. Loomis. New Jersey: Noyes Publications.

17-9. Schallamach, A., 1958. "Friction and Abrasion of Rubber." *Wear* 1: 384-471.

17-10. Kobayashi, M. et al. 1999. "Effect of Transferred Surface Films on the Behavior of Disc Brake Pads in Humidity Environment." SAE Paper No. 1999-07-339. Warrendale, PA: Society of Automotive Engineers.

17-11. Buell, S. 1892. Brake shoe. United States Patent No. 484404, filed April 6, 1891, and issued October 18, 1892.

17-12. DeFrance, M. 1925. Brake shoe. United States Patent No. 1557668, filed November 10, 1924, and issued October 20, 1925.

17-13. DeFrance, M. 1925. Method of making brake shoes. United States Patent No. 1557670, filed November 10, 1924, and issued October 20, 1925.

17-14. DeFrance, M. 1925. Method of making brake shoes. United States Patent No. 1557671, filed November 10, 1924, and issued October 20, 1925.

17-15. Nakagawa, M., and F. Nitto. 1986. Friction material using iron powder. European Patent No. 181616, filed November 7, 1985, and issued May 21, 1986.

17-16. Wood, F., and E. Clark. 1968. Friction material. United States Patent No. 3402054, filed November 29, 1965, and issued September 17, 1968.

17-17. Clark, E., and R. Summerling. 1972. Friction materials. United States Patent No. 3660120, filed May 14, 1968, and issued May 2, 1972.

17-18. Lowey, F., and C. Tower. 1946. Friction composition product. United States Patent No. 2408430, filed April 4, 1944, and issued October 1, 1946.

17-19. Fedorchenko, I. 2002. "Antifriction and Friction Cermet Materials." *Powder Metallurgy and Metal Ceramics* 41: 489-497.

17-20. Zhou, T. et al. 2007. "A Novel Route for Improving Creep Resistance of Polymers Using Nanoparticles." *Composites Science and Technology* 67: 2297–2302.

17-21. Rice, F. et al. 2001. "Crystalline Silica Exposure and Lung Cancer Mortality in Diatomaceous Earth Industry Workers: A Quantitative Risk Assessment." *Occupational and Environmental Medicine* 58: 38-45.

17-22. Loftipour, M. 1994. Friction material. Great Britain Patent No. 2268502, filed July 3, 1992, and issued January 12, 1994.

17-23. Austin, G. "Tripoli and Special Silica Stone." U.S. Department of the Interior, U.S. Geological Survey. Accessed July 6, 2010. http://minerals.usgs.gov/ minerals/ pubs/ commodity/ silica/041494.pdf.

17-24. Ripley, G., and C. Dana, eds. 1862. *The New American Cyclopædia: A Popular Dictionary of Common Knowledge*, vol. 14. New York: Appleton and Co.

17-25. Dolley, T., and W. Bolen. " U.S. Department of the Interior, U. S. Geological Survey. Accessed July 6, 2010. http:// minerals. usgs.gov/minerals/pubs/commodity/silica/780400.pdf.

17-26. Longely, J. 1935. Brake compositions. United States Patent No. 2020791, filed August 28, 1931, and issued November 12, 1935.

17-27. Alundum. Registered trademark of Norton Company Corporation, 1 New Bond Street, Worcester, Massachusetts, Registration No. 1624288, November 27, 1990.

17-28. Bayer, K. 1887. Verfahren zur Darstellung von Thonerdehyrate und Alkalialuminat. German Patent No. 43977, issued July 17, 1887.

17-29. Bayer, K et al. 1902. Verfahren zur Darstellung von Thonerdehyrate und Alkalialuminat. German Patent No. 65604, issued July 31, 1902.

17-30. Totten, G., and D. MacKensie, eds. 2003. *Handbook of Aluminum - Volume 1: Physical Metallurgy and Processes*. New York: Marcel Dekker.

17-31. Ingalls, W., ed. 1907. *The Mineral Industry: Its Statistics, Technology, and Trade during 1906."* Vol. 15. New York: Hill Publishing Company.

17-32. Asano, H., and K. Iwata. 1988. Friction material composite. European Patent No. 270089, filed December 2, 1987, and issued June 8, 1988.

17-33. Satapathy, B., and J. Bijwe. 2002. "Analysis of Simultaneous Influence of Operating Variables on Abrasive Wear of Phenolic Composites." Wear 253: 787-794.

17-34. Amethyst Galleries, Inc. "Amethyst Galleries Mineral Gallery." Accessed July 12, 2010. http://www.galleries.com/minerals/silicate/kyanite/kyanite.htm.

17-35. Webmineral. "Mullite Mineral Data." Accessed July 11, 2010. http://webmineral.com/data/Mullite.shtml.

17-36. Schneider, H., and S. Komarneni, eds. 2005. "Mullite." Weinham: Wiley-VCH Verlag GmbH and Co., KGaA.

17-37. Stedman, F., and R. Pocock. 1957. Friction lining compositions. United States Patent No. 2784105, filed November 8, 1955, and issued March 5, 1957.

17-38. Luther, W., and R. Koehring. 1960. Friction material. United States Patent No. 2945292, filed November 28, 1958, and issued July 19, 1960.

17-39. Spokes, R., and J. Littlefield J. 1959. Friction elements. United States Patent No. 2901456, filed March 1, 1955, and issued August 25, 1959.

17-40. Cox, C. 1959. Brake material for extreme heavy duty. United States Patent No. 2899740, filed August 30, 1957, and issued August 18, 1959.

17-41. Matejka, V. et al. 2008. "Effects of Silicon Carbide in Semi-metallic Brake Materials on Friction Performance and Friction Layer Formation." Wear 265: 1121-1128.

17-42. Satapathy, B., and J. Bijwe. 2004. "Wear Data Analysis of Friction Materials to Investigate the Simultaneous Influence of Operating Parameters and Compositions." *Wear* 256: 797-804.

17-43. Horth, F. 1999. Stabilizing friction levels. United States Patent No. 5972090, filed July 9, 1998, and issued October 26, 1999.

17-44. Yamamoto, K. et al. 2007. Friction material. United States Patent No. 7297728, filed July 1, 2004, and issued November 20, 2007.

17-45. Jang, H. et al. 2001. "Compositional Effects of the Brake Friction Material on Creep Groan Phenomena." *Wear* 251: 1477-1483.

17-46. Kim, S., and H. Jang. 2000. "High Temperature Friction and Wear of Friction Material: The Effect of the Relative Amount of Graphite and Zirconium Silicate (ZrSiO4). *KSTLE International Journal* 16: 365-372 (Korean).

17-47. Reynolds, H. 1979. Friction material for railroad service. United States Patent No. 4178278, filed June 26, 1978, and issued December 11, 1979.

17-48. Eschen, F. 1982. Friction material for railroad brake shoes. United States Patent No. 4352750, filed August 3, 1981, and issued October 5, 1982.

17-49. Hunt, R., ed. 1867. *Ure's Dictionary of Arts, Manufactures, and Mines*. 6th Ed., vol. 1. London: , Longmans, Green, and Co.

17-50. Mayrhofer, G. 2008. Sintered brake lining. United States Patent Application No. 2009/0236191, filed October 10, 2008.

17-51. Chugh, Y. 2007. Friction materials comprising coal combustion and coal gasification byproducts. United States Patent Application No. 2008/0121474, filed October 29, 2007.

17-52. Bush, H. et al. "The Application of Neutron Activation Analysis to the Measurement of the Wear of a Friction Material." *Wear* 20: 211-225.

17-53. Pollard, J. 1966. Cast railroad brake shoe and method of making same. United States Patent No. 3239319, filed May 10, 1963, and issued March 8, 1966.

17-54. Harold, J. 1979. Friction materials. Great Britain Patent No. 2016027, filed February 7, 1979, and issued September 19, 1979.

17-55. Headson, F. 1911. Brake band liner. United States Patent No. 997498, filed September 14, 1909, and issued July 11, 1911.

17-56. Brahs, G. 1951. Method of making friction elements. United States Patent No. 2553698, filed June 17, 1947, and issued May 22, 1951.

17-57. Pogosian, A., and N. Lambarian. 1977. "Estimation of Wear of Some Asbestos Reinforced Friction Materials." *Wear of Materials*. New York: American Society of Mechanical Engineers.

17-58. Suzuki, S. et al. 2005. Friction material. United States Patent No. 6887549, filed June 24, 2004, and issued May 3, 2005.

17-59. Roth, D. 1969. Copper base friction material with dispersed spinels. United States Patent No. 3428440, filed October 15, 1965, and issued February 18, 1969.

17-60. Baskin, Y. 1976. Friction material containing spinel as friction modifier. United States Patent No. 3959208, filed February 7, 1975, and issued May 25, 1976.

17-61. Fisher, W. 1923. Unwoven brakeband facing. United States Patent No. 1465389, Filed August 23, 1919, and issued August 21, 1923.

17-62. Nanfeldt, W. 1936. Molded brake lining and process for making the same. United States Patent No. 2061918, filed February 2, 1931, and issued November 24, 1936.

17-63. Miyoshi, K. 1998. *Solid Lubrication Fundamentals and Applications: Characterization of Solid Surfaces*." TM-1998-107249-CH2. Washington, DC: National Aeronautics and Space Administration.

17-64. Bowden, F., and A. Hanwell. "The Friction of Clean Crystal Surfaces." Proceedings of the Royal Society of London, A, 295, pp. 233-243.

17-65. Brady, G. 1940. *NAPA Materials Handbook*. New York: McGraw-Hill.

17-66. Long, H.1985. Basic Compounding and Processing of Rubber. Akron, OH: University of Akron American Chemical Society.

17-67. Knop, A., and L. Pilato. 1985. *Phenolic Resins: Chemistry, Applications, and Performance*." 1st ed. Berlin: Springer-Verlag.

17-68. Kercher, A. 1933. Electrical heating element and method of manufacture. United States Patent No. 1905232, filed December 17, 1928, and issued April 25, 1933.

17-69. Double, D. et al. 1988. Cementitious composite friction compositions.United States Patent No. 4792361, filed August 8, 1986, and issued December 20, 1988.

17-70. Kesavan, S., and D. Saiter, 1999. Coated friction pad for brake assembly. United States Patent No. 5971113, filed March 10, 1997, and issued October 26, 1999.

17-71. Torpey, M. et al. 2001. Abrasive coating on brake pads. United States Patent No. 6167992, filed July 14, 1998, and issued January 2, 2001.

17-72. Shao, X. et al. 2003. Brake friction material including titanium dioxide particles. United States Patent Application No. 2004/0140167, filed January 16, 2003.

17-73. Eisenhardt, F. 1932. Process of making brake lining and the like. United States Patent No. 1877651, filed May 24, 1924, and issued September 13, 1932.

17-74. Suzuki, S. et al. 2004. Friction material. United States Patent Application No. 2004/0241429, filed May 27, 2004.

17-75. Batchelor, C., and W. Jensen. 1965. Friction devices. United States Patent No. 3221853, filed August 29, 1962, and issued December 7, 1965.

17-76. Birchall, J. 1975. Friction material. United States Patent No. 3891595, filed August 24, 1973, and issued June 24, 1975.

17-77. Nakajima, O., and A. Hikichi. 2003. Friction material. United States Patent No. 6632857, filed April 5, 2000, and issued October 14, 2003.

17-78. Honma, T. et al. 1988. Friction material composition for brake lining. United States Patent No. 4785029, filed March 9, 1987, and issued November 15, 1988.

Chapter 18
Solid Lubricants

Enthusiasm is very good lubrication for the mind.
—Source Unknown

SOLID LUBRICANT ADDITIVES ARE ESSENTIAL for controlling the wear and friction of tribological composites. The most common solid lubricants are naturally occurring minerals that are locally available in a given geographic region. For instance, Miyoshi [3-36] described the role of molybdenum disulfide in the lubrication of the axles of Conestoga wagons travelling through the Climax, Colorado region during the exploration of the continental United States. In 1883, Dimick [18-1] patented another commonly used solid lubricant known as "black lead," an archaic name for graphite.

The role of solid lubricants in composite materials predates Dimick by more than two decades and was patented as a method for improving the lubrication of journal bearings. The improvement resulted from additions of plumbago (another outdated name for vein graphite) to a paper-based composite bound with gum-shellac. This composition was molded into shape and was a self-lubricating tribological composite in use as early as 1860 [18-2].

Today, friction material designers rely on graphite, molybdenum disulfide, and a number of other solid lubricant compounds as well as complicated mixtures to provide the necessary balance of properties. The general principle behind the use of solid lubricants is not new and can be seen in the Dixon Graphite Lubricant Laws [18-3] published in 1914:

> *It is generally recognized that there are two kinds of friction.*
>
> > *(a) Friction because of the contact of metal parts.*
> >
> > *(b) Friction because of the lubricant itself.*

Joseph Dixon considered friction to be a physical interaction between surfaces due to the irregularities at the surface. In addition, wear processes at the interface of sliding surfaces could only be prevented if the lubricant was highly viscous. In other words, the lubricant must possess a high degree of internal friction. In cases where the lubricant did not possess the necessary viscosity, the surfaces would eventually come into contact and interact, resulting in wear (which was therefore considered an inevitable product of sliding systems).

Today, the use of solid lubricants within a given composite might be considered a very challenging multiple criterion puzzle: one that desires to simultaneously increase the friction level and yet provide an effective means of reducing the wear rates of the final product. This diametrically opposed viewpoint is constrained by the fact that many tribological composites cannot operate with low coefficients of friction, which implies that the designer must select solid lubricants that will increase the friction output, provide protection against wear, and yet remain affordable.

As contradictory as these ideas may seem, this balance of properties can be achieved by artificially generating a stable and highly viscous third body layer at the interface. The most widely used and generally effective approach is to allow the friction material to become a reservoir for retaining reactive and non-reactive solid lubricants that are compatible with both contact surfaces.

Graphite

Graphite is a readily available and very effective solid lubricant found in numerous natural deposits around the globe. Naturally occurring graphite can be classified into three forms: amorphous, flake, or vein graphite. Each of these forms has a unique morphology, and purity levels vary greatly. The combination of purity and morphology affect the anti-friction and wear properties of these types of natural graphite.

Graphite can also be manufactured synthetically, typically in the form of an electrode for metal smelting operations. This form of synthetic graphite is commonly referred to as electrode graphite and is typically very pure. The manufacturing process strives to attain the highest possible purity level because electrode users require highly conductive electrodes that use less electrical power. Therefore, a common process control involves the measurement of the electrodes' resistivity. Conductivity, the reciprocal of resistivity, is a result of the alignment of the graphene layers of the graphite in the electrode. Therefore, highly aligned grades are higher in purity.

The structural aspects of graphite have been extensively studied by a number of researchers including Bernal [18-4] and Barrer [18-5], who studied it via adsorption methods, and the team of Lipson and Stokes [18-6]. These studies determined graphite to be composed of carbon atoms arranged in a hexagonal planar array where the electrons are delocalized above the surfaces of the plane. The planes, known as graphene layers, are weakly coupled together and capable of sliding relative to each other, especially under standard laboratory conditions. This structural characteristic enables graphite to function as an efficient and highly stable solid lubricant. However, it must be noted that graphite is not an intrinsically lubricating material and requires other materials to be present for effective lubrication.

The Mechanism of Graphite Lubrication

Savage [7-19] examined the lubrication mechanism of graphite and the formation of surface films at the sliding interface. The films, composed of graphene layers, adsorb low molecular weight species from the atmosphere (such as water vapor and oxygen) and decrease the force required to initiate relative motion. The reduction in the observed coefficient of sliding friction was validated by a series of *in vacuo* experiments that clearly documented the poor lubrication and wear characteristics of graphite under reduced atmospheric pressure. Additionally, Savage demonstrated the effect to be completely reversible by allowing atmospheric gases to reenter the test chamber, resulting in a rapid decrease in the force necessary to produce motion and also a concomitant reduction in wear rates. Savage's work was later confirmed by the team of Bowden and Young [18-7].

Deacon and Goodman [7-20] also examined graphite lubrication due its lamellar structure and compared it to other solid lubricants with similar structural properties. The transition from effective lubrication (i.e., low coefficient of friction) to higher friction was shown to be temperature dependent. At temperatures of approximately 752°F (400°C), the observed coefficient of friction for graphite was found to increase. Upon cooling, the coefficient decreased and provided further evidence for the adsorption/desorption of low molecular weight species as the underlying mechanism behind the lubricating ability of graphite.

Sourcing Graphite

Natural graphite can be obtained in various levels of purity, usually measured by its carbon content. All natural graphite is different, and all graphite sources must be carefully verified to ensure consistent product performance, especially with regard to the coefficient of friction. For this reason, the majority of friction material producers will seek out a reputable graphite supply company capable of providing a consistent product with an established history of exceptional quality. Often these companies offer an array of grades of carbon and graphite in many different particle sizes and purity levels.

Likewise, synthetic graphite is not immune from purity concerns. Anderson [12-27] recounts the effect of silicon carbide contamination in synthetic graphite. Concentrations of only a fraction of a percent were capable of generating excessive amounts of wear to mating members and produced heavy scoring across the surface. Because of this concern, the manufacture and methods for purification of synthetic graphite have emerged as a means to ensure effective lubrication under all standard operating conditions.

Natural Graphite

According to Mantell [18-8], natural graphite is typically available as amorphous graphite (also known as black lead). Graphite can also take the form of disseminated flakes (known as flake graphite). It can also be crystalline in nature, which is known as vein graphite (sometimes called plumbago).

The terms amorphous, flake, and vein are relatively ambiguous. However, amorphous forms are most likely to contain harsh abrasive impurities whereas vein graphite is generally regarded as the form with the highest purity. Flake can range in intermediate grades spanning the spectrum. Based on the type and concentration of impurities present, the lubricating ability of each of these types can give different performance attributes.

Amorphous Graphite

Amorphous graphite is a very fine particulate form of graphite with low degrees of crystallinity, which makes it appear amorphous [18-9]. However, it is not technically amorphous and has short range order. It is formed by metamorphosis of coal and generally has much lower graphite content than other forms. The most common measure of purity is to measure the ash content of the materials by heating in oxygen until the sample has been completely oxidized such as described in ASTM C561-91(2010)e1 [18-10]. Amorphous graphite is commonplace in friction materials, especially those produced in the local regions where it can be mined easily or by those that are insensitive to contamination concerns.

Guptill [18-11] is perhaps the first to describe a tribological composite that incorporates amorphous graphite. This metal matrix composite based on a matrix of soft bearing metal such as an alloy of lead and antimony contained additions of amorphous carbon. It was specifically noted that amorphous graphite was incorporated for reducing friction. Likewise, Nowalk [6-38] used the technique of saturating a ceramic body with a liquid solution containing amorphous graphite. This friction stabilizing body was capable of providing a heat-resistant ceramic matrix with a reservoir of lubricant present to prevent the occurrence of noise.

Naturally occurring amorphous graphite is not widely used in many modern resin-bonded friction materials primarily because of concerns with excessive abrasion and poor wear. These are typically attributed to the presence of harsh impurities. However, synthetic techniques have been developed that are currently being applied to the friction industry. For instance, Gibson and co-workers [18-12] developed an amorphous carbon coating on carbon fiber for use in wet friction material applications. Another technique by Dearnaley and Van Dyke [18-13] described a diamond-like film of amorphous carbon, which is an effective lubricant for coating a porous aluminum substrate.

Flake Graphite

Flake graphite is usually found in layered formations in metamorphic rock and is removed by crushing the surrounding rock and subsequent separation. Typical mined sources vary in purity, but flake graphite can be found in the 80–90% carbon range. Post-processing steps can further improve the purity level to approximately 99% [18-14], which is ideal for commercial lubricant applications.

The role of flake graphite in automotive tribological composites can be traced back to the earliest days of woven fabric linings. Johnson [18-15] incorporated flake graphite into a friction material by boiling the fabric in a lubricant solution containing finely divided graphite powder. The final product was exceedingly flexible and helped obviate chattering during use, which was deemed "disturbing to the operator." A similar composition based on this concept was also patented by Mead a few years later [18-16].

Spokes [18-17] developed an asbestos friction element containing agglomerates of rubber-bound graphite. The agglomerate was formed by mixing a rubber compound and graphite until it was suitably dispersed throughout the rubber matrix. This mixture was then comminuted into particles and incorporated into a composition friction element bound with a dual phase matrix of phenolic resin and rubber. Spokes noted that additions of fine flake graphite to the friction element composition were prone to drop the friction level significantly if the concentration exceeded 5% by weight. Additionally, it was reported that excessive concentrations of graphite resulted in a number of processing defects including extensive delamination cracks throughout the body of the composite.

In many friction material compositions, flake graphite provides effective lubrication when used in small concentrations and can also help to ensure that frictional heat generated at the interface is "well drained from the friction zone" [3-22]. The presence of this beneficial lubricant in small quantities can also aid in processing, especially in those operations involving compaction processes such as preforming or briquetting [18-18]. Flake graphite is considered to be the most widely used antioxidant in metal ceramic friction pairs [18-19] and might offset oxidative corrosion, which can lead to chemical wear concerns.

Flake graphite has been used in virtually all forms of friction material composite. Recent work by Gardner [18-20] incorporated flake graphite in a paper-based friction material. In this patent, the flake graphite was considerably larger in diameter—approximately 0.040–0.080 in. (1–2 mm). The benefit of the larger size was reportedly an improvement in particle flexibility and enhanced manufacturing.

Concerns with Flake Graphite

Despite the benefits of flake graphite, Zhan and Zhang [18-21] examined copper matrix composites containing graphite. The softness of the particles and the lamellar structure tended to form a solid lubricant film at the interface, which reduced friction and contributed to increasing the operational lifespan. However, graphite inclusions resulted in a drop in the mechanical strength of the composite. This was ascribed to crack nucleation by the formation of a cavity surrounding the graphite particles and might have been due to poor wetting of the particles by the copper matrix.

Lee [18-22] performed a series of experiments focused on the friction and wear performance of a sintered bronze friction material. It was found that flake graphite additions improved the wear resistance of the composite. However, the material containing graphite required higher compaction pressures during manufacturing to overcome low density and high porosity. It was further noted that graphite should be added last to the mixing cycle. Excessive mixing and the softness of the graphite flakes allowed microscopic particles of graphite to coat the surface of metal powder, which prevented effective bonding during the compaction and sintering phases of the manufacturing processes.

Vein Graphite

Vein graphite occurs in well-defined pockets or veins and is the most highly crystallized form of naturally occurring graphite. The carbon content of Sri Lankan vein graphite can reach levels of 97% and higher [18-23]. For over 140 years, vein graphite has been mined continually in Sri Lanka, formerly known as the Dominion of Ceylon. Therefore vein graphite is often called Sri Lankan or, in older references, it might be referred to as Ceylon graphite.

Vein graphite is the best naturally occurring graphitic lubricant available and possesses many highly desired characteristics that are well-suited for friction material development. Despite these benefits, vein graphite has typically been limited in use because of concerns about stability and availability of supply. Other sources of vein graphite have been located, with potential sites in the United States and India. However, virtually all commercially available vein graphite is obtained from mines in Sri Lanka, and supplies have been subject to disruption because of political unrest.

Synthetic Graphite

Crystalline forms of carbonaceous materials were first produced by Acheson in 1893 [18-24] in the form of carborundum (e.g., silicon carbide) for use in the abrasive materials industry. This manufacturing method was later modified and used to produce synthetic graphite [18-25]. This modified process has now become known as the Acheson method.

Acheson's method began by mixing carbon with a mixture of sand, sawdust, and salt such that the carbon, namely powdered coke, was present at roughly 50%. The final mixture was packed around a core of coke and heated in an electrical furnace. The formation process was described as heating by resistance, which increased the temperature of the core until radiant heat began to convert the packing material into silicon carbide in a similar manner to his previous patent.

Heating was then continued until the silicon carbide dissociation temperature was reach. According to Mantell [18-8], this was greater than 4000°F (approximately 2200°C) with "the production of good graphite" requiring temperatures of 4892°F (2700°C) and higher. At these temperatures, silicon carbide was destroyed, and silicon was driven off as vapor, leaving behind pure carbon that increased the size of the core.

When the resistivity of the electrode decreased to the desired level, the furnaces were allowed to cool and the electrodes removed for finishing. During the final machining operations, the graphite swarf was often collected and sold to graphite merchants who pulverize, classify, and sell it for other applications, including friction materials. Contamination from the packing materials and residual silicon carbide is an area of concern with graphite obtained from the Acheson process.

Modern Synthetic Graphite Production

Modern synthetic graphite electrodes are produced by the same fundamental principle of Acheson's process. However, many evolutionary changes have resulted in the production of higher purity electrodes with superior lubricating properties and a decreased, if not completely eliminated, concern of hard particle contamination. A recent patent application [18-26] exemplifies many of the features in today's graphite manufacturing process.

Daimler used a finely divided anode grade coke with low ash content (e.g., less than 5%). The coke was first mixed with a suitable carbonaceous binder such as pitch and formed into green electrodes. The green electrode was then baked to convert the pitch into carbon, typically by heating to temperatures between 1300–2000°F (approximately 700–1100°C). To increase the carbon content of the electrode, the baked electrode could be impregnated and re-baked, which incurred higher cost but can be repeated as necessary to ensure the electrode has achieved the desired carbon level prior to graphitization. Graphitization took place by heating to a preferred temperature range of 4000–4500°F (approximately 2200–2500°C).

Synthetic Graphite in Friction Materials

Synthetic graphite is commonly incorporated as a primary lubricant in the semi-metallic class of friction materials. It typically occupies a large portion of their compositional makeup—approximately 15% by weight [12-27]. However, synthetic graphite is also used in high-performance railway brake lining compositions [6-41], wet friction materials for automotive power train applications [18-27], and modern non-asbestos organic formulations [18-28].

Synthetic graphite forms a stable third body layer at the friction interface that is rich in graphite and prevents direct contact from occurring, resulting in a reduction in wear rates [18-29]. Lam and Yesnik [18-27] stated that synthetic graphite in their friction lining possessed high thermal conductivity, which assisted in the dissipation of heat away from the friction interface and offered improved performance.

Recent findings by Kolluri [18-30] examined thermal localization effects of graphite in friction materials and concluded graphite was beneficial for smoothing unwanted friction fluctuations. In addition, graphite suppressed the coefficient of friction, especially when used in finer particle sizes, and equalized thermal transport at the interface.

In this study, natural graphite was found to give higher friction levels than synthetic grades, a fact that could be attributed to source and purity of the natural graphite. It was also noted that physical properties of the friction material, such as softness, was capable of improving thermal localization. This finding coincides with reports of graphite addition decreasing the softness of non-asbestos brake linings for automobiles [18-31]. The ability of graphite to affect thermal transport might be the result of physical changes to the final product.

Polarized Graphite

Polarized graphite is synthesized by coating graphite with various inorganic compounds such as metallic molybdates, sulfates, or phosphates [18-32]. These compounds coat the graphene sheets, forming a polarized surface that improves adhesion to metal surfaces and aids in the formation of a stable third body layer. The incorporation of polarized graphite in a friction material composite resulted in a reduction in brake noise because of a reduction in the difference between the value for static friction and dynamic friction.

Based on the improved adhesion, polarized graphite offers the additional benefit of improving wear resistance above that provided by naturally occurring grades of graphite. Bonding of the polarized platelets to the surface protects metal surfaces and limits direct contact during sliding.

Despite the wear enhancement and stabilization of the coefficient of sliding friction, polarized graphite is more expensive than common grades of graphite. The higher cost appears to have effectively limited polarized graphite to a secondary function as an additive in a mixed lubricant system, such as those described by Holinski [18-33]. The use of polarized graphite in tribological composites is anticipated to remain low until a lower cost method is found.

Concerns with Graphite in Friction Materials

In many cases, graphite is a necessary [12-33] friction material additive capable of providing effective lubrication at the sliding interface. However, Rohatgi [18-34] limited graphite to concentrations of less than 20% by volume and cited drastic reductions of the coefficient of sliding friction, often to levels of approximately 0.2 at concentrations greater than 20%. Graphite must therefore be used at optimal concentrations to provide effective lubrication and yet maintain a safe level of friction.

Deacon and Goodman [7-20] cited the limitation of graphite's ability to lubricate at operating temperatures of approximately 752°F (400°C). Higher temperatures drive off the intercalated low molecular weight species, and lubrication is effectively diminished. At this point, graphite functions as a non-lubricating carbonaceous material, and other lubricating materials are needed to ensure effective operation, especially at higher temperatures.

Metal Sulfides

Metal sulfide compounds have been utilized as solid lubricants either singly or in combination with other materials in various tribological systems for well over a century. One of the earliest examples by Douglas [18-35] described a self-lubricating composition compounded with artificial galena and artificial stibnite in a paraffin mixture containing graphite. These important unctuous ingredients allowed for the development of an oil-free journal bearing design. This approach is similar to the rubber-based anti-friction concept illustrated by Mayall and Mayall [3-29].

Numerous beneficial properties of metallic sulfides have been reported by a number of groups including Bowden and Tabor [2-16], Thelen and Devine [18-36], and Samsonov [18-37]. In all cases, the general consensus stated the same conclusions regarding the function of metallic sulfides as a lubricating agent: metallic sulfides must be chemically active in the tribosystem to achieve the desired result.

Reactivity of Metallic Sulfides

Bowden and Tabor disclosed that metallic sulfides can provide effective lubrication only by reacting with the metallic surfaces to form a thin film at the interface. Third body layers resulting from chemical reactions with metallic sulfides [3-52] affected the dynamic friction characteristics, but did not necessarily imply a reduction in the friction level nor a change in the force necessary to induce motion [2-27].

Thelen and Devine concluded that wear behavior is a direct function of the reactivity of the lubricant. Wear reductions occurred only when the film was firmly attached to the surfaces, was readily sheared in cohesion, and could be regenerated or otherwise replenished as a result of abrasion or delamination.

Thelen and Divine also found that wear properties were correlated with the electronegativity of the metal atoms in the system. This was related to solubility where highly soluble sulfides tended to transfer to the more electronegative metal in the system forming a strongly adherent film and inhibited asperity interactions.

Similarly, Samsonov and colleagues compared metallic sulfides with metallic selenides of the same base metals. Generally, metallic sulfides exhibited lower friction levels than selenides, which was attributed to chemical properties of the materials.

For example, selenides tended to have higher cohesion between the lamella of the sulfide layers and therefore exhibited higher friction levels than the sulfides. Sulfides tended to have weaker cohesion between the lamella and transferred to the mating metals, forming the third body layer. In other words, the bond strength between sulfides and metallic surfaces was stronger than the cohesive bonds between the lamella.

Filip and colleagues [18-37] examined the friction layer formation processes that occurred during the operation of polymeric matrix composite utilized as friction materials for automotive brake applications. During this study, metal sulfide compounds and elemental copper were found to develop metallic alloys on the surface of the brake components. This compared well to the findings of Holinski and Hesse [18-38]. An understanding of these reactions offers an effective means for tailoring composition and enhancing performance.

Molybdenum Disulfide

Molybdenum disulfide is perhaps the most widely studied of all metallic sulfide solid lubricant additives. There are hundreds of publications that explore various aspects of this naturally occurring solid lubricant (e.g., molybdenite). Many of these documents were examined in the comprehensive literature survey by Winer [18-40], which covered a time period of forty years and encompassed findings from more than 250 papers. From this review, the fundamental mechanisms proposed for lubrication by molybdenum disulfide included low friction because of cleavage along the grain boundaries of individual crystallites, low friction because of adsorbed species (similar to that of graphite), and an intrinsic lubricating ability resulting from the ease of separation of the laminar plates of the crystals because of to the separation distance of the sulfur atom layers, which was taken from work by Dickenson and Pauling [18-41].

Salomon [18-42] explored the mechano-chemical factors associated with thin film lubrication and determined that molybdenum disulfide was an outstanding solid lubricant based on its ability to perform over a wide range of operating temperatures. Thin films of molybdenum disulfide were effective at extending the lifespan of sliding members. Additionally, the effect of burnish time was reported to be a significant contributor, and longer burnish times gave marked improvements in wear resistance.

Generally, molybdenum disulfide can be considered an intrinsic lubricant with excellent film forming capabilities. Film formation occurs by deposition of layers of thin platelets onto the mating metal surface. Favorable adhesion to the metal and low cohesion between the planes can drop the coefficient of sliding friction to very low values [18-43]. Third body layers based on molybdenum disulfide are stable to temperatures of approximately 662–752°F (350–400°C) in air. At higher temperatures, the conversion to molybdenum oxide occurs as a result of the loss of sulfur [7-20].

According to Salomon [18-42]. molybdenum disulfide can tolerate small amounts of soft impurities. A similar conclusion by Gardos [18-44] resulted from the examination of synergistic mixtures of solid lubricants and determined that oxides of molybdenum were effective in enhancing the friction and wear properties of thin film lubricants based on molybdenum disulfide.

However, and perhaps more importantly, McCabe [18-45] determined that pure molybdenum trioxide was not an abrasive oxide. Based on a standard ASTM abrasion method of the time, the abrasion of surfaces by pure molybdenum trioxide was actually less than the abrasion levels exhibited by technical fine grades of molybdenum disulfide or tungsten sulfide. In this case, the process of oxidative conversion of the metallic sulfide to the oxide can lead to the development of a superior lubricant.

Molybdenum Disulfide in Friction Materials

Molybdenum disulfide has been included in friction materials of all types. Lowey and Tower [17-18] described a sintered metallic friction element comprised of copper, tin, iron, graphite, and silica to which molybdenum disulfide could be added to provide an increased friction level especially under fade conditions. It should be noted that grade of molybdenum disulfide sourced by the authors contained approximately 10% silica as an impurity.

A similar concept of using impure molybdenum disulfide was reported by Okubo [18-46] in an asbestos-free friction material based on mixtures of wollastonite and glass fiber. This composite was bound in an organic matrix of resin and acrylonitrile rubber with 11.8 parts by weight molybdenum disulfide, which was only 50% pure.

To contrast the role of impure forms in friction materials, the work of Suzuki and colleagues at Nisshinbo [17-58] focused on corrosion induced by metallic sulfides including molybdenum disulfide. Sulfide-generated corrosion can lead to thickness variation of the transfer films deposited on the discs, resulting in unacceptable vibration (i.e., judder). A blend of metallic sulfides was selected with a low sulfate ion concentration (e.g., a pH of more than 4). The blend of sulfides relied on a balance of third body development and effective friction generation without the introduction of negative performance characteristics associated with acidic reactions induced from atmospheric moisture.

Lead Sulfide

Galena is the naturally occurring mineral form of lead sulfide. It is silver-gray in color with a hardness of approximately 2.5 on the Mohs scale. Galena is a high-density material with a specific gravity of approximately 7.5 relative to water. Lead sulfide, in the form of artificial galena, has been an additive in tribosystems for nearly a century and was first used by Douglas [18-35]. in his self-lubricating journal box design.

Additions of lead sulfide were also commonplace in early friction material composites but were intentionally added to the composition for a separate reason: to accelerate the oxidation process of drying oils by acting as a dryer. The drying action of lead sulfide enabled faster processing of the drying oil friction facings [18-47] and provided other benefits to the friction material. Most notably, lead sulfide is generally regarded as an effective lubricant [3-45] that also serves to stabilize friction levels [17-39].

Friction stabilization provided by lead sulfide can be attributable to chemical reactions that occur at the interface during braking. According to Weintraub [18-48], reactions because of elevated temperatures at the friction interface convert lead sulfide into lead oxide, especially in the presence of copper.

The conversion to lead oxide, Mohs hardness of approximately 2, was not universally accepted as a desirable trait of lead sulfide additions. Lidkea and Spokes [18-49] considered elemental lead to be a beneficial addition in friction material compositions. However, the rapid conversion to lead oxide was not acceptable. Lead oxide caused acceleration of oxidation in drying oil binder systems, perhaps uncontrollably. In this example, lead sulfide was the preferred drying agent and also provided friction stabilizing effects during use.

Lead sulfide might still be in use as a friction stabilizing agent but, based on toxicity, its use is expected to see a rapid decline in years to come. Miyaki and Kobayashi [18-28] developed a modern non-asbestos friction material that replaced lead sulfide with a composition containing milled electrode graphite and spherical bronze powder. The combination was found to provide similar performance characteristics, but without the toxicity concerns of lead.

Antimony Sulfide

Antimony sulfide can generally include both antimony trisulfide (Sb_2S_3), and antimony pentasulfide (Sb_3S_5). Although antimony pentasulfide has been used as a red pigment and incorporated into lubricant mixtures [18-50], its role as a tribological additive in friction material composites has been vastly overshadowed by the more commonly used antimony trisulfide.

Antimony trisulfide, which also occurs naturally as stibnite, has a hardness of 2 on the Mohs scale and has been included in friction materials for many decades. Antimony trisulfide is not an intrinsic lubricant like molybdenum disulfide, but is an active lubricant that enhances the frictional response of composites, especially during high temperature operation. This has been attributed to the oxidative conversion from antimony trisulfide to antimony trioxide.

According to Cho [18-51], the conversion introduces a slight increase in hardness of the particles that was shown to give improved friction, but that also created differential thickness of the third body layer because of abrasion from the more aggressive oxide. A similar finding was noted by Wirth and Whitaker [18-52] who also described increases in the friction of an automotive brake lining because of additions of antimony trisulfide. However, excess amounts of sulfides developed a thick third body layer that would shear easily at elevated temperatures and induced fade.

Antimony Trisulfide in Friction Materials

Antimony sulfide has been included in non-asbestos organic linings [15-53], low-steel friction materials [6-18], and semi-metallic linings [18-53] at concentrations of less than 10%, with many composition containing less than 5%. Typically these include antimony trisulfide as a friction enhancing solid lubricant.

However, Lumb and Edwards [18-54] described a friction material comprising phenolic resin, nickel powder, graphite, asbestos fiber, and antimony sulfide. The composition was mixed and molded under heat and pressure to form a rigid composite, which was subsequently clamped under a load and baked at temperatures sufficient to oxidize the phenolic resin binder. The resulting product was an asbestos fiber composite in a nickel sulfide matrix developed by reaction of the sulfur, from antimony trisulfide, with nickel powder. This product was shown to have exceptional thermal resistance and was ideally suited for heavy duty friction operation.

Antimony sulfide is widely used in modern non-asbestos friction materials today, especially in combination with other solid lubricants such as graphite [7-25]. However, concerns with the toxicity of antimony compounds are beginning to force tribological composite designers toward solid lubricants that are free of all antimony compounds.

Tin Sulfide

Tin sulfide can include tin (II) sulfide (SnS) and tin (IV) sulfide (SnS_2). According to Faullant [18-55] tin sulfides are effective replacement options for antimony trisulfide and provide a unique blend of favorable tribological properties including friction stabilization and wear reduction. These attributes were found in low-steel, semi-metallic and non-asbestos organic friction materials. The examination, which focused on the particle size effects of both tin (II) and tin (IV) sulfide in a model brake formulation, concluded that particle size effects were generally minimal.

It was noted that larger particle sizes of tin (II) sulfide gave higher coefficient of sliding friction relative to the larger particle sizes of tin (IV) sulfide. There were no significant differences in the reactivity of the tin sulfides. Heating of all samples in air resulted in the formation of tin dioxide (SnO_2) with no evidence of tin monoxide (SnO) being formed.

Suzuki [18-56] described the use of tin sulfide in combination with a mixture of tin and tin compounds in a non-asbestos friction material for automotive brake applications. The mixture containing tin sulfide provided a marked reduction in the occurrence of "metal pickup" (which can be viewed as an accumulation of reduced metal at the surface of the friction material that hardens forming a cutting edge and often results in severe abrasion to both surfaces).

Yamamoto [17-44] also determined that additions of tin sulfide to friction materials was an effective means for inhibiting the development of metal pickup at the friction interface. Tin sulfide was limited to a range of 3.5–75% as concentrations below this level were reportedly unable to prevent metal pickup, and excessive amounts were stated to adversely affect friction levels.

Generally, tin sulfides appear to have been used either singly or as a part of a mixture of solid lubricants. Lancaster [18-57] described a blend of tin (IV) sulfide and molybdenum disulfide with improved frictional performance and enhanced wear resistance when applied to smooth surfaces. This combinatorial approach is widespread, and blends with tin sulfides are very common.

Iron Sulfide

Iron sulfide has a number of natural and synthetic varieties in various morphological forms including Troilite (FeS), Greigite (Fe_3S_4), Marcasite (FeS_2- orthorhombic), and Pyrite (FeS_2 - cubic), which is commonly known as Fool's Gold. Several of these forms are found in nature. Each has uniquely different properties, and hardness values can range from 3.5 for Troilite to 6.5 for Marcasite.

Tribological studies of iron sulfide are very common and were at one time examined by the National Advisory Committee for Aeronautics [18-58 and 3-31]. Chemical reactions between ferrous sulfide (FeS) and steel were found to form a stable film at the sliding interface that caused increased friction at higher sliding speeds. Ferrous sulfide did not give indications of surface welding or seizure. However, it was not deemed to be beneficial for wear reduction. Additionally, sulfur-containing constituents in the tribosystem reacted with the steel surfaces and formed iron sulfide, which consequently affected the static friction levels as well.

In friction material composites, iron sulfide has been used across a wide range of products. Clark and Summerling [17-17] developed a friction material bound in a matrix of iron sulfide by reacting iron and sulfur together with asbestos, silica, and aluminum oxide to form a brake lining composite. Fedorchenko [17-19] reported that additions of iron sulfide to cermet friction materials (i.e., sintered metallic composition with ceramic additives) would greatly reduce the run-in period or burnish time. However, it was not specified if the additions provided wear resistance.

In a recent examination of the wear surfaces of automotive brake discs and friction material wear surfaces via Mossbauer spectroscopy, Vasconcellos and colleagues [18-59] found that iron pyrite concentrations would increase at the interface. The increased concentration in third body layers was associated with the decomposition of metal sulfide lubricants and other sulfur-containing species present in the composition.

Iron Sulfide in Friction Materials

Iron sulfide is listed in a number of patented friction material examples of potentially suitable lubricants that generally provide improvement in the wear resistance of the friction material. In metallic friction materials, the use of sulfur to alloy iron and form iron sulfide has been described by Larson and Biswas [18-60] as a means of wear reduction in a cast iron brake shoe. In resin-bonded composites, iron sulfide is often used in conjunction with other metal sulfide lubricants.

Suzuki and colleagues at Nisshinbo [18-61, 17-58 and 17-74] described the use of lubricating mixtures containing synthetic iron sulfide (FeS) with zinc and molybdenum sulfides. The most critical area of concern associated with the selection of metal sulfides was to limit the sulfate ion concentration of these compounds to concentrations less than 0.2 mg/g. Additionally, the pH of the final composite should be greater than 10, but less than 13, which could be achieved by the incorporation of alkaline additives such as slaked lime. Both aspects were necessary to offset corrosion induced by sulfur containing additives with excessive sulfate ion concentrations.

Iron sulfides will oxidize to the more thermodynamically stable iron (III) oxide, Fe_2O_3. Iron (III) oxide also occurs naturally in mineral form as Hematite. Iron (III) oxide is reddish in color, often associated with corrosion or rust, and has been used as a red pigment and polishing agent. The polishing action of iron (III) oxide can lead to increases in friction level, destruction of third body layers, and elevated wear rates. Care must be taken when using iron sulfide compounds in commercial friction material compositions.

Copper Sulfide

Copper sulfide has several forms including CuS_2, CuS, Cu_2S, and many more. The variety of copper sulfide compounds exists because of the unique bonding characteristics of elemental copper that allow for non-stoichiometric forms to exist. The more commonly found mineral forms include Cu_2S with a hardness of approximately 3 on the Mohs scale, and the rarer form CuS with a hardness value of approximately 2.

Copper sulfides have seen limited use in friction materials, most notably sintered iron-based friction materials. Schlömer [18-62] described a composition containing metal powders, mullite, graphite, silicon carbide, and copper sulfide, at a concentration of 2%. Copper sulfide was regarded as a friction supporting material and the concentration was limited to less than 5% of the total.

Fedorchenko [12-33 and 17-19] describes metallic sulfides in sintered iron composites. Copper sulfide and zinc sulfide were reported to increase the load bearing capacity of the materials greatly. In this example, copper sulfide was tested at a concentration of 3.4%.

Keller [18-63] systematically examined cupric sulfide (CuS) and cuprous sulfide (Cu_2S) in comparison to cupric oxide (CuO) in a rubber-bonded asbestos reinforced friction element. Cupric sulfide improved fade and recovery of the friction composition relative to both cuprous sulfide and cupric oxide. Additionally, cupric oxide generated higher levels of mating member wear and less friction stability than cupric sulfide. Cuprous sulfide was found to have similar friction stability concerns but generated an impressive 87-fold increase in mating member wear as compared to cupric sulfide.

Concerns with copper sulfide stability can also be seen in a novel method for developing a low friction coating. Shyne and Scheible [18-64] electrophoretically applied a mixture of copper sulfide and molybdenum disulfide to a sheet of stainless steel. This was then heated to temperatures of 1400°F (760°C) in an atmosphere of hydrogen. The reduction of copper sulfide formed hydrogen sulfide gas and elemental copper, which retained the molybdenum disulfide particles intact and provided for a low-friction surface on a substrate of stainless steel.

Copper-Iron Sulfide (Chalcopyrite)

Chalcopyrite is copper-iron sulfide with a hardness of roughly 3.5 on the Mohs scale. The use of chalcopyrite in friction material compositions is extremely limited and, although it has been mentioned in many patents, there is little definitive data regarding its use.

Lem [18-65] described the use of electromagnetic radiation in the form of microwaves as a curing method for friction materials and suggested that chalcopyrite, along with many other compounds, can be beneficial for enhancing cure rates during this process.

Vasconcellos and co-workers [18-59] concluded that copper iron sulfide additives in composite friction materials decomposed because of frictional heat. This liberated sulfur reacted with the hot iron of the brake disc to form a layer of iron pyrite at the surface and generally affected the overall tribological performance of the system. More work is needed to fully understand the true nature of chalcopyrite in friction materials.

Bismuth Sulfide

Bismuth sulfide occurs naturally as Bismuthinite, a very soft (Mohs of 2) mineral with a metallic luster, perfect cleavage, and high specific gravity. Its use in friction materials has been limited to a few citations (the majority in the last decade or so) as an additive with other lubricants.

For instance, Kesavan [18-66] used a mixture of bismuth sulfide in a non-asbestos organic friction material as a method for reducing brake noise. Ohe and Namba [18-67] stated that "metal catch" can be prevented by incorporating antimony oxide, antimony trisulfide, or lead sulfide into the composition. However, concerns with hazardous exposure required an effective solution that did not include antimony or lead compounds. The proposed solution was a mixture of metallic sulfides with examples containing tin, iron, and bismuth sulfides. The importance of bismuth sulfide was not fully illuminated in any case. Therefore, more work is needed in this area.

Zinc Sulfide

Zinc sulfide occurs naturally as wurzite and sphalerite, both of which have hardness values of approximately 3.5–4 on the Mohs scale. Synthetic versions such as Sachtolith [18-68], are prized for their low abrasion and non-toxic properties. The stability of these products allows them operate up to temperatures of approximately 660°F (approximately 350°C) after which they are converted to zinc oxide.

Holinski and Hesse [18-38] and Fedorchenko [12-33] described the benefits of zinc sulfide in sintered metallic and composition friction materials. Concentrations are typically limited to below 5% and zinc sulfide is generally used with other lubricants in a mixed lubricant system. As toxicology concerns continue to rise and the need for effective lubrication continues, zinc sulfide will most likely see more attention in the years to come

Synthetic Mixed Metal Sulfides

Metallic sulfides can also be produced synthetically by heating metals in the presence of sulfur without the presence of oxygen. This sulfidizing step converts metals into metal sulfides and allows for the use of metallic mixtures. For instance, Geringer [18-69] described the development of solid lubricants specifically targeted for use in automotive brake linings using a process where bi-metallic and tri-metallic mixtures could be used to produce mixed metal sulfide compositions with the general formula: $M^1M^2_xM^3_yS_z$. Mixtures of metals such as titanium, vanadium, manganese, iron, copper, zinc, molybdenum, tungsten, antimony, tin, and bismuth are all options.

Geringer disclosed an example of a bi-metallic sulfide produced by sulfidizing brass powder to form a copper-zinc sulfide. A comparison between simple mixtures of zinc sulfide and copper sulfide to the product obtained by sulfidizing brass demonstrated that the copper-zinc sulfide gave better friction stability and less wear than the composition containing the simple mixture of metal sulfides.

In addition, Hüner [18-70] developed a solid lubricant composition comprised of tin sulfide and carbon that is ideally suited for use in automotive friction materials. This composition was manufactured by heating elemental tin in the presence of carbon and sulfur to elevated temperatures under an inert atmosphere. The final compound is predominantly tin (IV) sulfide with small amounts of carbon and residual tin (II) sulfide. Further evaluation of this composition deemed it to be an acceptable replacement product for antimony sulfide.

The development of synthetic sulfide mixtures has opened an entirely new field of exploration for the development of sulfide lubricant systems. With increasing concern about toxicity and exposure, this method will most likely continue to see widespread success and growth.

Lubricant Mixtures

During the development of modern friction materials, designers often needed to expand the operating window across a wider range of temperature and pressure. This demanded that lubricant mixtures be used to achieve the desired balance of friction stability and low wear. Examples of this can be seen in the work by Centers [7-17], Lee and Jang [7-24], and Jang [18-71].

Combinations of graphite and metal sulfides have been shown to provide improved frictional performance (especially fade) hen compared to products containing only graphite. Metal sulfide compensates for the poor friction of graphite at elevated temperature and enables the development of a stable third body layer that ultimately controls the rate of wear.

Concerns with Metal Sulfides

But metal sulfides have a few areas of concern. Most notably is the corrosive nature of metal sulfides in a humid environment, as previously described by Suzuki. This can be compensated by selecting metallic sulfides that do not generate low pH solutions in the presence of water, or by the use of pH regulating agents such as calcium hydroxide.

The second area of concern is unwanted abrasion. Börjesson [18-72] studied the nature of third body layers in automotive brakes during operation at low contact forces and determined that metal oxides, produced by oxidation of metal sulfides in the composition, were capable of generating abrasion. The level of abrasion at low pressure was sufficient to remove deposits and led to unwanted wear during non-braking conditions. The development of unwanted vibration in the form of cold judder was described by Vikulov [18-73]. Cho [18-51] confirmed this result during his study of the tribological properties of solid lubricants and further supported the loss of lubrication and aggression because of conversion from sulfide to oxide that has been described by Buckley [18-74] and Leslie [18-75].

Hexagonal Boron Nitride

Hexagonal boron nitride is a commercial synthetic lubricant sometimes called white graphite. Hexagonal boron nitride has exceptional thermal stability up to temperatures of 1800°F (approximately 1000°C) in air [18-76] and is an intrinsic lubricant that does not require an adjuvant such as moisture to operate effectively. The use of hexagonal boron nitride has been limited by the extreme cost of this material, but future applications requiring extreme temperature lubrication might have a need for hexagonal boron nitride as a primary lubricant. This is an area where more work is needed.

Polytetrafluoroethylene (PTFE)

Polytetrafluoroethylene (PTFE) is a synthetic thermoplastic fluorocarbon polymer with an exceptionally low coefficient of friction. The incorporation of PTFE into a friction material would therefore be expected to provide an effective means for suppressing the friction level. However, Ono [18-77] described PTFE, used in the form of powder, as a friction regulator that enhanced friction at higher speeds, loads, and temperatures. The concentration was limited to a range of 3–7%.

PTFE is a soft solid and tends to form a third body layer in a different fashion from conventional metal sulfide lubricants. Lancaster [18-78] discussed the role of PTFE in composite tribology and reported the benefits of blending PTFE with a metallic sulfide such as molybdenum disulfide. The sulfide interfered with the third body formation process of PTFE and prevented the PTFE layer from growing to an excessive thickness. This once again demonstrates the benefit of using a lubricant mixture in tribological composites.

References

18-1. Dimick, G. 1883. Lubricating compound. United States Patent No. 287643, filed July 2, 1883, and issued October 30, 1883.

18-2. Devlan, P. 1860. Improvement in journal boxes. United States Patent No. 30128, issued September 25, 1860.

18-3. Joseph Dixon Crucible Co. 1914. *Graphite:* 16, No. 9: 3780. Jersey City, NJ: Dixon Graphite Productions.

18-4. Bernal, J. 1924. "The Structure of Graphite." Proceedings of Royal Society of London, A, 106, pp. 749-773

18-5. Barrer, R. 1937. "An Analysis by Adsorption of the Surface Structure of Graphite." Proceeding of Royal Society of London, A., 161, pp. 476-493.

18-6. Lipson, H., and A. Stokes. 1942. "The Structure of Graphite." Proceedings of Royal Society of London, A 181, pp. 101-105.

18-7. Bowden, F. and J. Young. "Friction of Diamond, Graphite, and Carbon and the Influence of Surface Films." Proceedings of Royal Society of London, A 208, pp.444-455.

18-8. Mantell, C. 1979. *Carbon and Graphite Handbook.* New York: Robert E. Kreiger Publishing Co.

18-9. Olson, D. 2003. "Graphite." *U.S. Geological Survey Minerals Yearbook.* Washington, DC: U.S. Department of the Interior.

18-10. ASTM International. 2003. "Standard Test Method for Ash in a Graphite Sample." ASTM C561 - 91(2010)e1 . West Conshohocken, PA: ASTM International, doi: 10.1520/C0561-91R10E01, www.astm.org.

18-11. Guptill, R. 1923. Bearing material and method of making the same. United States Patent No. 1464569, filed April 14, 1922, and issued August 14, 1923.

18-12. Gibson, D. et al. 1999.Amorphous carbon-coated carbon fabric wet friction material. United States Patent No. 5952249, filed December 17, 1996, and issued September 14, 1999.

18-13. Dearnaley, G., and M. Van Dyke. 2002. Lubricious diamond-like carbon coatings. United States Patent No. 6410144, filed June 29, 1999, and issued June 24, 2002.

18-14. Xanthos, M. 2010. *Functional Fillers for Plastics*. KGaA, Weinheim: Wiley-VCH Verlag GmbH and Co.

18-15. Johnson, C., "1919. Process of treating transmission bands. United States Patent No. 1293107, filed June 1, 1918, and issued February 4, 1919.

18-16. Mead, H. 1926. Lining for transmission and brake bands. United States Patent No. 1572301, filed December 4, 1924, and issued February 9, 1926.

18-17. Spokes, R. 1942. Friction elements. United States Patent No. 2276070, filed August 31, 1939, and issued March 10, 1942.

18-18. Potamoshnev, A. et al. 1966. "Some Operational Characteristics of Sintered Friction Materials Working Conditions of Dry and Lubricated Friction." *Fiskio-Khimischeskaya Mekhanika Materialov* 2: 363-364.

18-19. Hooten, N. 1969. "Metal Ceramic Composites in High Energy Friction Applications." *Bendix Technical Journal* 2: 55-61.

18-20. Gardner, T. et al. 2003. Friction facing composition and method of manufacture. United States Patent No. 6534565, filed August 28, 2001, and issued March 18, 2003.

18-21. Zhan, Y., and G. Zhang. 2004. Friction and Wear Behavior of Copper Matrix Composites Reinforced with Sic and Graphite Particles." *Tribology Letter* 17:.91-98.

18-22. Lee, J. H. et al. 2004. "A Study on the Friction and Wear of Bronze Sintered Friction Materials." *J. Korean Soc. Mfg. Proc. Eng.* 3: 5-10.

18-23. Taylor, H. 2006. "Graphite." In *Industrial Minerals and Rocks*, 7th ed. Edited by J. Kogel et al. Littleton, CO: Society for Mining, Metallurgy, and Exploration.

18-24. Acheson, E. 1893. Production of artificial crystalline carbonaceous materials. United States Patent No. 492767, filed May 10, 1892, and issued February 28, 1893.

18-25. Acheson, E. 1896. Manufacture of graphite. United States Patent No. 568323, filed December 27, 1895, and issued September 29, 1896.

18-26. Daimer, J. 2005. Graphite electrode for electrothermic reduction furnaces, electrode column, and method of producing graphite electrodes. United States Patent Application No. 2005/0254545, filed May 5, 2005.

18-27. Lam, R., and M. Yesnik. 1998. Friction lining comprising less fibrillated aramid fibers and synthetic graphite. United States Patent No. 5753356, filed June 3, 1994, and issued May 19, 1998.

18-28. Miyaki, T., and M. Kobayahsi. 2002. Non-asbestos type friction material. United States Patent Application No. 2003/0092795, filed October 1, 2002.

18-29. Yang, H. et al. 2010. Effect of the ratio of graphite/pitch coke on the mechanical and tribological properties of copper-carbon composites." *Wear* 268: 1337-1341.

18-30. Kolluri, D. et al. 2010. "Effect of Natural Graphite Particle Size in Friction Materials on Thermal Localization Phenomenon during Stop-braking." *Wear* 268:1472-1482.

18-31. Mathur, R. et al. 2004. "Controlling the Hardness and Tribological Behavior of Non-asbestos Brake Lining Materials for Automobiles." *Carbon Sciences* 5: pp. 6-11.

18-32. McIntyre, G., and R. Holinski. 2003. "Improvements of Friction Brake Lining Performance." SAE Technical Paper No. 2003-01-3316. Warrendale, PA: SAE International.

18-33. Holinski, R. 1987. Solid lubricant combination and utilization thereof in friction linings. United States Patent No. 4663060, filed April 7, 1986, and issued May 5, 1987.

18-34. Rohatgi, P. et al. 1990. "Friction and Wear in Metal-matrix Graphite Particle Composites." In *Tribology of Composite Materials* edited by P. Rohatgi et al. Materials Park, OH: ASM International.

18-35. Douglas, J. 1888. Self-lubricating journal box. United States Patent No. 379531, filed March 12, 1867, and issued March 13, 1888.

18-36. Thelen, E., and M. Devine. 1967. "Chemical Reactivity Factor for Metallic Sulfide Lubricants." *Tribology Translations* 10: 386-389.

18-37. Samsonov, G. et al. 1973. Lubricating Action of Refractory Metal Sulfides and Selenides. *Fiziko-Khimicheskaya Mekhanika Materialov* 9: 58-61 (Russian).

18-38. Filip, P. et al. 2002. "On Friction Layer Formation in Polymer Matrix Composites Materials for Brake Applications." *Wear* 252: 189-198.

18-39. Holinski, R., and D. Hesse. 2003. "Changes at Interfaces of Friction Components during Braking." *Proc. Inst. Mech. Eng*, Part D 217: 765-770.

18-40. Winer, W., 1967. "Molybdenum Disulfide as a Lubricant: A Review of the Fundamental Knowledge." *Wear* 10: 422-452.

18-41. Dickenson, R., and L. Pauling. 1923. "The Crystal Structure of Molybdenite." *Journal of the American Chemical Society* 45: 1466-1471.

18-42. Salomon, G. et al. 1964. "Mechano-chemical Factors in MoS2-film Lubrication." *Wear* 7: 87-101.

18-43. Bowden, F. et al. 1968. "Lubrication at High Temperatures of Refractory Solids." Proceeding of the Royal Society of London, A., 304, pp. 157-169.

18-44. Gardos, M. 1988. "The Synergistic Effects of Graphite on the Friction and Wear of MoS2 Films in Air." *Tribology Transactions* 31: 214-227.

18-45. McCabe, J. 1965. "Molybdenum Disulphide – Its Role in Lubrication." Proceedings of the Industrial Lubrication Symposium, London. Rowsley: Scientific Publications Ltd.

18-46. Okubo, H. et al. 1984 Asbestos-free friction materials. United States Patent No. 4477605, filed March 21, 1983, and issued October 16, 1984.

18-47. Stanley, F. 1922. Process of making friction facings. United States Patent No. 1420882, filed September 24, 1919, and issued June 27, 1922.

18-48. Weintraub, M. et al. 1974. "Wear of Resin Asbestos Friction Materials." In *Advances in Polymer Friction and Wear* edited by L. H. Lee. New York: Plenum Press.

18-49. Lidkea, H., and R. Spokes. 1936. Friction element composition. United States Patent No. 2052779, filed December 5, 1932, and issued September 1, 1936.

18-50. Soulen, J. 1976. Complex antimony sulfide as lubricant additives. United States Patent No. 3965016, filed May 27, 1975, and issued June 22, 1976.

18-51. Cho, M. et al. 2006. "Tribological Properties of Solid Lubricants (graphite, Sb2S3, MoS2) for Automotive Brake Friction Materials." *Wear* 260: 855-860.

18-52. Wirth, A., and R. Whitaker. 1992. "The Role of Transfer Film Chemistry in Automotive Friction Couples." Proceedings of the. International Congress on X-ray Optics and Microanalysis, Institute of Physics Conference Series 130, pp. 455-460.

18-53. Miyake, S. 1989. Brake frictional material. Japanese Patent No. 01-295035, filed May 20, 1988, and issued November 28, 1989.

18-54. Lumb, J., and M. Edwards. 1974. Friction material. United States Patent No. 3842031, filed November 28, 1972, and issued October 15, 1974.

18-55. Faullant, P. 2002. "Particle Size Effects of Tin Sulfides in Disc Brake Pads." SAE Paper No. 2002-01-2591. Warrendale, PA: Society of Automotive Engineers.

18-56. Suzuki, S. et al. 2007. Friction material. United States Patent No. 7297728, filed July 1, 2004, and issued November 20, 2007.

18-57. Lancaster, J. 1965. "Lubrication by Transferred Films of Solid Lubricants." Tribology Transactions 8: 146-155.

18-58. Levine, E., and M. Peterson. 1951. "Formation of Sulfide Films on Steel and Effect of Such Films on Static Friction." Technical Note No. 2460. Washington, DC: National Advisory Committee for Aeronautics.

18-59. Vasconcellos, M. et al. 2010. "Mossbauer Spectroscopy Characterization of Automotive Brake Disc and Polymer Matrix Composite (PMC) Pad Surfaces." *Wear* 268: 715-720.

18-60. Larson, H., and C. Biswas. 1982. Cast iron railroad brake shoes. United States Patent No. 4352416, filed December 12, 1977, and issued October 5, 1982.

18-61. Suzuki, S. et al. 2005. Friction material. United States Patent No. 6887549, filed June 24, 2004, and issued May 3, 2005.

18-62. Schlömer, J. et al. 1967. Friction material composition. United States Patent No. 3306715, filed July 7, 1965, and issued February 28, 1967.

18-63. Keller, E. 1969. Friction elements. United States Patent No. 3477983, filed June 23, 1966, and issued November 11, 1969.

18-64. Shyne, J., and H. Scheible. 1958. Electrophoretic method of applying a low friction coating. United States Patent No. 2826542, filed June 16, 1954, and issued March 11, 1958.

18-65. Lem, K. et al. 1996. Composition for use in friction materials and articles formed therefrom. United States Patent No. 5576358, filed February 3, 1995, and issued November 19, 1996.

18-66. Kesavan, S. et al. 2001. Solid lubricants containing bismuth sulfide for use in friction lining. United States Patent No. 6228815, filed June 29, 1999, and issued May 8, 2001.

18-67. Ohe, T., and M. Namba. 2003. Non-asbestos friction material. United States Patent Application No. 2003/0200898, filed April 24, 2003.

18-68. Sachtolith. Registered trademark of Sachtleben Chemie GmbH, Duisburg, Germany, Registration No. 1263885, January 17, 1984.

18-69. Geringer, M. 1999. Solid lubricant especially for friction linings, friction linings mixtures. United States Patent No. 5958846, filed November 3, 1997, and issued September 28, 1999.

18-70. Hüner ,R. et al. 2001. Solid lubricants with a tin sulphide and carbon base. United States Patent No. 6303545, filed October 6, 2000, and issued October 16, 2001.

18-71. Jang, H. et al. 1998. "The Effect of Solid Lubricants on Friction Characteristics." Technical Paper No. 982235. Warrendale, PA: Society of Automotive Engineers.

18-72. Börjesson, M. et al. 1993. "The Role of Friction Films in Automotive Brakes Subjected to Low Contact Forces," Proceedings of the Braking of Road Vehicles, I. Mech. E. C444/026, pp.259-268.

18-73. Vikulov, K. et al. 1998. "Brake Vibration and Disc Thickness Variation (DTV). In *Automotive Braking* edited by D. Barton and M. Haigh. London: Professional Engineering Publications.

18-74. Buckley, D. 1985. "Tribology: The Story of Lubrication and Wear." TM-101430. Washington, DC: National Aeronautics and Space Administration.

18-75. Leslie, A. 2004. "Mathematical Model of Brake Caliper to Determine Brake Torque Variation Associated with Disc Thickness Variation (DTV) Input." Technical Paper No. 2004-01-2777. Warrendale, PA: SAE International.

18-76. Lower Friction. "Hexagonal Boron Nitride (hBN) Powder." Accessed August 2, 2010. http://www.lowerfriction.com/pdf/11.pdf .

18-77. Ono, M. et al. 2004. Friction material composition and friction material therefrom. United States Patent Application No. 2006/0162259, filed February 5, 2004.

18-78. Lancaster, J. 1985. "Composites for Increased Wear Resistance: Current Achievements and Future Prospects." In *New Directions in Lubrication, Materials, Wear, and Surface Interactions: Tribology in the 80s,* edited by W. Loomis. NJ: Noyes Publications.

Chapter 19
Organic Filler Materials

Everything is organic and living, and therefore the whole world appears to be a living organism.
—Seneca

ORGANIC FILLER MATERIALS ARE FUNCTIONAL additives that are organic in the chemical sense (i.e., carbon-based compounds). Historically, these materials were derived from naturally occurring sources, which also makes them organic in the biological sense of the word. Common examples include a number of plant materials such as cork and wood, as well as animal-derived organic materials like leather. Also, advances in organic and polymer chemistry have now made it possible for a number of synthetic organic materials to find their way into friction facings.

Organic filler materials are typically classified as low temperature functional additives. They are added to a tribological composite to impart unique physical attributes to the structure of material and most often are present in the form of granular powders. The granules are dispersed throughout the matrix and affect tribological performance at low speeds under low-pressure applications. Occasionally, certain forms of organic filler materials have been found to aid in operator comfort by reducing vibration and noise. This is typically attributed to be the result of enhanced compliance at the sliding surface and an overall reduction in the modulus of the composite.

Early Organic Filler Materials

The history of organic filler materials began, like many other aspects of the friction industry, with the selection of readily available materials. The use of organic filler materials was stated to begin more than 4500 years ago [18-48] when wooden and leather friction materials were installed on quarry carts in the Sumerian region of Ancient Mesopotamia. Leather and wooden friction materials saw continued use until the early part of 20th century.

Early machinist textbooks stated that wood was a suitable friction material for general purpose use [19-1]. It was concluded that various species exhibited different performance characteristics. For example, poplar wood (from the genus *Populus*) was considered the best choice for use against iron mating members; oak wood (from the genus *Quercus*) and beech wood (from the family *Fagaceae*) were excellent secondary choices. Elm (from the genus *Ulmus*) was the ranked as a lower third choice because of the rapid drop in friction that would occur if oil contaminated the friction surfaces.

Wood and leather was ultimately replaced by simple saturated fabrics [4-1] and, as the development of woven asbestos friction materials began to emerge, the compositions began to rely heavily on many of the readily available organic materials which also included cork and wood.

Naturally Occurring Organic Filler Materials

Naturally occurring organic filler materials have typically been limited to additions of cork, wood (without regard to the specific form or species), and leather. A few of the more interesting studies that have been reported in recent literature were studied, but each shared the same recurring theme: the selected organic filler is available locally, moderately low in cost, and deliberately added with a desire to enhance a specific aspect of performance.

Cork

Cork is the outer bark of the cork oak tree (*Quercus suber*). It is indigenous to Europe and Africa with the majority of production in Portugal. Cork is a slow growing outer bark that can be removed with no damage to the tree. However, the harvesting of cork is regulated in many countries and requires extensive technical expertise to ensure survival of these beautiful evergreen trees [19-2].

Cork has been used in friction material development since the earliest part of the last century. Whitcomb [19-3] described a rubberized fabric design that incorporated cork in the form of machined inserts, which were embedded into the matrix. During vulcanization, the layers of fabric were bonded together while simultaneously forming an intimate bond with the cork plugs. The strong mechanical attachment between rubber and cork held the inserts in place during sliding. In this example, cork was intended to enhance the braking action by providing a large compliant region (i.e., a soft macro-asperity) at the interface [19-4 and 1-7].

Cork could also be ground into different ranges of particle sizes and formed into sheets of flexible, compliant material ideal for gaskets and wet friction applications. For instance, Bell [19-5] developed a laminated cork clutch facing targeted for replacing leather and asbestos materials. The laminated cork had a fabric facing that provided strength, while the cork gave the necessary compliance to ensure smooth engagement.

The compliance inherent in cork granules made it a favorite product for development of early clutch facings [19-6]. It was incorporated into mill board materials to improve the overall frictional response [19-7]. Other friction material manufacturers attempted to exploit the beneficial compression of cork in various compositions including a sulfur-treated cork product to ensure vulcanization of the rubber binder [19-8], cork coated with phenolic resins and mixed with plasticizers to ensure compliance [19-9, 19-10], and rubber-coated cork particles for use in laminated friction materials [19-11] and sheeter mixes [19-12]. Because of the importance of cork in resilient friction materials [19-13], attempts were also made to use it in molded friction materials.

Schultz [19-14] combined phenolic resin-coated cotton fibers with coated cork granules and granules of polymerized cashew nut shell liquid (CNSL). This mixture was molded to form a porous friction material for use in clutches and wet brakes. A similar example by Maierson and Todd [19-15] contained 8.7% cork in a rubber modified phenolic resin matrix reinforced with asbestos that had been specifically formulated to operate in a multi-disc, wet friction brake system.

Cork is not as widely used today as in previous years. However, but Asano and Iwata [17-32] developed a friction material with a porosity level of 15–30% that incorporated a blend of equal amounts of cork and granules of polymerized CNSL. The cork was necessary to convey the desired level of compression, but could not be used in concentrations exceeding 30%.

Wood

Wood, usually in the form of sawdust, was a common organic filler material in composition railroad brake shoes prior to 1940. For example, Kinzer [19-16] used sawdust and specifically stated that the addition was intended to improve the friction level of his composition. Allen and Morrison [19-17] combined sawdust with asphaltum, comminuted iron, and pulverized rock to form a castable brake shoe composition. However, they did not specify the reason for incorporating sawdust. Laeufer [12-2, 12-3] and DeFrance [17-14] also used sawdust as a portion of their cement-based friction compounds.

As an organic filler material, wood has some very interesting tribological characteristics dependent on the deformation and adhesion aspects of the selected material [19-18]. Atack and Tabor examined the frictional characteristics of wood and concluded that the mechanism for generating friction was a result of changes in the mechanical properties of wood influenced by humidity. These changes altered the deformation behavior under a given load, and the overall surface behavior of the wood dictated adhesional responses between sliding bodies.

It is also of interest to note that wet wood generated higher friction levels (i.e., more deformable) than when dry, which could account for its widespread success in early railroad friction applications that are sensitive to the difference between dry and wet friction levels.

In more recent years, the addition of wood granules or sawdust to modern friction materials has virtually disappeared. However, Takujui [19-19] described a method for the production of a friction material based on additions of wood particles. The process involved forming a friction material by conventional means and then subjecting it to thermal treatment under inert atmospheres. The thermal treatment converted wood into carbon and provided the automotive friction material with improved friction responsiveness and reduced rates of wear.

Leather

Leather is derived from tanning animal skins, most commonly that of cattle. Early uses of leather involved simply attaching a sheet of leather material to a wooden brake block. Wieterhold and Oak [19-20] developed a method for attaching leather to the block by use of fastening clamps. A similar idea was patented by Bayley in 1876 [19-21] and did not require the use of nails or screws as the method of attachment.

Leather is a soft, compliant, flexible material that could be formed into brakes [19-22], especially band and drum brake linings. It could also be shaped into conical sections for use in cone clutch devices [19-23]. Despite the compliance and excellent friction characteristics, issues with leather as a friction material became evident as early as 1892 [13-4].

Generally, leather tends to become smooth during use. Under severe operating conditions, it was prone to char badly [16-7]. These factors resulted in a friction level that was unsuitable (i.e., lower than desired). Therefore, leather friction materials were either replaced with a fresh piece of leather or treated with a friction enhancing solution such as those described by Newman [19-24]. Pulverized leather was sometimes incorporated into casting compositions for brake shoes similar to the use of sawdust.

Leather has also been incorporated into non-asbestos friction material in the form of "leather fibers" [19-25]. The fibers were pretreated with a drying oil and were beneficial in the pursuit of a quiet friction material that did not squeak during use.

Aside from the occasional esoteric resurgence of leather in friction materials, the use of leather is not expected to continue because of poor thermal resistance coupled with erratic performance attributes. Both have virtually eliminated it from consideration in all modern design circles.

Miscellaneous Natural Filler Materials

Virtually any known organic filler material can be obtained from naturally occurring sources. For instance, Laeufer [19-26] developed a wood substitute by combining cornstalks and pig's blood in a matrix of Portland cement. Formolo [13-14] described the use of nut shells in a composition friction material for railroads. Ibhadode and Dagwa [19-27] developed an asbestos-free friction lining from palm kernel shells. Mutlu [19-28, 19-29] investigated the use of organic filler materials derived from rice and cedar pines in a brake pad composition.

In the majority of these examples, the materials are plant-derived carbon-based compounds and will degrade in a similar manner to that of other cellulosic compounds. However, Mutlu found that rice husks would enhance friction because of the siliceous content of the ash generated by the heat of sliding friction. In addition, cedar pine cone dust was found to induce fade, but boric acid additions were capable of offsetting this effect and could be used in conjunction to provide a suitable friction material.

Synthetic Organic Filler Materials

Synthetic organic filler materials are typically polymeric materials that have been developed by modification of a naturally available product or by synthesis techniques from simple starting monomers.

Synthetic organic filler materials can be subdivided into two large families of materials: CNSL derivatives and synthetic elastomeric materials, which can also include chemical modifications to natural rubber compounds. The compositions of many modern friction materials often employ a blend of organic filler materials to achieve the desired mixture of compliance and thermal resistance, but a vast majority relies heavily on the addition of polymerized CNSL granules.

Polymerized CNSL Granules

Polymerization of CNSL with a curing agent, such as formaldehyde was described by Harvey [8-29]. The reaction yielded an amorphous material that could be pulverized into granules. The granular form of polymerized CNSL was used in early friction materials for "maintenance of uniformity of friction characteristics at high temperature." This ability to affect the friction level of a brake lining composition earned polymerized CNSL granules the name friction dust.

Much like phenolic resins, the polymerization process can be tailored in a number of ways by using various aldehyde compounds in the presence of different catalysts and with different levels of curing agent. For instance, Bruce [8-31] polymerized CNSL with acidic compounds and later adjusted the pH of the friction material to prevent corrosion of the brake drums.

Kuzmick and Whitelaw [19-30] described a partially cured friction dust that had the consistency of rubber. This particle was used in concentrations of nearly 20% in early molded friction materials and could also be produced by polymerization of marking nut oil (*Semecarpus anacardium*), a similar plant-derived phenolic mixture.

The dominance of polymerized CNSL dust in friction materials has continued to modern times. Yasushi and Kunio [15-24] reported that polymerized CNSL dusts were effective friction regulating additives. The ability of these polymeric granules to regulate friction is likely a combination of properties including compatibility with the resin matrix and an ability to conform to the surface under high normal loads, which serves to increase the hysteretic component of friction generation. In addition, the particles typically are moderately thermally stable and have a high carbon yield when exposed to elevated temperatures.

There are a number of commercial CNSL particles available today, and many have found widespread use in modern NAO formulations. Of the various forms, the two most common families are formaldehyde condensed particles (often called brown particles because of the amber color of the granules) and black particles (which are condensed with furfuraldehyde and have a dark, nearly black color).

Formaldehyde Condensed Particles

The condensation reaction between formaldehyde and CNSL has been fully described by Harvey [19-31]. The reaction can be accelerated by elevated temperature and chemical catalysts or driers such as manganese salts. Formaldehyde can be present in gaseous form, as aqueous solutions of formalin, or in solid form as paraform. In each case, the reaction profiles are adjusted to ensure that the reaction is conducted in the most efficient manner and follows a typical phenolic resin condensation reaction profile.

Furfuraldehyde Condensed Particles

Condensation of CNSL with furfuraldehyde was described by Caplan [19-32, 19-33]. The reaction was different in that it began with a mixture of an organic sulfate such as diethyl sulfate, furfuraldehyde, and CNSL. The mixture was combined through agitation and heating. When it reached a prescribed temperature, the product was dispensed into shallow pans and baked at a temperature of approximately 275°F (135°C) for an extended period of time. The fully cured product was then ground into powder and sold under the trademark Cardolite [19-35]. There are a number of suppliers for this product today, and each grade has its own unique chemistry that provides a wealth of options for further modification.

Friction Dust in Friction Materials

Kuzmick and Whitelaw [19-30] were perhaps the first group to employ a polymeric CNSL granule in friction compositions. From this simple beginning, the market has expanded to include a number of particles with multiple compositions in a number of sizes. CNSL particles have been used in virtually all types of friction materials including rubber-bound low-steel compositions [17-54], injection-molded asbestos-filled drum lining compositions [19-35], metallic friction materials [19-36], railroad composition brake blocks [17-47], and modern non-asbestos organic friction materials 18-28].

The most commonly cited justification for the use of polymerized CNSL granules in friction materials is for the enhancement of the friction level [6-2], where the addition of friction dust was deemed to be specifically beneficial for improving "cold" friction levels [6-48] of brake linings.

In addition to enhancing friction, the use of these materials has been found to assist in the manufacture of certain composite materials [7-6] and also to improve wear resistance of automotive brake linings. The reduction in wear was often noted in the cold temperature sections where the effects of thermal degradation were less likely to occur [17-45].

Concerns with CNSL Dust in Friction Materials

The primary concern with using polymerized CNSL particles in a friction material is an inherent characteristic of all organic materials. This class of materials has a moderately low capacity to withstand elevated temperatures for prolonged periods of time.

Jacko [3-41] described the destruction of the surface layer of friction materials as a result of operation at elevated temperature. It was observed that as the surface temperature increased, the friction surfaces lost organic content, specifically noting the loss of CNSL dust. The friction surface therefore became more porous, which might partially account for the reductions in brake noise noted by Cho [19-37].

The phenolic resin character of CNSL implies that CNSL particles will experience degradation mechanisms similar to that of phenolic resins [9-18, 9-19, 9-20]. Compositional studies by Tanaka confirmed the chemical similarity between CNSL and phenolic resins in a resin-based friction material [7-31]. It was noted that thermal degradation processes not only destroyed the methylene bridges between phenolic moieties, but also destroyed any cross-links that formed along the fifteen carbon side chain. This led to a large number of possible degradation products that, if trapped at the friction interface, were associated with the induction of fade [13-6]. For this reason, many researchers limit the concentration of CNSL dust in friction materials to maintain high temperature friction stability [19-38].

Today, chemical modifications to CNSL and chemical modifications with distillates of CNSL have resulted in the development of a number of different products [19-39, 19-40] with many unique and novel properties. This type of research is expected to continue and will hopefully lead to future CNSL-based derivatives that provide enhanced physical and thermal attributes for the next generation of polymerized CNSL friction dusts.

Rubber Particles

Particles of rubber, also known as rubber crumb and tire peel, are common synthetic organic filler materials. This family of additives is generally limited to only a few possible candidates. Fiction material designers typically will use only small amounts to adjust specific friction properties. High concentrations of rubber particles are expected to introduce similar concerns to those of any organic particle, namely poor friction stability at elevated operating temperatures.

Of the variety of rubber particles available, the more commonly selected materials are tire peel (a form of recycled rubber obtained from commercial tire sources), styrene-butadiene (SBR) particles, and acrylonitrile-butadiene (NBR) particles.

Tire Peel

Tire peel is scrap rubber that has been ground into powder, typically by cryogenic means [19-41]. Its composition will vary depending on the type of tire or base rubber compound used in the reclamation process. In addition, recycling companies might co-grind a number of different sources, which further convolutes a thorough analysis of the composition of the particles.

The performance of rubber particles is often compared to polymerized CNSL dust [19-42]. The benefits associated with tire peel additions are also expected to mirror those of CNSL dust and include a generalized effect on the cold friction levels with reduced wear under these conditions. The list of negative attributes associated with tire peel can result in poor performance at elevated temperatures, but might also affect the manufacturing process. Typically, this involves difficulty in preforming because of rebound of the particles upon release of the compressive force. Rubber powders that are easily compacted with little or no rebound can also be considered beneficial for preforming. However, the concentration must be carefully controlled so that the desired friction level at elevated temperatures is maintained.

SBR Crumb

SBR crumb is similar to tire peel and in some cases the terms are used interchangeably. SBR is typically not compatible with phenolic resins and therefore does not bond well with resinous matrices. This can affect wear rates as well as the physical properties of the product. There is little information in the literature that specifically describes the use of SBR crumb. However, DeGaugue [19-43] patented a composition brake block for railcars from a polymeric matrix comprised of SBR and CNSL. To ensure that the friction levels were acceptable under wet conditions, a "long wear friction particle" such as a vulcanized hard rubber crumb was included along with asbestos and cast iron particles.

NBR Crumb

NBR rubber is perhaps the most common form of rubber crumb used in the friction material industry today. Traditionally, acrylonitrile-butadiene copolymers such as the one described by Konrad and Tschunkur [19-44] have been used as physical property modifiers for phenolic resin bound materials [19-45], often with combinations of oil and phenolic resins [8-32]. A similar method was described by Sarbach [19-46] who developed a synthetic rubber composition based on mixtures of NBR and CNSL.

Modern researchers have also used NBR as a toughening agent for phenolic resins. Lu [19-47] modified phenolic triazine and benzoxazine resins compounds by milling with NBR prior to using in automotive friction materials. Matejka [17-41] and Izyumova [15-60] also used NBR modified resins in friction material composites to enhance the physical performance of their compositions. Likewise, Lu [19-48] compared NBR modified phenolic resins with a resin modified with boric acid and with an unmodified resin. NBR modified resins improved wear within a specified ratio of modification.

Today the use of NBR crumb is similar to that of CNSL dust. The NBR crumb can be considered a particle that readily deforms under load. This generates friction at cold operating conditions because of hysteresis of the elastomer. However, physical blends of rubber with phenolic resin will not give the same level of performance as a chemically modified resin [10-15].

Recycled NBR

Recycled grades of NBR offer the design engineer an opportunity to include a compliant particle at a moderately lower cost and to impact the environment by reducing waste from entering landfills. These materials have been used since 1947 [19-49] and have typically been reinforced with carbon black with other fillers such as silica. Some grades can retain processing oils that are not capable of being polymerized and will be liberated at the interface during sliding. This can lead to frictional variation, especially at higher temperatures. Additionally, the composite designer should make every effort to understand the composition of degradation products, which can contain metal oxides that might adversely affect wear rates.

Virgin Grades

Virgin NBR is not vulcanized and can be added to the composition in either latex or granular form. Granular forms are usually produced by spray drying latex to form particles. They are combined with an anti-caking agent. The particles can be dry mixed into a resin-bonded system or solvated [19-50] to form the primary binder for a rubber based friction material. Additionally, the particles do not have the resiliency of a fully vulcanized recycled material and will flow under compressive loads (i.e., cold flow). This can be beneficial in manufacturing processes where effective preforming is necessary. Virgin grades are generally several times more expensive than recycled grades and are therefore used only in applications that can justify the higher cost.

Fluoroelastomer Crumb

Fluoroelastomer crumb is typically available as a recycled granular product obtained from the production of fluoroelastomer compounds, such as Viton [19-51] or other suitable FKM polymers. The use of FKM polymers (those that are approximately 80% fluoroelastomer content), in friction materials does not appear to have become commonplace. However, fluoroelastomers offer superior heat and chemical resistance in comparison to other common types of rubber compounds. Additionally, they are significantly more expensive. Until the price disparity is reduced, or a clear justification is found, the role of fluoroelastomer crumb is expected to remain limited.

Polychloroprene

Polychloroprene rubber is a chlorinated elastomer with an extended history of use in the friction materials industry. Driscoll [19-52] and Bruce [19-53] both used chloroprene rubber as a binder for asbestos-based friction materials more than 70 years ago. Today, chloroprene might still be in use, especially for railway friction applications [19-54], but little information is available regarding its selection and potential benefits. This is an area where more work is needed.

References

19-1. Oberg, E., and F. Jones, eds. 1915. *Machinery's Handbook for Machine Shop and Drafting Room*, 5th ed. New York: The Industry Press.

19-2. Pereira, H. 2007. *Cork: Biology, Production and Uses.* Amsterdam: Elsevier.

19-3. Whitcomb, W. 1907. Composite sheet or layer. United States Patent No. 841294, filed October 10, 1906, and issued January 15, 1907.

19-4. Whitcomb, L. 1913. Friction device. United States Patent No. 1082834, filed July 11, 1913, and issued December 30, 1913.

19-5. Bell, H. 1917. Clutch facing. United States Patent No. 1239864, filed December 8, 1916, and issued September 11, 1917.

19-6. Tresslar, G. 1922. Friction clutch member. United States Patent No. 1414490, filed April 3, 1918, and issued May 2, 1922.

19-7. Fisher, W. 1922. Clutch facing and process of making same. United States Patent No. 1436158, filed February 25, 1921, and issued November 21, 1922.

19-8. Short, C. 1925. Clutch disc for motor vehicle clutches. United States Patent No. 1524214, filed May 28, 1917, and issued January 27, 1925.

19-9. Weisenburg, A. 1936. Clutch plate. United States Patent No. 2054210, filed May 22, 1933, and issued September 15, 1936.

19-10. Denman, H. 1940. Method of making brake blocks. United States Patent No. 2185333, filed August 1, 1936, and issued January 2, 1940.

19-11. Smith, C. 1957. Friction material. United States Patent No. 2787569, filed November 15, 1952, and issued April 2, 1957.

19-12. Yocom, A. et al. 1952. Cork containing composite sheet material. United States Patent No. 2584959, filed September 14, 1948, and issued February 5, 1952.

19-13. Batchelor, C., and E. Carey. 1954. "Friction Facts, Fortes, and Foibles." Technical Paper No. 540030. Warrendale, PA: Society of Automotive Engineers.

19-14. Schultz, H. 1951. Friction material. United States Patent No. 2553215, filed November 25, 1946, and issued May 15, 1951.

19-15. Maierson, T., and R. Todd. 1960. Friction material for use in brakes. United States Patent No. 2954853, filed November 30, 1955, and issued October 4, 1960.

19-16. Kinzer, J. 1895. Brake shoe. United States Patent No. 542414, filed May 9, 1895, and issued July 9, 1895.

19-17. Allen, A., and J. Morrison. 1897. Brake shoe. United States Patent No. 596132, filed May 29, 1897, and issued December 28, 1897.

19-18. Atack, D., and D. Tabor. 1958. "The Friction of Wood," Proceedings of the Royal Society of London, A 246, pp. 539-555.

19-19. Takuji, I. et al. 1993. Method for producing friction material. Japanese Patent No. 05-070610, filed September 17, 1991, and issued March 23, 1993.

19-20. Wieterhold, V., and V. Oak. 1875. Brake block for wagons. United States Patent No. 164891, filed April 1, 1875, and issued June 22, 1875.

19-21. Bayley, J. 1876. Wagon brake. United States Patent No. 174183, filed September 21, 1875, and issued February 29, 1876.

19-22. Frood, H. 1905. Brakes. Great Britain Patent No. 190416222, filed July 22, 1904, and issued May 25, 1905.

19-23. Snider, A. 1916. Cone clutch facing and process of making the same. United States Patent No. 1180885, filed November 26, 1915, and issued April 25, 1916.

19-24. Newman, J. 1912. Composition of matter. United States Patent No. 1019989, filed June 1, 1911, and issued March 12, 1912.

19-25. Sugita, Y. and Y. Fujita Y.1981. Friction material. Japanese Patent No. 56-057869, filed October 17, 1979, and issued May 20, 1981.

19-26. Laeufer, J. 1906. Artificial wood or substitute material. United States Patent No. 818833, filed January 31, 1906, and issued April 24, 1906.

19-27. Ibhadode, A., and I. Dagwa. 2008. "Development of Asbestos-free Friction Linings Material from Palm Kernel Shell." *Journal of Brazilian Society of Mechanical Sciences* 30: 166-173.

19-28. Mutlu, I. 2009. "Investigation of Tribological Properties of Brake Pads by Using Rice Straw and Rice Husk Dust." *Journal of Applied Science* 9: 377-381.

19-29. Mutlu, I. 2009. "Investigation of Using Cedar Pine Cone Dust and Boric Acid in Brake Pads." *Journal of Applied Science* 9: 382-386.

19-30. Kuzmick, J., and A. Whitelaw. 1941. Molded composition friction body. United States Patent No. 2256646, filed February 23, 1938, and issued September 23, 1941.

19-31. Harvey, M. 1929. Cashew-nut-shell condenser product and process for making same. United States Patent No. 1725797, filed June 15, 1928, and issued August 27, 1929.

19-32. Caplan, S. 1939. Treatment of cashew nut shell liquid. United States Patent No. 2176059, filed November 6, 1936, and issued October 17, 1939.

19-33. Caplan, S. 1943. Polymerization-furfuraldehyde condensation product of cashew nut shell liquid distillation residue and method. United States Patent No. 2317587, filed July 17, 1940, and issued April 27, 1943.

19-34 Cardolite. Registered trademark of Cardolite Corporation, Newark, New Jersey, Registration No. 0391140, October 28, 1941.

19-35. Beare, C. 1948. Injection of brake linings. United States Patent No. 2452284, filed January 11, 1946, and issued October 26, 1948.

19-36. Schultz, H. 1951. Friction material. United States Patent No. 2569539, filed October 30, 1947, and issued October 2, 1951.

19-37. Cho, M. et al. 2005. "Effects of Ingredients on Tribological Characteristics of a Brake Lining: An Experimental Case Study." *Wear* 258: 1682-1687.

19-38. Armour Research Foundation of Illinois Institute of Technology. 1957. "Aircraft Landing Gear Brake Friction Materials Development." ARF Project No. G016, AD669388, July 31, 1957.

19-39. Sato, S. et al. 2010. Process for the production of cardanol. United States Patent No. 7825284, filed April 5, 2006, and issued November 2, 2010.

19-40. Tsap, G. et al. 2007. Melamine ring-containing co-polymers; methods of making and using the same. United States Patent Application No. 2008/0103283, filed December 20, 2007.

19-41. Liang, S. B. et al. 1998. "Transient Heat Transfer of Scrap Tire Peel and Its Fine Powder in the Cryogenic Grinding Process." *Chemical Engineering Communications* 163: 133-144.

19-42. Cecere, J. 1997. Bonding a friction material brake lining element to a metallic backing plate element. United States Patent No. 5595266, filed June 26, 1995, and issued January 21, 1997.

19-43. DeGaugue, C. 1954. Composition brake block. United States Patent No. 2686140, filed March 29, 1952, and issued August 10, 1954.

19-44. Konrad, E., and E. Tschunkur. 1934. Rubber like masses from butadiene hydrocarbons and polymerizable nitriles. United States Patent No. 1973000, filed April 20, 1931, and issued September 11, 1934.

19-45. Keller, E., and R. Spokes. 1946. Friction elements. United States Patent No. 2394783, filed May 9, 1942, and issued February 12, 1946.

19-46. Sarbach, D. 1943. Synthetic rubber composition. United States Patent No. 2325981, filed August 26, 1941, and issued August 3, 1943.

19-47. Lu, Y. et al. 2002. "Modeling Wear Traces of Automotive Friction Materials by Cantor Set." *Tribology Transactions* 45: 258-262.

19-48. Lu, Y. 2002. "Friction Performance of Phenolic Binders," In *Proceedings of the International SAMPE Symposium and Exhibition* 47: 236-241.

19-49. Spokes, R. et al. 1947. Friction element. United States Patent No. 2428298, filed November 16, 1942, and issued September 30, 1947.

19-50. Twiss, S., and E. Sydor. 1961. Friction elements and method of making the same. United States Patent No. 3007890, filed May 7, 1959, and issued November 7, 1961.

19-51. Viton. Registered trademark of E. I. DuPont de Nemours and Company, Wilmington, Delaware, Registration No. 0675761, March 24, 1959.

19-52. Driscoll, J. 1936. Wear and heat resisting composition. United States Patent No. 2052610, filed September 23, 1932, and issued September 1, 1936.

19-53. Bruce, D. 1937. Friction material and method of making the same. United States Patent No. 2077669, filed June 1, 1934, and issued April 20, 1937.

19-54. Littlefield, J. 1982. Composition friction element for a railroad brake shoe. United States Patent No. 4313869, filed May 19, 1980, and issued February 2, 1982.

Chapter 20
Inorganic Filler Materials

The secret, which nobody realized at the time, was that with this sand, which was 100 percent inorganic, you can control exactly what goes in it.
—Anita Tucker

INORGANIC FILLER MATERIALS ARE AS COMMON as dirt, sand, or other minerals found in the vicinity of a given friction material manufacturer. The role of an inorganic filler material is often considered to be a low-cost, space filling substance that reduces the amount of binder necessary to consolidate the composite. Inorganic filler materials are typically used in higher concentrations, especially in lower cost composites. Additionally, inorganic filler materials occur in many shapes and sizes with a plethora of possible combinations that can be explored by the composite designer. However, in many cases, the importance of these materials is commonly overlooked. Despite this oversight, inorganic filler materials are actually critical to the success of many friction producing composites.

The Role of a Filler in Friction Materials

The role of inorganic filler materials in modern friction materials is to provide the designer with alternative materials of lower cost that fill available space while maintaining a high degree of thermal resistance. Common inorganic filler materials typically do not affect the wear and friction behavior significantly.

In other words, the inorganic filler should not function as an abrasive or a lubricant and, therefore, should serve to support the performance of the product as a secondary component to the other constituents. This general class of materials is not easily defined, but usually consists of heat-resistant minerals with a moderate to low hardness values.

Like abrasive materials, inorganic filler materials must be selected with the same considerations regarding size, shape, and particle friability. These factors will affect the development or destruction of beneficial third body layers at the friction interface. Particle friability for a given inorganic filler material might be less of a concern than for harder abrasive additives. However, poor adhesion between binder and filler particles, when coupled with easily fractured particles, can result in rapid wear, and the final product will often exhibit frictional instability.

Finally, as with all minerals that possess an inherent level of natural variance, knowledge of the impurities in the material must be determined by comprehensive analysis and extensive testing. Slight variations in the concentration of impurities (e.g., quartz) in moderately soft filler materials such as talc can result in a number of issues, including excessive abrasion and poor mating member compatibility. Purity is an essential prerequisite for all inorganic filler materials.

Carbonaceous Filler Materials

Carbonaceous filler materials are typified by forms of coal, coke, and carbon black. These materials are not truly inert and do not lubricate as efficiently as graphite. Therefore, they can be considered non-lubricating carbon filler materials. Although their use in automotive friction materials varies greatly because of philosophical differences of the designers, several forms of carbon have promising thermal characteristics that make them suitable for high temperature operation.

Coal

Coal is a combustible sedimentary rock that occurs in many forms. Anthracite, bituminous, and lignite (or brown coal) are three of the better known varieties. Coal was first used in friction materials in 1906 [12-1] in the form of coal cinders, a porous thermally altered carbonaceous residue resulting from the combustion of coal. Coal cinders were included as an ingredient in a molded friction compound for railroad friction materials and were deemed beneficial because of their hardness and porosity.

Several years later, Thompson [20-1] described an asbestos-based friction material comprised of equal parts asbestos and bituminous coal in a double-bodied linseed oil binder system. Johnson attempted to use coal as a source of continued lubrication during braking. The mechanism proposed was essentially a form of thermal cracking whereby frictional heat generated during operation would liberate a viscous substance (i.e., tar) and increase the coefficient of friction (i.e., a frictioning agent) while also providing improvements in wear.

Spokes [20-2] described a friction material bound with a rubbery mixture of polymerized linseed oil and phenolic resin. This product contained asbestos and a pyrobitumenous substance that could be manufactured by rolling into long strips suitable for drum brake applications. Likewise, Bruce [20-3] used brass and asbestos in a chloroprene rubber mixture as a sheeter friction material. In both cases, the use of bituminous materials was added to the compositions as carbonaceous filler.

As recently as 1984, coal was listed as a potential additive in friction materials [20-4]. However, in this specific instance, anthracite coal was listed. No specific information regarding its use was reported, and it might have been present only as a space filling material. Coal is typically considered a combustible material today. Its use in friction material appears to be limited to a few low-energy industrial applications that do not present the hazards of ignition during use.

Petroleum Coke

Petroleum coke, also called pet coke or petrol coke, is a carbon-rich residue derived from petroleum refining. There are many grades of petroleum coke, which vary depending on the coking process. Many forms of coke can be used as fuel or in the production of electrodes for smelting operations. However, friction material producers have generally adopted petroleum coke in the form of either needle coke or sponge coke, with the occasional use of shot coke.

The nomenclature for these types of coke serves as a rough descriptor. Needle coke is best described as a material with a layered structure and fractures in a somewhat linear fashion along lines that are parallel to the layers producing needle-like particles. Sponge coke has a cellular pore structure that typically does not possess short range order and that resembles, for all intents and purposes, a sponge. Shot coke is generally spherical in appearance and does not share the aspects of either needle or sponge coke.

According to Mantell [18-8], petroleum coke was at one time considered a waste product that was used in the fueling of the stills at refineries. Today petroleum coke is being produced at a volume of approximately 60–70 million tons per year [20-5] and is finding widespread use as an alternative fuel that competes with coal. The friction material industry consumes only a tiny portion of the available supply and generally uses a precisely specified range of particle sizes of a few basic types.

Pet Coke in Friction Materials

DeFrance [17-12, 7-13, 17-14] was perhaps one of the earliest users of coke. His patents described a molded composition brake shoe for railway cars. The compositions were simple asbestos-based compounds in a metallic shell with coke listed at a level of 8–13 parts by weight. However, no mention of the benefits of coke was reported.

Several years later, Spokes [20-6] described a friction element comparing the use of petroleum coke to bituminous coal. Bituminous coal is effective for improving cold friction output of brake linings. When operating under low-energy conditions, bituminous coal was found to give reduced wear. However, at higher energy operating conditions, coal decomposes with a significant degradation in braking performance. This was further compounded by concerns of coal converting to an excessively abrasive carbon residue (i.e., ash) during operation.

Petroleum coke, like bituminous coal, was also found to provide improved friction and wear, but acted by a distinctly different mechanism. The thermal energy of braking acted to soften petroleum residues in the coke, which would exude out at the interface, forming an effective hydrocarbon lubricant. The petroleum residue (called asphalt content) was present at approximately 3% by weight. When this type of coke was added to the composition as a replacement for coal, the overall tribological performance was greatly enhanced.

Nanfeldt [20-7] stated that brake linings should contain density elements, which are primarily for resisting normal pressure (i.e., load bearing particles), and should also contain shear resisting elements that will fracture with a jagged edge to eliminate glazing of the surfaces. In addition, the composition must also contain surface elements that maintain friction resistance. One exemplary agent that fills all of these requirements was described as large particles of petroleum coke with a nominal diameter of approximately 40 mesh (approximately 400 microns). Similarly, Spokes and Littlefield [17-39] used petroleum coke in their composition as a convenient means for increasing the friction output at low pressure.

Petroleum coke is often described as a friction enhancing agent or a performance modifier [4-5 and 5-10] and is still widely used in friction materials. Petroleum grades are sold as calcined varieties with little or no hydrocarbon residue. However, green (non-calcined) versions might also be available from various suppliers. The use of green coke in today's friction material market is considered unorthodox.

Modern friction materials typically use additions of petroleum coke combined with other forms of carbon such as graphite. One notable example of this approach [20-8] described a synergistic blend of graphite and coke in a range of combinational ratios from 2:1 to 7:1. This range was considered ideal for ensuring that the final product had effective lubrication and exhibited the desired levels of high temperature wear. However, the overall concentration of the mixture was limited to 25% by volume.

Metallurgical Coke

Metallurgical coke is produced by the destructive distillation of coal. It has an open porous structure but, the process of distillation tends to concentrate impurities native in the coal. Therefore, metallurgical coke often contains more impurities (i.e., higher ash content) and can be more abrasive than a comparably sized grade of petrol coke.

Despite this fact, Seiz and Eldrige [20-8] described a heat-resistant paper material made from synthetic organic fibers (such as aramid fibers) that is coated with a layer of carbon granules and bonded together with a thermosetting resin. The preferred carbon granules were listed as petroleum coke, but metallurgical coke was also listed as a possible alternative. The study cited the use of this paper in wet friction systems such as automatic transmission friction elements.

Shepley and Carter [20-9] described a friction material containing an iron powder-metal sulfide alloy such as manganese sulfide. This unique composition gives an extensive list of raw materials, including metallurgical coke, as one of many possible carbon filler materials. The primary interest was focused on the ability of iron-metal sulfide alloy to reduce noise and improve wear performance. Little information was provided regarding the benefits of metallurgical coke.

In light of the limited information available, metallurgical coke can be anticipated to be an effective replacement for petroleum coke in many tribological applications. The concentration and composition of the ash from this material should be identified and fully understood to prevent the development of surprising friction transitions that can occur when metallurgical coke is used in high concentrations.

Carbon Black

Carbon black is elemental carbon in the form of colloidal particles. Carbon black was first produced more than 3500 years ago [20-10] as "lamp black." Today, carbon black is produced by the highly controlled decomposition of hydrocarbons, which can be in either liquid or gas phase. The decomposition process yields a finely divided, high purity form of carbon. Modern production processes yield either thermal black or furnace black (which is the more common of the two). Thermal black and furnace black differ in their feedstock: furnace black typically results from pyrolysis of heavy oils whereas thermal black is produced by incomplete combustion of natural gas.

Carbon black is chemically and physically different from soot, which is typically less than 60% carbon. Carbon black is often greater than 97% carbon. Nearly 90% of all carbon black is used in the manufacture of rubber and elastomer compounds such as tires, belts, and hoses. The second largest industrial application uses only 9%, typically as a pigment. The remaining 1% is considered an essential ingredient in a vast number of other applications, which would include its use in engineered tribological composites.

Carbon Black in Rubber-Based Friction Materials

Carbon black has been used in rubber-based friction materials for nearly a century. Novak [5-16], was perhaps one of the first to incorporate carbon black into an asbestos reinforced rubber friction material. This product was solvated and extruded by conventional means to form a rubber-bonded brake lining.

Tseng [20-11] described a molded asbestos friction material with a blend of phenolic resin and rubber, together with carbon black and the necessary curing agent (sulfur). The composition was mixed in an intensive rubber mixer, calendared to size, cut into strips, and formed into linings prior to curing in an oven. It was noted that composition contained a relatively large amount (7.7% by volume) of carbon black, which was considered necessary to reduce the wear rate of the final product.

Novak [20-12], on the other hand, described a rubber composite produced by bonding layers of friction material together with an adhesive. This compound was a typical solvated asbestos-rubber mixture containing only a small amount (three parts) of carbon black, and is similar to many of the early rubber-based friction products.

Nanfeldt's [20-13] description of a novel molded friction material bound with sulfurized linseed oil expanded on the behavior of many of the additives used in friction materials of this time. For instance, blood proteins and lead oxide were combined to provide a method of in situ generation of elemental lead at the interface.

Mica was added to prevent the formation of a glazed surface, which could imply removal of third body layers from the interface. Along a similar line of thinking, carbon black was added to stabilize the coefficient of friction by also serving to prevent the formation of a glazed surface at the friction interface.

Composition friction elements for railway applications also rely heavily on rubber as a basis of their binder system. This logically implies a need to rely on conventional rubber curing systems and on carbon black as a reinforcing filler. Compositions such as those described by Spokes [20-14] and DeGaugue [20-15] use carbon black at levels from 10 parts to 10% of their example compositions. The use of carbon black in rubber-based friction materials has continued to today. Patents by Chester [20-16] and Harold [17-54] are two of the many resources that described formulations containing carbon black in modern rubber-based friction materials.

Carbon Black in Other Types of Friction Material

Carbon black has also been used widely in other forms of friction material. The majority of this research was reported by workers in Japan. For instance, Yasuhiro [20-17, 20-18] used carbon black in the binder system of a carbon-based friction material designed for use in clutches and brakes. The impregnating solution of thermosetting resin contained up to 50% carbon black by weight and enhanced the load carrying ability of the final product.

Yasusaburo [20-19] described a wet friction material bound with a fluorine containing elastomer. The product was a blend of fiber such as potassium titanate, a suitable peroxide curing agent, and carbon black at levels ranging from 15–50% by weight. The wide range of carbon black additions were noted to be an effective means for adjusting the hardness of the final product.

Carbon black has also been included in the development of non-asbestos organic friction materials. Masanori [20-20] patented a friction material bound together in a matrix that that contained at least 3% carbon black, but that could contain up to 20% by volume. The unique grade of carbon black was specified to have very precise iodine and dibutyl phthalate (DBPA) absorption characteristics. The benefit of this composition was a marked reduction in low-frequency brake noise during operation. A similar patent by Takashi and Shinichiro [20-21] also described how carbon black additions to friction materials might limit the development of judder because of degradation of the constituents by frictionally induced thermal energy (i.e., carbon black might limit deposition at high operating temperature).

Barytes

Barytes, a pulverized form of barite (the mineral form of barium sulfate), is perhaps the most commonly used friction material additive in the world. Barytes has a moderately low hardness value (approximately 3–3.5 on the Mohs scale), high specific gravity (approximately 4.5), very low solubility in water, and a high melting point (greater than 2732°F/1500°C). Additionally, the cost of barytes is very low in comparison to other raw material additives. This blend of properties makes barytes ideally suited for use in friction materials.

Headson [20-22] recorded the use of barytes in his patent from 1921 and was perhaps the first to use barytes in a composition friction material. However, the role of barytes as a filler material in phenolic plastics dates back to the work of Baekeland as early as 1909 [20-23]. Despite this prolonged period of use, the role of barytes is generally defined as inert filler and is commonly used to lower the cost of the final product. However, evidence suggests barium sulfate can function as a powerful friction augmenting agent based on its chemical reactivity.

The Reactivity of Barytes

The reactivity of barytes has been examined for more than a century. Howe [20-24] cited Mostowitsch [20-25] in 1909 and described the decomposition reactions of barium sulfate in the presence of carbon. The reaction proceeded by extraction of the oxygen from the sulfate, which liberated carbon dioxide and formed barium sulfide. The reaction temperature began at 1112°F (600°C), which was well below the decomposition point of pure barium sulfate. However, this can vary depending on the concentration of carbon present.

Recently, McCormick [10-15] examined the interactions of polymeric matrix materials in friction materials. The examination concluded that phenolic resins provided a carbon source capable of reacting with barytes at high temperatures. The proposed mechanism suggested the reduction of barium sulfate into the sulfide, which confirmed the work of Mostowitsch and provided additional support for the reactivity of this inert filler at the friction interface during operation.

The Use of Barytes in Brake Linings

After sliding, the surfaces of a brake couple were examined by Ma and colleagues [20-26] using x-ray diffraction methods. It was determined that barytes had formed a film on the discs at operating temperatures of 572°F (300°C) and higher. The film formation process was viewed as a heavy debris filling mechanism where high density barytes particles filled the interstitial voids and provided protection from wear. This was also confirmed by Handa and Kato [6-16].

The film forming ability of barytes was also reported in the work of Österle [2-23]. The presence of barytes crystals at the friction interface were clearly identified, but at a greatly reduced size with particles in the nanometer size range. This indicated that the sliding action fractured barite particles into extremely small fragments.

Batchelor [6-24] and the team of Pogosian and Lambarian [17-57] classified barytes as a friction modifier and not as a filler material. The incorporation of barytes in asbestos reinforced friction materials was found to enhance friction stability during fade conditions. Similar results were reported by Kragelskii [20-27] who confirmed the stabilizing effects of barytes in asbestos reinforced friction materials and also classified barium sulfate as an anti-seizure additive in metallic friction materials (e.g., sintered metallic compositions).

Despite the numerous benefits of barium sulfate in conventional friction materials, there are also concerns with its use, especially at higher concentrations. Kim [20-28, 20-29] reported that barytes can adversely affect the friction levels of the brake linings at high speeds and high concentrations. This can potentially lead to stick-slip during sliding. This has been widely associated with noise generation in brake lining operation and should be avoided by limiting the concentration of barytes to acceptable levels.

Calcium Carbonate

Calcium carbonate, also known as whiting, is a white mineral solid with several polymorphs including limestone, chalk, marble, travertine, and calcite. These forms have variable chemical characteristics. The general properties for calcium carbonate include a moderately low specific gravity (approximately 2.7) and a hardness of only 3 on the Mohs scale. The melting point varies depending on the source of the specific polymorph, but temperatures in excess of 1544°C (840°C) will decompose calcium carbonate to calcium oxide and release gaseous carbon dioxide. Calcium carbonate is widely used because of its low density and color. Most commercially supplied grades take the form of a very fine powder.

Headson [20-22] was one of the first to include whiting in his list of acceptable filler materials. Tucker [20-30] followed a similar pattern and also included whiting as one of many acceptable filler materials, a practice that continues in the patent literature of today. However, one early patent by Denman [20-31] describes a vulcanized rubber backing material with whiting as a constituent. His backing material listed whiting at a concentration of 51% by weight, which seems excessive. However, this may have been justified based on comments from Brady [17-65] who describes whiting as filler that imparts crack and tear resistance when added to rubber compounds.

Many modern friction materials still rely heavily on calcium carbonate additions. Miyaki and Kobayashi [18-28] made use of small amounts of calcium carbonate (approximately 5%) in a lead sulfide free non-asbestos friction material for disc brake applications. This low level is in direct contrast to the formulation by Horiguichi [13-20] for drum brake friction materials that included calcium carbonate at a level of 57%.

Generally, calcium carbonate can provide resistance to shrinkage in polymeric composites [20-32] and might have potential applications in lubricating compositions [20-33], but the majority of use appears to be as a low-cost filler. Whiting is moderately heat-resistant with very low cost, and pure versions do not contribute to excessive abrasion. However, silica contamination should be carefully understood prior to selection.

Calcium Hydroxide and Calcium Oxide

Calcium hydroxide and calcium oxide are generally classified together under the category of pH modifying filler materials. The general terminology for this class of materials in convoluted and includes names such as quicklime, slaked lime, hydrated lime, and simply, lime.

Chemically, quicklime is calcium hydroxide (CaO). It is often manufactured by roasting calcium carbonate to drive off carbon dioxide. Quicklime can react with water to form calcium hydroxide ($Ca(OH_2)$), which is also known as slaked lime, hydrated lime, hydraulic lime, pickling lime, and milk of lime. To avoid confusion regarding names, only the terms calcium hydroxide and calcium oxide are used whenever possible in this discussion.

Laughlin [8-13] used calcium hydroxide in an oil-rubber bound asbestos brake lining to saponify fatty acids and produce an in situ soap compound. The process of saponification uses a strong base that, in the presence of water, will hydrolyze the ester bonds of fatty acid forming a glycerol and a calcium salt of the fatty acid. A similar saponification process was also recorded by Blume [20-34] in his description of a linseed oil-bound asbestos friction material. Blume also noted that calcium hydroxide additions provided the additional benefit of shrinkage control during cure.

One of the more interesting early applications of calcium hydroxide was patented by Driscoll and Bruce [20-35] in 1936. This process began with a woven asbestos fabric that had been treated with a suspension of hydrated lime (calcium hydroxide) in water. The lime treated fabric was subsequently coated with sodium silicate and formed into a brake lining with exceptional levels of heat and water resistance. In this example, calcium hydroxide improved adhesion between sodium silicate and asbestos fiber.

Aside from these early applications, calcium hydroxide is most often added to modern friction material compositions to control pH of the composite when wet. One of the first groups to positively identify this application of hydrated line (calcium hydroxide) was Laher and Kelleher [20-36]. They used 1.9% in a molded brake lining composition to offset the effects of residual acids present in the resin (because strong mineral acids were often used as catalysts during the condensation process of many phenolic resins). It was also noted that calcium hydroxide additions increased the hardness of the final product.

Several years later, Miller [20-37] reported an increase in physical strength during his description of a steel fiber reinforced friction material, which incorporated calcium hydroxide at a level ranging from 2–10% by weight. In this unique composition, calcium hydroxide was found to react with phenolic resin, increasing the strength of the friction material matrix. The enhanced strength allowed the composite to withstand higher dynamic loads experienced during braking.

Modern friction materials use a more conservative approach and typically employ calcium hydroxide in small concentrations. For example, Iwata and Asano [7-8] developed a friction particle for use in a conventional friction material that contained a small amount of calcium hydroxide (approximately 2%). Tadokoro and co-workers at Nisshinbo [20-38] developed a non-asbestos friction material using only 3% calcium hydroxide. Chinda [15-50] followed a similar pattern and incorporated calcium hydroxide at 2% of the composition. In these cases, it can only be assumed that calcium hydroxide is present as a method of regulating pH as described by Lem [20-39] and, therefore, higher concentrations are likely not justified.

Concerns with Calcium Hydroxide and Calcium Oxide

Calcium hydroxide and calcium oxide are both alkaline (basic) materials. This provides a means for imparting corrosion resistance in automotive brake applications by ensuring that water at the surface does not promote oxidation. Like all chemically active ingredients, they can also introduce other problems. For instance, one of the most noted concerns is the potential to accelerate the polymerization of certain polymer systems.

Brady [17-65] reported that lime (most likely calcium oxide) can stiffen rubber compounds and accelerate the vulcanization process. Long [17-66] also described lime (again most likely calcium oxide) as an effective accelerator for reducing the cure times by increasing the speed of vulcanization in rubber compounds. Long stated the critical need for the concentration of accelerators to be controlled accurately.

The effect can also be found in non-elastomeric systems. Reflecting back to the work of Miller [20-37], the behavior of electropositive bivalent metal ions can accelerate the cure of phenolic resins, most notably describing the acceleration of resole resin cure rates in the presence of calcium hydroxide as described by Park [20-40].

Another prominent concern centers on the use of calcium oxide in friction materials. Calcium oxide will react with water in the atmosphere and liberate heat. This process, known as slaking, forms calcium hydroxide and will vary with the grade of calcium oxide and particle size. For this reason, additions of calcium oxide must be carefully controlled. Mixtures, as well as the raw materials, must be fastidiously protected against moisture. Because of the difficulties inherent in controlling these factors, friction material designers generally incorporate the lower reactivity calcium hydroxide in dry friction materials in lieu of quicklime.

Mica

Mica is not a single material, but is actually a class of phyllosilicate minerals with more than 30 members. The most common forms include biotite, lepidolite, phlogopite, and muscovite. All forms of mica are sheets of silicate tetrahedrons interconnected with six member rings and can be easily divided along their basal planes. Therefore, they are ideal for preparing ultra smooth surfaces for the study of monomolecular layers of boundary lubricants [20-41] and other aspects of tribological interactions.

Mica compounds are typically soft minerals (2.5–3.0 on the Mohs scale) with moderately low specific gravity (approximately 3.0 depending on variety). In addition, the members of the mica group exhibit a unique blend of properties including exceptional flexibility (because of the lamellar structure), a high degree of heat resistance, and transparency. The transparent nature of thin sheets of mica, (e.g., muscovite) at one time made it ideal for making thin plates suitable for furnace windows as described in 1770 by D'Auteroche [20-42].

It appears that mica was used first in the manufacture of sintered metallic friction elements, where it was considered a lubricant [6-44]. This perspective might have merit when it is considered that mica exhibits perfect shear in the basal planes and would have been more readily sheared in comparison to the sintered metallic welds present between the particles of the matrix (i.e., the mica platelets would have been the weakest link when exposed to shear forces during sliding).

In contrast, Nanfeldt [20-13] described mica as shear resisting element in his sulfur linseed oil based binder system and included it to prevent formation of glaze at the friction interface. Nanfeldt did not describe the mechanism, but mica flakes in a soft matrix might have generated friction because of plowing action at the interface. This would tend to scour the surface, removing deposits generated by higher braking temperatures and effectively stabilizing the coefficient of friction.

Mica is also a common additive intended to reduce brake noise. It has been used for this purpose for nearly a century. One of the first examples patented was by Nash [20-43] who used small amounts of mica powder in a brake dressing compound to help alleviate brake "squeaking." The dressing compound was applied to the friction surfaces of brake linings that had worn and become glazed.

Modern friction materials designers such as Yamashita [20-44] and Nakagawa [20-45] incorporated mica into molded friction materials specifically for reducing low-frequency noise. Concentrations were limited to an upper limit of approximately 30%, but significant reductions in brake noise were found at much lower concentrations of mica. Mitsuhiko [20-46] expanded on the role of mica in friction materials by specifically noting the beneficial ability of mica to suppress stick-slip, which can be considered the underlying mechanism for the noted reductions in brake noise. It must be noted that mica was reportedly difficult to mix into the composition effectively, implying that care must be taken to ensure consistent dispersion.

Poor dispersion of the mica flakes can lead to a number of other quality concerns, and Mitsuhiko incorporated a pre-mix strategy to overcome this issue. This technique is similar to the use of surface coatings described by Yang [20-47] and has been shown to improve wear rates relative to untreated grades of mica, especially when used in elastomeric friction compounds.

Vermiculite

Vermiculite is a weathered form of mica [20-48] composed of hydrated laminar sheets of silicates. With the application of sufficient heat, water vapor is dislodged from between the lamella in a process known as exfoliation. This produces an expanded type of vermiculite in the form of hinged sheets of material with an accordion-like shape.

According to the United States Environmental Protection Agency (EPA), the use of expanded vermiculite is common in insulation and has been used as a loose insulation in houses since 1919 [20-50]. Vermiculite was found to match many of the properties of asbestos including low conductivity, good abrasion resistance, and ready availability at low cost [6-35]. This combination of properties not only makes vermiculite an ideal insulation, but also an effective additive for replacing asbestos in friction materials.

As early as 1932, vermiculite was incorporated into friction materials in the form of Jeffersite. Watson and Redfield [8-18] used a small amount (2% by weight) to improve the wear resistance of their rubber friction compound. More recently, Kaufman [12-10 and 15-23] described a friction lining containing a larger concentration of vermiculite (from 25–50%) as a replacement for asbestos. Kimura [20-50] used vermiculite in a friction material for dampers as a means of increasing the porosity of the friction material. Filip [18-38] described vermiculite as a lubricant in his composition.

In virtually all cases, mica and vermiculite share a number of similarities and can be used in similar concentrations with similar expectations of performance. However, there is one significant concern with vermiculite. Vermiculite can be contaminated with asbestos, as indicated in Kaufman [20-4] and EPA [20-50]. The presence of even small amounts of asbestos in vermiculite can result in reductions in use in the future.

Calcium Fluoride

Calcium fluoride is more commonly known as the mineral fluorspar or fluorite. By definition, it is a 4 on the Mohs scale, and can range in specific gravity from 3.1–3.6 depending on composition. Calcium fluoride has a melting point of nearly 2600°F (1427°C) and is commonly used as a fluxing agent in the manufacture of metals—a process that dates back to 1855 [20-51].

Calcium fluoride has also been classified as an effective high temperature lubricant, especially for drawing wire composed of aluminum and aluminum alloys [20-52]. The beneficial lubrication aspects were described by Buckley [18-74]. Friction material researchers typically include it among the list of acceptable filler materials, but with two noteworthy exceptions.

Fedorchenko [20-53] described the addition of metal fluorides such as calcium fluoride to a sintered iron-based friction material. At temperatures of 977°F (525°C) metallic fluoride became an effective lubricant. At this temperature, molten fluorides formed a film at the interface and dropped the friction level. The drop in friction level was necessary for prevent seizure between the sliding surfaces. It was also pointed out that fluorides can weaken the structure of a sintered iron material, but alloying agents such as molybdenum can offset the drop in strength. Fedorchenko suggested a range of 6–9% for calcium fluoride in the iron-based composition.

In composition brake shoes, Formolo [20-54] described calcium fluoride in his self-lubricating brake shoe, but did not expand on the benefits gained from its use. It is anticipated that calcium fluoride can contribute to high temperature lubrication, but more work is definitely needed in this area.

Cryolite

Cryolite is a halide mineral, chemically known as sodium hexafluoroaluminate (Na_3AlF_6), with a hardness and specific gravity of approximately 3. Cryolite was generally considered a single source mineral with the largest commercial deposit isolated on the west coast of Greenland. The principal mine site, Ivittuut; was depleted in 1987 [20-55]. Other natural sources are available, but synthetic production methods, such as those described by Kadotani and Isobe [20-56], are capable of generating a consistent supply with higher quality that have essentially replaced natural grades.

Synthetic grades of cryolite have effectively met the needs of consumers in aluminum smelting operations. They have also replaced cryolite in grinding wheel production, where cryolite plays an important role in improving the surface finish. In grinding operations, cryolite is thought to act as a flux at the friction interface and to lubricate the surface [20-57].

In friction materials, cryolite first appeared as a pigment in an adhesive composition combined with phenolic resin and asbestos fiber. This specific adhesive composition was used by the American Brake Shoe Company [20-58] to bond of drum brake linings to the brake shoe. In the patent by Kuzmick [20-59], a number of asbestos-based compositions for railway brake systems were illustrated. Thermal softening of the polymeric matrix, which was typically rubber, allowed hard particles to be depressed back into the matrix, preventing abrasion of the wheel at high temperatures. This also affected friction levels. Cryolite, used in concentrations of approximately 10% with an organic bond content of 20%, was reported to melt at the interface, forming an amorphous glassy layer that acted as a flux to prevent abrasion, but that also aided in retaining particles at the interface. The retention of particles at the interface can be seen as a deliberate attempt to control the third body layer and thus stabilize wear rates.

Several years later, Honma [17-78] patented a friction material with a limitation on organic binder content at 17–20%, where concentrations less than 17% were deemed unacceptably low in strength and concentrations higher than 20% exhibited a greatly reduced porosity level and suffered from poor performance. Additions of cryolite were made to improve the fade performance of this compound and were limited to between 3–20% of the total.

Kaolin

Kaolin is an aluminum silicate clay that is soft (Mohs of 2.5) with a specific gravity of approximately 2.6. This clay is often white in color and has widespread applications in the rubber industry. Typically, kaolin and other similar forms of clay were used in early flexible rubber-based drum brake lining compositions such as those described by Nanfeldt [20-60,17-62], often at concentrations of approximately 10%.

Modern research on kaolin is still underway. It is typically studied in comparison to other mineral fillers such as mica. One example by Yamashita [20-44] described many combinations of mineral fillers and included kaolin, in the form of kaolinite, as a means for the reduction of brake noise in automotive brake lining compositions.

Kaolin is another example of a common filler material that is not truly inert. It begins to lose water at temperatures above 1022°F (550°C). The resulting dehydrated aluminum silicate is more disordered and exhibits different tribological characteristics than hydrated forms. Further heating can result in the nucleation of aluminum silicate (e.g., formation of mullite) in a mixture with cristobalite silica, both of which are considered to be highly abrasive. Despite these concerns, more work is needed to fully appreciate the potential benefits of kaolin in non-asbestos tribological composites.

Feldspar

Feldspar is another member of the alumino-silicate mineral family that includes several minerals such as Anorthite [6-18]. Feldspar is generally harder (6 on the Mohs scale) and typically has limited use in friction materials. There is one exception in work by DeLukacsevics [6-37] who used a combination of feldspar with kaolin and a number of hard ceramic additives in his "fired" friction element for high temperature operation. It is interesting to note that in this example, kaolin acted as a ceramic binder instead of as an additive.

On the other hand, Sanders [17-5] used feldspar in an organically bound friction material composition. This simple formulation relied on small additions of copper for conductivity, asbestos for reinforcement ,and hard refractory materials such as feldspar, which was listed at a concentration of 50%, to generate the necessary friction level and resist fade. Sanders also pointed out that silica could be used in place of feldspar, but both should be sized very finely to prevent excessive abrasion from occurring.

References

20-1. Thompson, J., 1930. Composition of matter. United States Patent No. 1761057, filed August 26, 1927, and issued June 3, 1930.

20-2. Spokes, R. 1934. Composition friction element. United States Patent No. 1961177, filed June 13, 1932, and issued June 5, 1934.

20-3. Bruce, D. 1941. Friction element. United States Patent No. 2253608, filed December 2, 1936, and issued August 26, 1941.

20-4. Kaufman, E. Friction lining material. United States Patent No. 4487729, filed March 15, 1983, and issued December 11, 1984.

20-5. Pet Coke Consulting. "Advice for Buyers and Sellers of Petroleum Coke. Accessed August 23, 2010. http://www.petcokeconsulting.com/primer/index.html.

20-6. Spokes, R. 1941. Friction element. United States Patent No. 2267803, filed August 16, 1938, and issued December 30, 1941.

20-7. Nanfeldt, W. 1942. Process of making brake lining. United States Patent No. 2273770, filed August 16, 1938, and issued February 17, 1942.

20-8. Seiz, D., and L. Eldridge. 1992. Friction material having heat-resistant paper support bearing resin-bonded carbon particles. United States Patent No. 5083650, filed May 24, 1991, and issued January 28, 1992.

20-9. Shepley, C., and D. Carter. Friction material. United States Patent No. 5501728, filed July 22, 1994, and issued March 26, 1996.

20-10. International Carbon Black Association. "What is Carbon Black?" Accessed August 26, 2010. http://www.carbon-black.org/what_is.html.

20-11. Tseng, A. et al. 1934. Molded brake lining material. United States Patent No. 1963511, filed May 7, 1932, and issued June 19, 1934.

20-12. Novak, I. 1935. Friction element. United States Patent No. 1997308, filed February 15, 1932, and issued April 9, 1935.

20-13. Nanfeldt, W. 1939. Molded friction material. United States Patent No. 2175480, filed July 2, 1936, and issued October 10, 1939.

20-14. Spokes, R. 1940. Composition material for friction elements or the like. United States Patent No. 2211354, filed February 18, 1937, and issued August 13, 1940.

20-15. DeGaugue, G. 1964. Composition brake block. United States Patent No. 3152099, filed June 3, 1960, and issued October 6, 1964.

20-16. Chester, J. 1879. Improvements in or relating to friction materials. European Patent No. 839, filed August 8, 1978, and issued February 21, 1979.

20-17. Yasuhiro, A. 1979. Friction material. Japanese Patent No. 54-138879, filed April 20, 1978, and issued October 27, 1979.

20-18. Yasuhiro, A. 1979. Friction material. Japanese Patent No. 54-138880, filed April 20, 1978, and issued October 27, 1979.

20-19. Yasusaburo, S. 1983. Wet friction material. Japanese Patent No. 58-213077, filed June 3, 1982, and issued December 10, 1983.

20-20. Masanori, I. 1993. Friction material. Japanese Patent No. 05-117633, filed October 24, 1991, and issued May 14, 1993.

20-21. Takashi, N., and Y. Shinichiro Y. 1993. Friction material. Japanese Patent No. 05-222353, filed February 12, 1992, and issued August 31, 1993.

20-22. Headson, F. 1923. Lining or friction member for brake bands, brake shoes, or other articles. United States Patent No. 1468634, filed May 9, 1921, and issued September 25, 1923.

20-23. Baekeland, L. 1918. Composition containing phenolic condensation products. United States Patent No. 1259472, filed December 2, 1909, and issued March 19, 1918.

20-24. Howe, J. 1909. Recent work in inorganic chemistry." Proc. Amer. Chem. Soc. Eschenbach Printing, Easton, Pennsylvania, 1909.

20-25. Mostowitsch, W. 1909. "On the Behavior of Heavy Spar with Temperature and Its Reactions." *Metallurgie* 6: 450.

20-26. Ma, Y. et al. 2008. "Effects of ZrSiO4 in Non-metallic Brake Friction Materials on Friction Performance." *Tribology International* 41:166-174.

20-27. Kragelskii, I. 1965. "Friction and Wear." Washington, DC: Butterworths.

20-28. Kim, S. et al. 2004. "Tribological Properties of Polymer Composites Containing Barite (BaSO4) or Potassium Titanate (K2O-6(TiO2)." *Tribology Letters* 17: 655-661.

20-29. Kim, S. et al. 2001. "Effect of Humidity on Friction Characteristics of Automotive Friction Materials." *KSTLE International Journal* 2: 150-153 (Korean).

20-30. Tucker, G. 1934. Rubber-bonded asbestos product and method of making. United States Patent No. 1956053, filed April 17, 1931, and issued April 4, 1934.

20-31. Denman, H. 1932. Brake lining. United States Patent No. 1851087, filed August 15, 1931, and issued March 29, 1932.

20-32. Bhattacharya, S. 1986. *Metal-filled Polymers: Properties and Applications*. New York: Marcel Dekker.

20-33. Khan, M., and S. Jha. 2009. Friction material compositions and additives. United States Patent No. 7588609, filed July 30, 2004, and issued September 15, 2009.

20-34. Blume, W. 1933. Friction element. United States Patent No. 1924021, filed January 5, 1931, and issued August 22, 1933.

20-35. Driscoll, J., and D. Bruce. 1936. Treated fabric and process of making the same. United States Patent No. 2033928, filed December 23, 1931, and issued March 17, 1936.

20-36. Laher, F., and C. Kelleher. 1950. Molded brake lining. United States Patent No. 2534607, filed March 7, 1944, and issued December 19, 1950.

20-37. Miller, M. 1988. Friction material. United States Patent No. 4722950, filed July 7, 1986, and issued February 2, 1988.

20-38. Tadokoro, F. et al. 2005. Non-asbestos friction material. United States Patent No. 6838161, filed July 2, 2002, and issued January 4, 2005.

20-39. Lem, K. et al. 1996. Composition for use in friction materials and articles formed therefrom. United States Patent No. 5576358, filed February 3, 1995, and issued November 19, 1996.

20-40. Park, B. et al. 2002. "Effect of Synthesis Parameters on Thermal Behavior of Phenol-Formaldehyde Resol Resin." *Journal of Applied Polymer Science* 83: 1415-1424.

20-41. Bailey, A., and J. Courtney-Pratt. 1955. "The Area of Real Contact and the Shear Strength of Monomolecular Layers of a Boundary Lubricant," Proceeding of the Royal Society of London, A 227, pp. 500-515.

20-42. D'Auteroche, J. 1770. "A Journey into Siberia Made by order of the King of France." London.

20-43. Nash, G. 1927. Brake dressing compound and method of making same. United States Patent No. 1639611, filed September 4, 1926, and issued August 16, 1927.

20-44. Yamashita, Y. et al. 1993. Friction material for making brake pads. United States Patent No. 5266395, filed January 17, 1992, and issued November 30, 1993.

20-45. Nakagawa, M. et al. 1993. Friction material for making brake pads. United States Patent No. 5266395, filed January 17, 1992, and issued November 30, 1993.

20-46. Mitsuhiko, N. et al. 1993. Friction material and method of manufacturing such material. United States Patent No. 5268398, filed August 17, 1992, and issued December 7, 1993.

20-47. Yang, A. et al. 1991. "Abrasive Wear in Filled Elastomers." *Journal of Materials Science* 26: 5823-5837.

20-48. Hindman, J. 2006. "Vermiculite." In *Industrial Minerals and Rocks*, 7th ed. Edited by J. Kogel et al. Littleton, CO: Society for Mining, Metallurgy, and Exploration.

20-49. Environmental Protection Agency. "Asbestos." Accessed August 30, 2010. http://www.epa.gov/asbestos/pubs/verm.html.

20-50. Kimura, K. et al. 1998. Friction material for dampers and process for producing the same. United States Patent No. 5830566, filed October 13, 1995, and issued November 3, 1998.

20-51. Lawton, B. 1855. Improvement in journal box alloys. United States Patent No. 13466, issued August 21, 1855.

20-52. Buehler, W., and J. Nachman. 1960. Elevated temperature working of metals and alloys. United States Patent No. 2950526, filed April 22, 1957, and issued August 30, 1960.

20-53. Fedorchenko, I. 1979. "Modern Theories of the Mechanism of Friction and Wear and the Main Trends in the Development of Composite Frictional and Bearing Materials – A Review." *Porosh. Metall.* 196: 53-65 (Russian).

20-54. Formolo, J. 1998. Self-lubricating brake shoe material. United States Patent No. 5821282, filed October 26, 1995, and issued October 13, 1998.

20-55. Minerals Zone. "Cryolite." Accessed August 30, 2010. http://www.mineralszone.com/minerals/cryolite.html.

20-56. Kadotani, M., and S. Isobe. 1972. Process for the production of high quality synthetic cryolite. United States Patent No. 3656894, filed May 8, 1970, and issued April 18, 1972.

20-57. Kaaber, H. 1999. Grinding aid material, method of producing it, use of it, and abrasive material containing it. United States Patent No. 5888258, filed April 18, 1997, and issued March 30, 1999.

20-58. Spokes, R. 1951. Bonding element for friction element assemblies. United States Patent No. 2554128, filed September 4, 1947, and issued May 22, 1951.

20-59. Kuzmick, J. et al. 1996. Molded composition brake shoe comprising organic bond, hard mineral filler, and cryolite. United States Patent No. 3227249, filed June 25, 1959, and issued January 4, 1966.

20-60. Nanfeldt, W. 1936. Brake lining. United States Patent No. 2061919, filed October 31, 1931, and issued November 24, 1936.

Chapter 21
Metallic Filler Materials

L'injure se grave en métal; et le bienfait s'escrit en l'onde.

—Jean Bertaut (ca. 1611)

THE PERMANENCE OF METALLIC FILLER materials endows modern friction materials with many unique benefits. Metallic fibers were discussed previously in this book. Generally, metallic fillers offer similar benefits without enhancing strength characteristics as effectively as a fibrous form. Many modern tribological composites are based on the benefits gained by incorporating various metallic powders into the composite—a process that historically began with the addition of cast iron grit.

Cast Iron Grit

Cast iron grit is a hard metallic filler material that has usually been limited to use in railway friction materials, most notably as an ingredient in composite brake shoe compounds. According to Walsh and Spurr [21-1], cast iron grit acted more like an abrasive and less like a metal in this application. This was actually beneficial for promoting abrasion of the wheel surface and for preventing cracking at the friction surface, which can lead to excessive wear by delamination mechanisms.

The first use of cast iron grit in a friction material was patented in 1900 by Morrison and Allen [17-1] as an ingredient in a cast asphalt brake shoe. One-half century later, DeGaugue [19-43, 20-15] used cast iron grit at a concentration of 30% by weight in his composition brake block. Cast iron grit in this example was intended to provide friction stabilizing effects where the wet and dry friction values were similar, which coincided with the reports by Wise and Lewis [12-22]. The similar values for both wet and dry friction levels are one of the most noted beneficial characteristics of cast iron grit additions in friction compositions.

On the other hand, Spokes [17-39] concurred with the beneficial friction generating properties of cast iron grit, but expressed concerns based on the potential for scoring and excessive abrasion. Like all aspects of formulation design, raw material concentration can be critical.

Modern asbestos-free composition brake shoes can utilize different fibers for structural reinforcement, but they continue to rely on cast iron grit [17-47] as an additive. It is anticipated that cast iron grit will continue to be used for years to come. However, in light of the abrasion and hardness of these particles, its role in non-rail tribological composites is limited and growth is not expected to occur.

Sponge Iron

Sponge iron, also known as reduced iron powder, has been produced since 1872 [21-2] by one of two methods: carbon reduction and hydrogen reduction. The carbon reduction method began by combining simple mixtures of iron oxides, most notably hematite or magnetite, with carbon. The mixture was heated in the absence of oxygen until the carbon abstracted oxygen from the iron oxide and liberated gaseous carbon oxides. The loss of the gaseous oxides leaf a residual mass that, upon cooling, was a highly purified form of elemental iron, which could be used in the manufacture of steel or other ferrous metal alloys. Years later, Sieurin [21-3, 21-4] described a similar method, which refined the process by including lime. This became known as the Sieurin process [21-5].

Shimer and Shimer [21-6] described a reduction method that began by melting oxides of iron. Upon reaching the liquid phase, the oxides were reduced with a blast of heated hydrogen gas. Hydrogen reacted with oxygen present in the melt and left iron behind. It is interesting to note that a previous patent also used a similar concept of superheated gas to remove the oxygen [21-7].

A subtly different approach, using a reducing mixture of hydrogen previously combined with atmospheric oxygen, was recorded in an English patent [21-8] by Gurlt (ca. 1856). Regardless of the method used, the production of sponge iron powder is a fundamental process that has been refined throughout the years and now yields iron powders that are very soft, highly pure, and capable of being granulated to produce a powder ideally suited for friction materials.

Sponge Iron in Friction Materials

Despite the long history of manufacture, sponge iron is a relatively new entrant onto the scene of friction material development. Wellman [21-9] recorded that conventional sintered metallic friction elements were based on copper with small additions of other metallic elements such as tin, lead, and iron to the mixture. These multi-metal compositions were capable of being replaced with a mixture of graphite and iron, especially if the iron powder was a soft spongy iron powder with low apparent density.

The attributes of this particulate form of soft iron made it easily compacted into the desired shape. Its ductility provided a measure of green strength to the composite that enhanced manufacturing. These ideal characteristics were quickly realized by manufacturers of sintered metallic friction elements. However, despite the many benefits, the wear behavior of these materials was not ideal and modifications to the composition were necessary.

Ankeny [21-10] and Luther and Koehring [17-38], combined iron with conventional solid lubricants such as graphite and molybdenum disulfide; but also incorporated elemental sulfur into their compositions. The concentration of the sulfur was limited to only 1% of the iron content and was reported to enhance the wear resistance of the compound.

Weinman [21-11] also used sponge iron with 1% sulfur additions, but further refined the role of sponge iron in sintered metallic friction materials by adding an iron-aluminum alloy. The use of this specific alloy was an attempt to offset the low coefficient of friction during cold operation, a phenomenon known as cold pedal. A similar approach was also taken by Sampson [21-12] using an iron-nickel alloy.

Sponge Iron in Semi-Metallic Friction Materials

Modern semi-metallic friction materials are often based on the example composition recorded by Rhee and Kwolek [6-9]. Their resin-bonded friction element used a concentration of 20% by volume iron powder and 10% steel fiber, which was needed for structural reinforcement. These were bound in a matrix of phenolic resin and molded by conventional means. A coarse grade of sponge iron was preferred because the unique sponge morphology, specifically attributed to the minute holes in the granules, was reported to assist in the reduction of noise during braking.

Newell [21-13] described a baseline semi-metallic friction material, similar to that of Rhee and Kwolek, and replaced sponge iron with mill scale, a mixture of iron oxides that forms on the surface of hot rolled steel. By adjusting the ratio of sponge iron and mill scale in the composition, Newell obtained improved friction output, especially when the brake linings were cold. This recalls the previous cold pedal issues associated with sintered friction materials.

Sponge iron is still being used in friction materials today. Efforts to improve the friction level are generally focused on the use of alloying agents in combination with iron. These are typically carbon, sulfur, or phosphorous [21-14, 21-15]. In general, these attempts also record favorable changes in the wear properties of the friction couple.

Because sponge iron is essentially pure iron, the dominant negative characteristic associated with its use is corrosion. Compositional modifications can introduce materials that act as sacrificial anodes that will preferentially corrode. Other options include the use of materials intended to push the pH to highly alkaline ranges.

Ogiwara [21-16] described a rust preventative treatment that prevented corrosion at the bond line between the backing plate and the friction material. This coating composition was applied to the friction material exterior and inhibited the destructive de-bonding of the lining from the carrier plate because of the ingress of excessive corrosion (a process sometimes referred to as rust jacking).

Carbonyl Iron

Carbonyl iron powder is a highly pure form of iron powder produced by the decomposition of iron pentacarbonyl. The powder has the form of spherical particles that are usually in the micron size range. The use of iron carbonyl powder in friction material is limited because of the high cost of manufacture, but a few records of its use have been found.

For instance, Rossez [21-17] incorporated iron carbonyl powder in a polytetrafluoroethylene matrix to form a magnetic friction material for use in magnetic clutches. Carlson [21-18] used iron carbonyl powder as the friction material in a novel magneto-rheological brake. This device used a layer of iron carbonyl powder in a precisely controlled magnetic field to control the yield stress of the powder, thus producing induced resistance to motion as a function of an applied magnetic field. The spherical nature of the particles and their purity enable this type of brake to function repeatedly without generating excessive wear to the system. Aside from these uses, the role of carbonyl iron is an area that remains relatively unexplored in the arena of tribological composites.

Aluminum Powder

Elemental aluminum in friction materials has taken the form of a fibrous additive for replacing asbestos [6-6], as a foil or ribbon woven into the matrix and compressed to form the body of the composite [21-19, 21-20, and 16-17], or as a matrix [5-38].

Likewise, aluminum can be obtained in powder and flake form with various sizes and shapes for a number of applications in the friction materials industry. Aluminum powder has been used in modern composition brake shoes for trains [21-21], in paper-based friction materials [13-42], and in non-asbestos organic friction materials [13-21, 18-28]. Typically the concentration is limited to less than 10–15% of the composition.

As we have already seen, aluminum can also be incorporated in sintered metallic friction materials as an alloy with iron as described by Weinman [21-11], and as a part of the matrix in aluminum bronze composites. The aluminum bronze class of sintered metals was studied by Kryachek and Shvedkov [6-49, 12-30] and was found to exhibit increases in the coefficient of friction as the aluminum content increased. Increases in aluminum concentration changes the microstructure of the alloy resulting in enhanced wear resistance, especially at concentrations of about 10%.

In conventional resin-bonded friction materials, the benefits of aluminum are generally classified as a means of enhancing frictional performance at elevated temperature such as during fade conditions [6-25]. Furthermore, Satapathy and Bijwe listed aluminum as an effective ingredient for imparting recovery to a friction material after fade occurred [10-16]. These changes in friction associated with thermal conditioning are also a source of concern when using aluminum powder.

Concerns with Aluminum Powder

Bowden and Thomas [16-20] examined the surface temperature of sliding solid and reported that certain metals such as aluminum undergo an exothermic reaction during sliding. This can lead to the development of hot spots where temperatures can reach 3632°F (2000°C).

In addition to hot spot generation at the interface, which can lead to localized regions of excessive degradation, the use of aluminum in friction materials is generally not considered a favorable option because of the production of very hard abrasive oxides [21-22]. This concept was discussed previously in other chapters (i.e., Oehl and Paul [3-12]) and adds further support to this area of concern. The use of aluminum powder is therefore considered a "reactive metallic filler material" and should be used with care.

Copper Powder

Elemental copper has been a staple of the friction material industry with an extensive worldwide history of use dating back to 1913 [21-23]. Early uses of copper typically took the form of wire reinforcement in a woven friction material, which used a blend of asbestos and camel hair in this example. By 1918 [6-3], copper chips were being molded into rubber-bound brake linings for use in band brakes and shoes for drum brakes.

The use of copper in sintered metal friction materials is ubiquitous with this class of friction materials and a number of examples illustrated the manufacture of metallic friction elements based on copper [6-44]. Sintered copper materials were formed into clutch materials [21-24, 5-33] and brake linings. The fundamental characteristics of copper made it ideal for sintered friction materials. Its use as a matrix material continues to this day.

In conventional resin-bonded friction materials, copper powder can be beneficial as a heat conductor, scavenger, lubricant, or coolant because of the endothermic nature of phase changes associated with melting. These benefits, which were first pointed out by Walker [21-25], have made copper an ideal friction material additive. In addition, copper allows for a potentially infinite number of material combinations to be exercised during the research and development phase of composite development.

For instance, Schlomer [18-62] blended copper and iron together with tin to form a sintered iron friction material. Myers [6-13] described a mixture of iron, copper, tin, and lead with small amounts of phenolic resin to form a hybrid sintered material. And finally, Iwata and Asano [7-8] included small amounts of copper powder in a friction particle designed to augment the performance of a conventional non-asbestos organic friction material.

In recent years, copper has also been a subject of interest in the development of new friction material concepts. During studies of boric acid additions in phenolic composites, Mutlu [21-26] found copper to give increased friction output. In addition, finer particles of copper were shown to have decreased the wear resistance of the composite relative to larger particle sizes.

Copper additions can also benefit manufacturing by increasing the thermal conductivity of the composites and might provide a way to reduce molding times during production [16-27]. The effect on thermal conductivity is not limited to copper. Additions of brass or steel can have similar effects [16-11], but copper is a convenient agent that will affect the thermal characteristics of the composite during all aspects of its operational life.

Concerns with copper powder are typically focused on toxicological issues, resulting in restrictions on the amount of copper in friction materials. These restrictions are expected to limit the use of copper in all of its forms. Therefore, a race to find a suitable replacement in underway.

Zinc Powder

Much like copper, zinc has been continually present in the friction material industry since the earliest days of its infancy. Allen [13-9] described a compressed paper and lead foil composition useful as a friction material for wheel brakes of trains. It was noted that zinc could be an effective replacement for lead in the composition. Nanfeldt [7-2] reported his invention of sulfurized oil as a suitable binder for woven friction materials, especially those reinforced with wire, which was preferably zinc.

Zinc powder, sized less than 100 mesh (approximately 150 microns) was used by Driscoll [21-27] to coat asbestos fibers prior to weaving. The zinc powder additions at the friction surface were reported to prevent scoring, and the proposed mechanism behind the ability of zinc to prevent excessive wear of the mating members involved the deposition of a thin film of zinc on the drum surface.

Likewise, Griffin [21-28] used zinc in a sheeter friction material to reduce scoring of the brake drums, but attributed the mechanism to a scavenging action that removed hard particles from the friction interface and thus prevented excessive abrasion of the mating member. Regardless of the mechanism for preventing scoring, zinc is a beneficial additive to the arsenal of the friction material designer.

Zinc is a moderately soft material with a hardness of approximately 2.5 on the Mohs scale and a moderately high melting point of approximately 780°F (415°C). Oxidation of zinc yields a white solid with a modest increase in hardness (4.5 on the Mohs scale) that can be effective for stabilizing friction at elevated temperatures. In addition, zinc is a very effective sacrificial anode that can be added to friction linings to protect them from corrosion. Also, because of the distribution of particles throughout the matrix, the protection afforded by zinc is continually renewed because of wear [21-29] thus providing a lifetime of corrosion protection for the product.

Lead Powder

Lead and lead powder have served the industry as solid lubricants for well over 100 hundred years. In 1883, Shimer and Shimer [16-16] patented a brake shoe design that used the softness of lead and its low melting point to cast lead sections into recesses in the body of an iron brake shoe. Lead ensured adequate contact, most likely by increasing the true area of contact, and also provided lubrication that prevented "granulation of the iron shoe surfaces."

Stenberg [4-6] and Blank [21-30] stated that lead wire in woven friction materials was an effective friction stabilizing agent. In addition, Stenberg claimed that lead is a lubricant that gives the added benefit of effectively preventing scoring of brake drums. This was reportedly accomplished by preventing the formation of abrasive slag particles at the surface. This concept was shared by Blank who confirmed that lead could prevent the formation of abrasive particles at the friction interface and recorded its ability to lubricate efficiently.

Lead powder was also in use during this time. Workers at General Motors [7-22] coated asbestos fiber with a phenolic resin, and bonded lead powder to the surface of the fiber. After compaction and molding, the asbestos friction material contained a well-dispersed solid lubricant in the form of finely divided lead particles.

Lead continued to be used in a number of applications, including wet friction materials for automatic transmissions [12-25] as well as in sintered metals and semi-metallic compositions such as those described by Batchelor and Carey [19-13]. In virtually all examples, lead was cited as a solid lubricant. It was used globally until concerns regarding toxicity and environmental exposure resulted in a decline in use. Today, lead is banned from consideration by many friction material consumers because of its toxicity.

Brass Chips

Brass is an alloy of copper and zinc that can vary extensively in composition. Brass can contain small amounts of arsenic or tin, which prevents dezincification. In some cases, lead might also be present and, in fact, was typically present to enhance the machining ability of brass. Brass wire was a common reinforcement in early woven friction materials [16-22]. Brass chips were introduced into friction materials as a result of recycling woven friction materials containing brass wire [21-31] or by the introduction of swarf from brass machining processes.

Alternatively, Spokes [21-32] describes a composition brake lining comprised of asbestos and chopped brass wire in a rubber mixture. This composition was pressed into the cavity of a brake shoe and retained by using barbs as a form of mechanical attachment in lieu of rivets or similar attachments. Similarly, Headson [6-3] used a rubber friction compound for band brakes comprised mostly of asbestos, but added 2% brass shavings instead of chopped wire.

Regardless of the method of manufacture, brass additions to a tribological composite are expected to increase the thermal properties of the material [7-12, 16-11]. Additionally, brass is widely considered an effective scavenger that removes deposits from the mating member and improves friction stability, especially during the recovery from high temperature operation [6-24 and 19-38].

Jacko and Rhee [6-25] also described the scavenging effect as the breaking up of surface films and suggested that brass can affect the stability of third body layers by selectively transferring to the mating member. The mechanism for this type of selective transfer has been reported by Kragelskii [21-3] as preferential diffusion of zinc out of brass and is similar to dezincification. The end result is a residual copper-rich layer at the interface that assists in the formation of lubricating film, which can reduce the coefficient of friction [20-27].

Xian [16-35] reported brass concentrations in polymer-matrix friction materials as a significant factor in the wear and friction performance. It was noted that as the concentration of brass increased, the concentration of copper at the interface also increased. This provided wear resistance with enhancements in the friction level.

Brass is also a favored metallic filler material because it does not promote catalytic degradation of the polymer matrix—a process that can be seen with copper and iron [9-24]. However, despite the many benefits, brass is not without its concerns.

One issue is based on the bond strength of the polymer to the metallic surface of the chips. This often assumed property is necessary to prevent the loss of brass at the interface, which can lead to excessive transfer. Similar findings were also noted by Kerridge and Lancaster [16-34] relating to the transformation toward severe wear processes that accompanies metallic transfer of brass to the harder mating member.

Halberstadt [6-29] viewed the transfer process as an alloying process where brass would react with certain rotor alloys at the friction interface. The resulting alloys were found to induce metal transfer and promoted excessive two body abrasion. The end result was severe surface damage in the form of extreme scoring.

Today, many companies rely on the benefits of brass and use small concentrations of brass chips and brass fibers in their compositions [21-34]. Lead-free versions of brass are becoming more common because of concerns with lead content. Additionally, regulations regarding copper content are expected to impact the use of brass in the future.

Bronze Powder

Bronze is an alloy of copper, usually with tin, and is generally harder and more brittle than brass or copper. Additional elements can be added to produce bronze that is easily formed into wire, which was used in early folded and compressed fabric friction materials [21-35] for clutch applications. Metallic bronze has also been used as a friction material in the form of a cast bronze shoe integrally molded onto the carrier shoe [21-36] and as an *in situ* alloy developed during sintering [21-37]. This type is typically used in wet friction systems.

Additions of bronze to friction materials are commonly found as shaved bronze fiber or as powder. Powder forms can be air-atomized or water-atomized grades and can possess a spherical morphology. Bronze additions are typically present at concentrations of less than 10% and often in the presence of elemental tin and metallic sulfide lubricants.

For instance, Miyaki and Kobayashi [18-28] described a modern, non-asbestos organic friction material with only 3% spherical bronze powder. Likewise, Suzuki and Hattori [17-58, 18-56] described the importance of bronze additions and incorporated bronze, in both fiber and powder forms, with additions of tin powder and a metallic sulfide lubricant. This unique alloy of bronze was specified to contain at least 8% tin and might be effective for reducing metal pickup. However, excessive amounts of tin can diminish the friction level of the product. A similar claim was also reported by Yamamoto [17-44].

Bronze additions are expected to provide many benefits of brass and copper. However, they but also face similar concerns with lead content and can experience restrictions because of the presence of copper in the alloy. Despite these concerns, bronze is an area of exciting possibility and more work is needed to fully appreciate its significance.

References

21-1. Walsh, P., and R. Spurr. 1979. "Design of Composite Friction Materials for Railway Brakes," Proceedings of the Institution of Mechanical Engineers, Railway Div. C161, pp.137-141.

21-2. Blair, T. 1872. Improvements in iron sponge. United States Patent No. 126924, issued May 21, 1872.

21-3. Sieurin, S. 1913. Reducing ore by means of solid carbonaceous material. United States Patent No. 1054873, filed April 1, 1911, and issued March 4, 1913.

21-4. Sieurin, S. 1929. Process for the production of iron sponge. United States Patent No. 1714280, filed December 17, 1927, and issued May 21, 1929.

21-5. Comyns, A. 2007, *Encyclopedic Dictionary of Named Processes in Chemical Technology*, 3rd ed. Boca Raton, FL: CRC Press.

21-6. Shimer, G., and S. Shimer. 1875. Improvement in processes and furnaces for manufacturing iron. United States Patent No. 165375, filed March 10, 1875, and issued July 6, 1875.

21-7. Smith, G. 1869. Improvement in the manufacture of wrought-iron and steel direct from the ore. United States Patent No. 92894, issued July 20, 1869.

21-8. Gurlt, A. 1856. Improvements in the manufacture of iron and steel. British Patent No. 1679, issued July 16, 1856.

21-9. Wellman, S. 1958. Friction assembly. United States Patent No. 2863211, filed November 29, 1955, and issued December 9, 1958.

21-10. Ankeny, T. et al. 1960. Friction material. United States Patent No. 2945291, filed November 28, 1958, and issued July 19, 1960.

21-11. Weinman, E. 1963. Friction material. United States Patent No. 3101527, filed September 2, 1960, and issued August 27, 1963.

21-12. Sampson, F. 1962. Friction material. United States Patent No. 3067493, filed July 27, 1959, and issued December 11, 1962.

21-13. Newell, C. 1989. Friction material containing mill scale. United States Patent No. 4855336, filed July 25, 1988, and issued August 8, 1989.

21-14. Ogiwara, O. 1981. Friction material comprising an iron powder having a carbon content of from 0.5-1 percent. United States Patent No. 4280935, filed August 31, 1979, and issued July 28, 1981.

21-15. Nakagawa, M., and F. Nitto. 1987. Friction material using iron powder. United States Patent No. 4665108, filed November 8, 1985, and issued May 12, 1987.

21-16. Ogiwara, O. 1985. Semi-metallic friction pad subjected to rust preventative treatment. United States Patent No. 4505963, filed April 12, 1983, and issued March 19, 1985.

21-17. Rossez, J. 1965. Magnetic clutch with a magnetic lining. United States Patent No. 3177998, filed January 15, 1963, and issued April 13, 1965.

21-18. Carlson, J. et al. 1998. Controllable brake. United States Patent No. 5842547, filed July 2, 1996, and issued December 1, 1998.

21-19. Crane, J. 1927, Brake. United States Patent No. 1637894, filed November 5, 1925, and issued August 2, 1927.

21-20. Leonard, J. 1930. Friction metal article. United States Patent No. 1745173, filed April 5, 1929, and issued January 28, 1930.

21-21. Ueda, Y. et al. 1966. Brake lining from graphite and metal particles in a thermosetting binder. United States Patent No. 3269976, filed October 11, 1965, and issued August 30, 1966.

21-22. Scieszka, S. 1980. "Tribological Phenomena in Steel-composite Brake Material Friction Pairs." *Wear* 64: 367-378.

21-23. Société anon, d'électricité et d'automobiles Delaunay-Belleville. 1913. Improvements in friction brakes. Great Britain Patent No. 191311456, filed August 1, 1912, and issued September 18, 1913.

21-24. Palm, J. 1939. Clutch. United States Patent No. 2163152, filed February 18, 1937, and issued June 20, 1939.

21-25. Walker, F. 1945. Friction material. United States Patent No. 2369502, filed March 7, 1940, and issued February 13, 1945.

21-26. Mutlu, I. et al. 2007. "Wear Performance of Some Phenolic Composites with Boric Acid." *Industrial Lubrication Tribology* 59: 38-45.

21-27. Driscoll, J. 1931. Friction element and method of making same. United States Patent No. 1812326, filed February 21, 1930, and issued June 30, 1931.

21-28. Griffin, L. 1937. Friction material and method of making the same. United States Patent No. 2087283, filed April 21, 1934, and issued July 20, 1937.

21-29. Hell, M. et al. 2002. Friction lining, especially for brakes and clutches, and a method for producing a friction lining. United States Patent No. 6481555, filed March 21, 2001, and issued November 19, 2002.

21-30. Blank, H. 1938. "Brake Linings." *Automotive Industries,* September 17, 1938, pp. 346-353.

21-31. Tseng, A. 1933. Molded brake lining. United States Patent No. 1896335, filed July 15, 1929, and issued February 7, 1933.

21-32. Spokes, R. 1917. Composition lining for brake shoes. United States Patent No. 1242069, filed November 18, 1915, and issued October 2, 1917.

21-33. Kragelskii, I. et al. 1978. "The Mechanism of the Initial Stage of Selective Transfer during Friction Contact." *Wear* 47: 133-138.

21-34. Kikuchi, K. et al. 2009.Friction modifier, method of producing friction modifier, and friction material. United States Patent Application No. 2009/0270284, filed April 27, 2009.

21-35. Slade, E. 1929.Clutch disk and method of making the same. United States Patent No. 1715105, filed May 9, 1923, and issued May 28, 1929.

21-36. Peterson, J. 1933. Brake shoe. United States Patent No. 1898310, filed August 2, 1930, and issued February 21, 1933.

21-37. Wellman, S. 1949. Improvements in sintered friction material. Great Britain Patent No. 628110, filed December 10, 1945, and issued August 23, 1949.

Chapter 22
Testing and Evaluation of Friction Materials

In every machine power is lost through friction; and in every man, the noblest and the purest, there is resistance to be overcome ...

—Alexander Maclaren

BECAUSE OF THE CRITICAL NEED FOR SAFE and effective operation under diverse operating conditions, friction materials have always been subjected to extensive empirical testing. Test methods commonly include the determination of the coefficient of sliding friction at various speeds and temperatures, the rate of wear, and many aspects for characterizing mechanical strength. In all cases, testing is focused on specific measurable attributes that have been found to be important to the end user of the composite.

An example of an early friction material test stand was constructed by Herbert Frood. His simple device was powered by waterwheel, built in his garden shed [22-1], and provided a convenient source for generating energy that allowed comparison of his early woven friction compounds under similar conditions. Using this data, he was able to screen materials and select those with superior performance.

From these humble beginnings, the art of friction material testing has evolved into a complex field, encompassing techniques that range from simple dimensional measurements to complex analytical and chemical methods. In an effort to limit the scope to a reasonable length, the fundamental methods of interest are limited to physical test methods, select chemical analysis techniques, and basic approaches for evaluating tribological behavior.

Physical Test Methods

Physical test methods for comparing tribological composites are similar to those employed in the examination of other composite materials. McLellan [22-2] provided a comprehensive list of methods for testing physical characteristics of brake linings that included ultimate tensile strength, assembly shear strength, compressibility, specific gravity, hardness, dimensional stability, and stiction.

Ultimate Tensile Strength

Ultimate tensile strength is the maximum stress that a material can withstand prior to failure under tension. In ductile materials, failure is usually accompanied by a change in the cross sectional area (e.g., necking), but in brittle materials failure occurs suddenly with little or no change in the cross section of the sample. The failure mode in friction materials will vary based on the type of material.

For example, a rubberized fabric material will experience a different mode of failure than a rigid phenolic resin-bonded composite. A sintered metallic friction material can exhibit limited ductile properties prior to failure and differ from both rubberized fabric or molded materials. Traditionally, the methods most commonly used are plastic composite methods such as the ASTM D638-10 [22-3] or ASTM D882-10 [22-4]. Many companies often modify these methods to meet their own individual needs, but results will typically fall within a broad region of compliance.

Because many commercial friction materials are polymeric composites, variations in the structure because of the occurrence of flaws, voids, and errors in bonding, as well as the potential for crack nucleation at stress centers, results in an increased potential for error in measurement of the ultimate tensile strength. The statistical nature of the measure of a distribution of flaws is known as the Weibull modulus. This is a dimensionless number corresponding to the variability in measured strength. It reflects the distribution of flaws in the material, especially in brittle glass-like materials [22-5, 22-6]. This requires investigators to perform multiple tests using identical samples to fully quantify the effects of a given additive on the ultimate tensile strength of the final product.

Assembly Shear Strength

Assembly shear strength is a measure of the structural integrity of the brake lining after it is attached to its carrier plate. One common method uses a rigid fixture to hold the brake assembly in place while a blade is forced against the friction material. The load is increased until separation occurs between the friction composite and carrier plate. Typically this method involves a variation of the SAE J840 procedure [22-7].

Recently a comprehensive examination of the procedure was conducted by a multi-company team [22-8] using a two-level, four-factor design of experiments. From this testing, the generalized conclusions were modifications to the original procedure that incorporated a normal force during the application of a shear load and a change in the placement of the blade. Ideally, the shear blade and the carrier plate should have a nominal separation distance of 0.0394 inches (1mm). In addition; it was recommended that a minimum of five samples be tested from each population.

Compressibility

The compressibility of a brake lining or a clutch facing allows the lining to conform to the mating surface more readily under a given application force. This improves engagement "feel" and can have an impact on the generation of noise, especially in brake linings. Two common methods for determining the compressibility of a brake lining are the Gogan Hardness test and an assembly compression test.

Gogan Hardness Test

In 1932, Gogan [22-9] filed a patent for a modified hardness test using a circular flat penetrator under a prescribed load. Based on the depth of penetration, the hardness of the material could be determined. This method has been used by friction material manufacturers as a quality control method since 1933 [8-19] and, according to Nicholson [5-1], is not a hardness test per se, but "really a compression test."

The test is sensitive to the thickness of the sample. Thicker materials will appear to be softer than thinner sections of the same composition. Typically, friction materials will have a Gogan Hardness value of 15–40 where a material with a value of 15 is harder (less penetration) than a material with a value of 40.

SAE International has a published procedure, SAE J379 [22-10] that provides a comprehensive description of the method as well as the four common Gogan scales. The procedure is not as widely used as in the past, but is still preferred by industrial material suppliers.

Assembly Compression Test

Because of concerns with excessive compressibility in a friction material and consumption of fluid reserves in the brake system, brake linings (especially disc brake pads) are tested using a sophisticated compression test machine, which operates in accordance with SAE 2468 [22-11]. This test applies a fixed load onto the carrier plate of the brake lining using a piston of the correct size and measures the deformation to within a few microns. This allows friction material designers and vehicle systems engineers to specify precisely the necessary compression allowance that brake linings might have during use.

Furthermore, the samples can also be heated to measure the compressibility changes as a function of temperature. Temperatures as high as 752°F (400°C) are sometimes used to examine the necessary fluid requirements in a vehicle operating under worst case fade conditions. Likewise, the test can also be performed under different application forces, allowing the designers to determine the compression values as a response of both temperature and pressure.

Specific Gravity

The specific gravity of a brake lining composition is a measure of its density relative to water using the Archimedes method of water displacement. In its simplest form, the sample is weighed in air and then suspended in water. The relative density of the sample can be determined from the amount of water displaced. The procedure was adopted by SAE International as SAE J380 [22-12] and provides a comprehensive description of the method.

Because each material is unique in composition and the specific gravity of a product can be altered by the process of manufacture, each organization must determine the acceptable specific gravity range for each material.

Hardness

Hardness is one of the most important parameters in the determination of abrasion resistance [3-1] and is a common metric that can be easily tested in a number of ways including scleroscopic methods [18-31]. However, the most common methods generally include hardness by indention such as Rockwell hardness. Typically, friction material producers use the superficial Rockwell scales such as L, M, R, and S to describe their materials.

In the past, common measurement techniques included the ASTM B347 [22-13] procedure specifically for testing of sintered metallic friction material, but this procedure has been withdrawn and is no longer currently active as a standard method.

One of the most recent standards is SAE J2654 [22-14], which uses a Rockwell method to determine the hardness of the brake lining compositions. There are several factors that affect brake lining hardness and compositional effects. For instance, an increase in the concentration of hard ceramic particles has been shown to dramatically affect the hardness of the final product [22-15] and thus can have an impact on both noise and wear performance.

Additionally, process conditions (most notably baking of the products) can be an effective technique for changing the hardness of the composite [22-16, 21-26]. Although, hardness is not typically correlated with frictional output [22-17] for a given friction material, hardness has been found to affect the wear rates of friction materials, which in some cases resulted in improvements of nearly 200 times better wear resistance [22-18].

When evaluating friction materials, the heterogeneous nature of the composite requires a number of locations to be tested per lining in order to gain an indication of statistical certainty. If the linings are heavily loaded with rubber compounds, or are based on an elastomeric matrix, the methods described by Scott [22-19] are considered an excellent method for determining basic physical attributes of the materials. Those methods should be followed when possible.

Finally, hardness testing is not always a simple task. As noted, compositional effects and processing changes introduce wide ranges of variation that can convolute the data and make hardness testing challenging. In addition, each test method should be precisely controlled because of environmental influences, including humidity [22-20], which has been shown to affect hardness of disc brake friction linings.

Dimensional Stability

Dimensional stability in brake linings has been described in detail by Joshi [22-21] and also by Nicholson [5-1]. It compares measurements taken at ambient condition to those taken from the same locations on a heated brake lining. Any dimensional changes that occur upon heating can be quantified as either swell or shrinkage of a composite. The sample is measured again upon cooling, revealing permanent dimensional changes that occurred during heating, which can most often be seen as growth (expansion) in the thickness of the lining.

The dimensional stability of a lining is important because of its potential for influencing wear measurements. Bush [17-52] found that brake linings had a typical wear value of only 2×10^{-5} inches (5×10^{-5} cm) per brake application. This small amount of wear is difficult to measure with conventional devices and requires several hundred stops to generate a wear value large enough to measure accurately. In addition, the presence of shrinkage or swell during use will only offset the true wear of the lining, making precise quantification difficult. The current method of measuring the swell and growth of a brake lining is found in SAE J160 [22-22].

Stiction

Stiction is a phenomenon where adhesive bonds are formed between the lining and the mating member. The most typical explanation for this effect is corrosion-induced bonding between the surfaces of the friction lining and the metallic mating member. At this time, the industry does not have a standard method for evaluation. However, a basic method would typically involve saturating a sample of lining with water or saline solution, and clamping it under a given load to the mating member (most often cast iron). After a defined period of time (usually approximately twenty-four hours) the force necessary to break the adhesive bond is measured. Despite the severity of this failure mode, a solution was provided by Shibata [10-19] whereby the addition of a pH augmenting additive to the friction composition can elevate the pH into an alkaline range, which inhibits the formation of corrosion at the interface and prevents stiction from occurring.

Chemical Test Methods

Chemical test methods include a wealth of possible analysis techniques ranging from simple solvent extraction, which is used to evaluate the state of cure of the phenolic resin, to vastly complex methods of elemental analysis such as x-ray fluorescence and x-ray diffraction. Many of the more advanced techniques are often too elaborate for effective quality control purposes and therefore are not widely used in day-to-day manufacturing environments.

A few of the fundamental methods include acetone extraction, pH, moisture, and ash content. In addition, there are a number of specific tests, such as iodine number used for classifying drying oils, or methods for characterizing phenolic resins by inclined plate flow or hot plate cure time. Generally, these methods are best developed from industry partners and will allow for direct lab-to-lab comparison of the raw materials.

Acetone Extraction

Perhaps the most common test is acetone extraction. This method seeks to determine the state of cure of a lining. This method has been used for more than 45 years [22-23] and is still in use in modern friction material laboratories [22-24]. It has been standardized [22-25] and is used to ensure that the final friction linings meet pre-established standards. The test is destructive but can be completed in only a few hours, allowing for a convenient method of process control.

pH

The acidic or basic nature of raw materials in a friction material can have profound effects on the final product. Acidic ingredients can accelerate the cure of the resins and result in defective products. Mildly acidic ingredients might not create issues during production but can lead to elevated degrees of surface corrosion when coupled with a cast iron mating member, which can lead to stiction or judder.

Likewise, basic (alkaline) ingredients are typically incorporated to inhibit corrosion but can retard the cure of resins and also generate defective parts. Highly reactive alkaline materials can also pull moisture from the environment and undergo chemical reactions if the mixtures are not contained and stored properly. Because pH is also a measure of purity, changes in basic raw materials should be monitored closely.

Generally, the most common technique for determining the pH of a friction material composite involves placing a sample in water and checking the change in pH over time. Many methods have been developed by vehicle manufacturers who provide specific guidelines for sample preparation and for reporting the pH values for a given friction material.

Moisture Content

Moisture can induce pH changes in raw materials, specifically metallic sulfides [18-40, 18-42], and can also affect the curing profile of the binder, especially phenolic resins. The effect on phenolic resin can retard or accelerate the resins; depending on the composite and the nature of other ingredients present [22-26, 22-27]. Because of the number of possible effects associated with moisture contamination, many companies have developed an analysis technique that meets their expectations. In general, these methods involve the determination of the moisture by difference where the mass of the sample is measured and after heating (usually to the boiling point of water) the final mass is taken. The difference in mass is attributed to moisture being driven off.

Ash Content

The ash content is a measure of the non-organic constituents present in a raw material and is typically used only for evaluation of organic compounds. The term "loss on ignition" can be used to describe this test method, which involves heating the raw material in a furnace to temperatures of 1472°F (800°C) and higher.

Heating is usually performed in the presence of oxygen. All organic compounds are destroyed, leaving behind a heavily oxidized residue that can be further analyzed by elemental methods to determine the inorganic fraction present in the lining. Using a simple mass difference approach, the amount of organic material can also be determined. Typically, these methods are shared between suppliers and producers to ensure consistent quality control and reporting of data.

TGA and DSC

A simple and convenient tool for both moisture and ash content is thermo-gravimetric analysis (TGA). This test uses a computer controlled system to heat the sample under precise conditions to provide changes in weight as a function of temperature. The systems can operated in different atmospheres and are highly flexible; however, the sample size is quiet small and requires a number of samples to be tested to ensure some measure of statistical significance. This method is widely used for preliminary testing of virtually all organic constituents including organic fillers, fibers, and binder materials.

In addition to TGA, another commonly used method for examining the cure rates of binders and reactive additives is differential scanning calorimetry (DSC). This method compares the heat rates of two samples, a control and a specimen, to determine the differential heat flow. From this information, the analyst can provide an indication of cure rates as well as temperatures for the onset of cure and thermal degradation. Both TGA and DSC are highly effective but require a number of samples and great deal of time in which to develop a solid understanding of the properties of the materials and systems. Therefore, these are best limited to areas of product development and are too time consuming for effective in-process quality control measures.

Tribological Test Methods

Friction and wear processes are affected by a number of variables including pressure, temperature, velocity, lubricants, and the composition of the mating members. To examine a tribosystem fully, these factors must be controlled within the limits of operational consideration. Therefore, tribological test methods are usually designed to simulate actual operating conditions of a given automotive system such as disc brakes or automatic transmissions.

The most common methods include simple tribometers (such as a pin-on-disc) or sample dynamometers. Dry friction material suppliers typically use small sample machines like the Chase machine or FAST machine to examine a new formulation quickly. Further testing usually includes single ended inertia dynamometers or drag dynamometers such as the Krauss machine. Additionally, test equipment can include dual-ended systems and even complex four-wheel inertia systems have all been used to simulate braking of an automobile. For wet friction systems, industry standard machines include the SAE #1 and SAE #2, but many others are also available. Finally, all of these methods are generally followed up with extensive vehicle and component testing, as in the case of automotive transmissions.

Chase Machine

The Chase machine was developed by Chase while working at General Motors [22-28] and is also known as the friction material test machine (FMTM). This test machine uses a small sample of friction lining against a brake drum with a precisely defined composition to determine basic characteristics of friction and wear.

Burkman and Highley [22-29] describe the use of a Chase machine during the evaluation of brake lining materials with several procedures specific to General Motors. These procedures examined fade and recovery performance, friction level under repeated drag conditions, and also the stability of friction over time. Muzechuk [22-30] reports a modification to the Chase machine using a scissor adapter and small friction samples along with a disc to evaluate early disc brake friction compositions.

Today, the Chase machine is used by many industrial friction material suppliers for comparison between friction material compositions; however, these results are not easily correlated to existing devices. For instance, Fox and Alvi [22-31] developed a hoist dynamometer to evaluate full-scale band brake materials for cranes because of a lack of correlation with the data obtained by the Chase machine. Likewise, automotive friction materials developers have also decreased use of this machine because of its lack of correlation with modern automotive braking systems.

One area where the Chase machine still finds use is for providing an edge code for friction output of commercial automotive brake linings. This process was established in 1967 by the American Association of Motor Vehicle Administrators (AAMVA) and was continued by the Automotive Manufacturers Equipment Compliance Agency, Inc. (AMECA) [22-32]. This organization provides a means for indentifying the manufacturer and rated friction output of a brake lining using the SAE J661a procedure [22-33]. Most North American friction manufacturers today strive to have a friction lining with a nominal coefficient of between 0.35–0.45 when using this procedure; however, many European formulations are often targeted to have a coefficient of 0.45–0.55. There are a number of exceptions to this general rule of thumb.

Additionally, the Chase machine is still an effective means for quickly examining friction performance between two compositions if provision is made for the testing of a suitable number of samples. McCuen [22-34] stated nearly fifty years ago that the Chase machine was to be used as a preliminary testing method to evaluate performance and it still remains an effective means for quickly examining lower cost raw materials.

Modern versions of this device are computer controlled with software that is more easily programmable. These advancements are coupled with mechanical improvements that allow Chase machines to operate at higher pressures and faster velocities, and under higher temperatures than the original designs. Each of these factors can be adjusted and varied to create a number of novel test methods for evaluating friction and wear performance at levels more intensive than previously described.

FAST Machine

The Friction Assessment Screening Test (FAST) machine was developed by workers at Ford Motor Company and has been thoroughly described by Anderson [22-35]. The FAST machine uses a small sample of friction material operating against a flat surface and is essentially a pin-on-flat type friction test. The device can be run in constant output mode by varying operational parameters that try to keep the friction level constant, or it can operate using constant load by adjusting input parameters. Tests are typically 90 minutes in length and perform the same amount of work per specimen allowing direct comparisons to be drawn from the data. For these reasons, Ford Motor Company made the test mandatory for all Ford passenger cars beginning with the 1967 model year [22-36].

Additionally, the FAST test machine has been widely used in academic studies of friction materials to evaluate the effects of raw material changes. Weintraub [22-37] reported a correlation between the degree of cure with phenolic resins and the initial friction levels of resin-bonded friction materials. In this study, higher cure states gave higher initial friction levels than compositions that were not fully cured.

Lu [22-38, 22-39] used a combinatorial approach and examined 23 different raw materials using the FAST machine. The data was subsequently used to optimize a commercial brake pad formulation for automotive use. Tang [22-40], along a similar line of reasoning, examined a number of raw materials using relational analysis methods and an approach to determine the effects of the raw material candidates in an industrial friction material composition, which also allowed the composition to be further refined.

Chapman [22-41] explored the wear resistance of aluminum-boron carbide cermets (i.e., ceramic-metallic composition) for potential application as automotive brake components. It was concluded that aluminum-boron carbide cermets exhibited an order of magnitude better wear resistance and did not fade as readily as conventional metallic mating members.

Much like the Chase machine, the FAST machine remains a favorite among academic researchers but is not widely used in the development of automotive friction materials. Correlation between the small scale device and actual vehicles is also difficult, but the FAST machine will most likely continue to be used for small scale quality control and raw material evaluation.

Sub-Scale Dynamometers

Sub-scale dynamometers are shrunken versions of full size test machines and rely on proper scaling to ensure that the sample of friction material experiences the same unit loading of pressure and operates under the same kinetic energy profiles.

Rusnak [22-42] examined several friction material test methods in 1970. At this time, the tests could be categorized into research test methods, developmental test methods, qualification test methods, and quality control verification test methods. Research methods were fundamental tests for elucidating basic aspects of friction and wear relative to materials. Developmental tests were examinations of friction couples under precise conditions to examine the effect of alloys or alternative friction materials. Qualification methods were designed to validate the product for commercial use. Quality control methods were to ensure compliance with previously established validation results.

During Rusnak's examination, the Girling sample dynamometer was compared to the FAST machine and to the Chase machine. The Girling sample dynamometer was found to have the disadvantage of a small rotor and small sample, which was affected by the heterogeneity in the composition, but was found to correlate more readily to an actual vehicle than either the FAST or Chase machines. A similar study performed by Preston and Forthofer [22-43] expressed similar concerns regarding the size of the components but also found the Girling dyno to compare reasonably well to a six-passenger sedan equipped with four piston caliper front disc brakes and duo-servo drum brakes.

The Krauss Machine

The Krauss machine was developed by Erick Krauss and is predominately used in Europe [22-36]. This machine uses full size components to test both disc and drum versions of automotive brakes. Originally this machine was a constant input machine, but newer versions have been modified and can operate under constant output mode. This allows for a number of different tests to be performed and is finding more widespread use in quality control verification as well as for the development of materials in industrial and academic circles.

Inertia Dynamometers

Inertia dynamometers are very expensive machines that can run full vehicle corner assemblies, in some cases including full wheel and tire assemblies complete with suspension components [22-44]. Inertia dynamometers rely on stored energy in a rotating flywheel to provide the required kinetic energy. Simplified versions were in use as early as 1935 [22-45]. The basic operation consists of accelerating a flywheel to a known speed and disengaging the drive source. The brake is applied immediately, and the kinetic energy stored in the flywheel is dissipated by friction. These tests can be controlled by braking speed, rate of deceleration, initial braking temperature, constant torque output, or constant pressure input. Modern brake dynamometers can control a number of factors simultaneously and are highly effective for evaluating both friction and wear performance of new compositions.

The number of procedures is extensive and includes methods for friction assessment, noise generation, wear resistance, and simulations under controlled environments. Procedures are often written based on data obtained from vehicle test methods, such as a Detroit City Traffic test [22-46], or as simulations of driving conditions, such as those found in the Los Angeles metropolitan area [22-47]. These methods rely on the repeatability and control of an inertia dynamometer to evaluate friction materials precisely for these rigorous applications and take out some of the error introduced by environmental and operator factors.

Inertia dynamometers can also be produced as dual-ended machines, as described by Preston and Forthofer [22-43], that allow for a partial brake system to be evaluated. This method is the current favorite for providing direct correlation to a vehicle and was widely used in the validation of materials prior to beginning vehicle test programs. Howard and Winge [22-48] took this process to its ultimate conclusion and computerized four single-ended dynamometers together to form a quadruple dynamometer for simulation of the entire vehicle.

Today, friction material producers rely on the control and repeatability of inertia dynamometers to ensure consistent friction and wear performance over the widest possible range of operation. Smaller companies and academic researchers can also utilize these systems by contacting local testing houses. However, the cost of these methods can easily exceed $10,000 for a single wear test [22-49], especially those that span the course of several days. Therefore, many companies prefer to rely on lower cost evaluation methods and use inertia dynamometers during the final phases of product validation.

Vehicle Testing

Vehicle testing methods are the final examination of the performance of a friction material. Testing of this magnitude takes several days or even weeks to complete, and requires a full-time staff of mechanics, drivers, and engineers. In addition, various environmental effects can delay the testing, thus amplifying the costs of a single test.

The number of vehicle test methods varies extensively and can range from simple city traffic procedures to extensive trials of durability. For example, McCuen [22-34] records that General Motors embarked on an extensive program beginning in October 1957 that examined brake lining performance. After twelve months of testing, General Motors had accumulated more than 1.68 million miles of travel (approximately 2.7 million kilometers). Other examples include a 7300 mile (nearly 12,000 km) by Radlinski [22-50] that encompassed a route covering most of the western half of the United States.

Workers in Europe have also conducted similar feats of testing endurance. Mackenzie [22-51] described a European proving trial that involved 23 drivers and a series of cars of various sizes to evaluate the wear rates of commercial friction materials under real world driving conditions. This test covered 3000 miles (almost 5000 km) per car and gives an extensive discussion of the local terrain and temperature profiles of the linings during use.

Vehicle testing is becoming more expensive and the need to find a lower cost option is always a priority. However, because of the immense number of operating variables, vehicle testing is the final and most decisive of all tests. Failure under this level of operation is simply not an option.

Wet Friction Material Testing

Wet friction materials operate in fluids and are also subject to the same concerns as dry friction. In this case, the system operates in a modified fluid with a number of additives, lubricants, and the like. Despite the differences, the need for effective testing and qualification is also very critical.

Often the test methods are based on the SAE # 1 and SAE # 2 machines. These machines have been well described by Smith [22-52] and are considered an effective means for testing friction materials for automotive powertrains. Modifications to the machinery are numerous and a host of different methods are also commonplace. One example by Watts and Sparrow [22-53] described a modification to the SAE # 2 machines that allows for the testing of wet band brakes used in automotive transmission systems.

In manual transmissions, the friction couples are usually coated cone-shaped rings that engage on the inside or outside surfaces of mating cones. Shift quality is a strong performance trait in these systems and requires friction material producers to find methods for evaluating the friction and wear, but also the subjective feel of the engagement. One novel method has been described by Seto [22-54] using an automotive transmission under precise computer control.

Like brake dynamometers, there are a number of smaller devices that have been built and used in the evaluation of wet friction materials. One early system by Olson [22-55] was specifically designed for evaluating wet friction systems of large earthmoving devices. This device, known as the FACT machine (i.e., friction assurance control test) allowed for testing various materials and fluids with precise control to determine the desired performance aspects of new combinations.

As systems have advanced, research continues and new systems are constantly emerging by leaders in the wet friction materials industry. Many of these systems are evaluated with a number of fluids to determine the effect of the material versus the effect of the fluid [22-56]. Growth and advancement in this area will continue for many years to come and is expected to reveal a number of new and interesting lubricant chemistries as well as compositional modifications to the friction facings.

Final Words on Testing

Regardless of the material, the liquid, or the system, friction is highly dependent on the surface structure of the mating bodies. The very action of sliding induces changes in the structure and friction varies throughout the test. Variation is always present.

This is critically important when testing small samples where the local variation in surface texture can affect the total friction force and dominate other effects. In addition, the number of variables involved in testing a given material pair across the widest possible range of operation necessitates that friction material producers must develop large test plans that encompass as many of these aspects as possible to be successful [22-57].

References

22-1. Technical Staff of Ferodo, Inc. "Friction Materials: A Combination of Properties." *Ferodo International Technical News*, F13 (Publication date unknown).

22-2. McLellan, R. 1988. "Requirements of Friction Materials in Commercial Vehicle Disc Brake Applications." In Proceedings of the International Conference on Disc Brakes for Commercial Vehicles, 1–2 November. London: The Institution of Mechanical Engineers pp. 9-13.

22-3. ASTM International. 2010. "Standard Test Method for Tensile Properties of Plastics." ASTM D638 – 10. West Conshohocken, PA: ASTM International.

22-4. ASTM International. 2010. "Standard Test Method for Tensile Properties of Thin Plastic Sheeting." ASTM D882 – 10. West Conshohocken, PA: ASTM International.

22-5. Petrovic, J., and G. Hurley. 1990. "Vapor-liquid-solid (VLS) SiC Whiskers: Synthesis and Mechanical Properties." In *Fiber Reinforced Ceramic Composites: Materials, Processing and Technology*, edited by K. Mazdiyasni. New Jersey: Noyes Publications.

22-6. Klein, C. A. 2009. "Characteristic Strength, Weibull Modulus, and Failure Probability of Fused Silica Glass." *Optical Engineering* 48: 2009.

22-7. SAE International. 2007. "Test Procedures for Shear Strength of Automotive Brake Pads and Brake Lining Assemblies." SAE J840. Warrendale, PA: SAE International.

22-8. Basch, R. et al. 2006. "Evaluation of SAE J840 Lining Shear Test Procedure." SAE Paper No. 2006-01-3191. Warrendale, PA: SAE International.

22-9. Gogan, J. 1934. Strength of material testing device." United States Patent No. 1978302, filed June 1, 1932, and issued October 23, 1934.

22-10. SAE International. 2009. "Gogan Hardness of Brake Lining." SAE J379. Warrendale, PA: SAE International.

22-11. SAE International. 2006. "Road Vehicles - Brake Linings - Compressibility Test Procedure." SAE J2468. Warrendale, PA: SAE International.

22-12. SAE International. 2009. "Specific Gravity of Friction Material." SAE J380. Warrendale, PA: SAE International.

22-13. ASTM International. 1900. "Test Method for Hardness of Sintered Metal Friction Materials." ASTM B347. West Conshohocken, PA: ASTM International (withdrawn 1991).

22-14. SAE International. 2004. """Hardness of Brake Lining." SAE J2654. Warrendale, PA: SAE International.

22-15. Cerit, A. et al. 2008. "Effect of Reinforcement Particle Size and Volume Fraction on Wear Behavior of Metal Matrix Composites." *Tribology Industry* 30: 31-36.

22-16. Baumgardt, W. 1991. "Post-baking of Phenolics." In *Best of Thermosets*, vol 1. Newton, CT: Society of Plastics Engineers.

22-17. Green, A. 1955. "Friction Between Unlubricated Metals - A Theoretical Analysis of the Junction Model," Proceedings of the Royal Society of London, A 228, pp.191-204.

22-18. Axén, N., and S. Jacobson. 1994. "A Model for the Abrasive Wear Resistance Of Multiphase Materials." *Wear* 174: 187-199.

22-19. Scott, J. 1955. "Rubber Hardness Testing." *Rubber Chemistry and Technology* 28: 1071-1081.

22-20. Kobayashi, M., and N. Odani. 1997. "Study of Stabilization Friction Coefficient of Disc Brake Pads in Cold Conditions." SAE Paper No. 973030. Warrendale, PA: Society of Automotive Engineers.

22-21. Joshi, M. 1980. "Disc Brake Linings - Analytical Study and Selection Criteria." SAE Paper No. 800782. Warrendale, PA: Society of Automotive Engineers.

22-22. SAE International. 2007. "Dimensional Stability of Friction Materials When Exposed to Elevated Temperatures." SAE J160. Warrendale, PA: SAE International (Method under review).

22-23. Russell, G. 1975. "The Manufacture of Disc Brake Linings." SAE Paper No. 750228. Warrendale, PA: Society of Automotive Engineers.

22-24. Bijwe, J. et al. 2008. "Optimization of Brass Contents for Best Combination of Tribo-performance and Thermal Conductivity of Non-asbestos Organic (NAO) Friction Composites." *Wear* 265: 699-712.

22-25. ASTM International. 2004. "Standard Test Method for Acetone Extraction of Phenolic Molded or Laminated Products." ASTM D494 – 04. West Conshohocken, PA: ASTM International.

22-26. Park, B. D. et al. 1999. "Differential Scanning Calorimetry of Phenol-Formaldehyde Resins Cure-Accelerated by Carbonates." *Polymer* 40: 1689-1699.

22-27. Geimer, R., and A. Christiansen. 1996. "Critical Values in the Rapid Cure and Bonding of Phenolic Resins." *Forest Products Journal* 46: 67-72.

22-28. ASME Expert Panel on Alternatives to Asbestos in Brakes. 1988. "Analysis of the Feasibility of Replacing Asbestos in Automobile and Truck Brakes." New York: American Society of Mechanical Engineers.

22-29. Burkman, A., and F. Highley. 1967. "Laboratory Evaluation of Brake Lining Materials." SAE Paper No. 670510. Warrendale, PA: Society of Automotive Engineers.

22-30. Muzechuk, R. 1967. "New Devices to Assist in Friction Material Development." SAE Paper No. 670512. Warrendale, PA: Society of Automotive Engineers.

22-31. Fox, J., and L. Alvi. 1972. "The Hoist Dynamometer - An Instrument for Full Scale Evaluation of Band Brake Materials." SAE Paper No. 720786. Warrendale, PA: Society of Automotive Engineers.

22-32. Ameca. "About Ameca." Accessed October 19, 2010. http://www.ameca.org.

22-33. Society of Automotive Engineers. 1997. "Brake Lining Quality Control Procedure." SAE J661a. Warrendale, PA: Society of Automotive Engineers.

22-34. McCuen, N. 1959. "What's New in Brake Linings." SAE Paper No. 590212. Warrendale, PA: Society of Automotive Engineers.

22-35. Anderson, A. et al. 1967. "A New Laboratory Friction and Wear Test for the Characterization Of Brake Linings." SAE Technical Paper No. 670079. Warrendale, PA: Society of Automotive Engineers.

22-36. Rhee, S., and H. Schwartz. 1979. "Test Methods for Automotive and Truck Friction Materials." In *Wear Test for Plastics: Selection and Use*, edited by R. Bayer. ASTM STP 701, pp. 39-49. West Conshohocken, PA: American Society for Testing and Materials

22-37. Weintraub, M. et al. 1972. "The Effect of Cure Variation on Chemical and Functional Properties of Resin Bonded Friction Materials." SAE Paper No. 720054. Warrendale, PA: Society of Automotive Engineers.

22-38. Lu, Y. et al. 2002. "Optimization of a Commercial Brake Pad Formulation." *Journal of Applied Polymer Science*. 84: 2498-2504.

22-39. Lu, Y. 2006., "A Combinatorial Approach for Automotive Friction Materials: Effects of Ingredients on Friction Performance." *Composites Science and Technology* 66: 591-598.

22-40. Tang, C. F., and Y. Lu. "Optimization of an Industrial Friction Material Formulation." *Tribology Transactions* 46: 19-23.

22-41. Chapman, T. et al. 1999. "Wear Resistant Aluminum-boron Carbide Cermets for Automotive Brake Applications." *Wear* 236: 81-87.

22-42. Rusnak, R. et al. 1970. "Friction Material Testing." SAE Paper No. 700517. Warrendale, PA; Society of Automotive Engineers.

22-43. Preston, J., and R. Forthofer. 1971. "Correlation of Vehicle, Dynamometer and Other Laboratory Tests for Brake Friction Materials." SAE Paper No. 710250. Warrendale, PA: Society of Automotive Engineers.

22-44. Anderson, A. et al. 1973. "Asbestos Emissions from Brake Dynamometer Tests." SAE Paper No. 730549. Warrendale, PA: Society of Automotive Engineers.

22-45. Bockius, C., and J. Hunt. 1935. "Brake Drum and Lining Development." SAE Paper No. 350103. Warrendale, PA: Society of Automotive Engineers.

22-46. Schwartz, H., and S. Rhee. 1977. "Brake Dynamometer Evaluation of the Detroit Traffic Road Test." SAE Paper No. 770026. Warrendale, PA: Society of Automotive Engineers.

22-47. Hass, G. et al. 1966. "Laboratory Simulation Of Driving Conditions in the Los Angeles Area." SAE Paper No. 660546. Warrendale, PA: Society of Automotive Engineers.

22-48. Howard, D., and J. Winge. 1967. "An Automatically Programmed Quadruple Dynamometer for Vehicle Brake Testing." SAE Paper No. 670144. Warrendale, PA: Society of Automotive Engineers.

22-49. Blau, P. and B. Jolly. 2005. "Wear of Truck Brake Lining Materials Using Three Different Methods." *Wear* 259: 1022-1030.

22-50. Radlinski, R. 1972. "Evaluation of the Use of Automotive Braking Systems during a 7300 Mile Cross-Country Trip." SAE Paper No. 720032. Warrendale, PA: Society of Automotive Engineers.

22-51. Mackenzie, D. et al. 1966. "A European Proving Trial." *Auto. Eng*. 56:18-25.

22-52. Smith, G. et al. 1967. "Putting Automatic Transmission Clutch Friction Researchers on Speaking Terms." SAE Paper No. 670051. Warrendale, PA: Society of Automotive Engineers.

22-53. Watts, R., and S. Sparrow. 1988. "Development of a THM 700-R4 Band Friction Test for the SAE 2 Machine." SAE Paper No. 881675. Warrendale, PA: Society of Automotive Engineers.

22-54. Seto, N. 1989. Method and apparatus for evaluating a feeling of a gear shifting operation in a manual transmission. United States Patent No. 4849888, filed March 10, 1988, and issued July 18, 1989.

22-55. Olson, A. 1972. "Evaluation of Friction Materials." SAE Paper No. 720366. Warrendale, PA: Society of Automotive Engineers.

22-56. Hoerbiger. "Welcome to Hoerbiger." Accessed October 19, 2010. http://www.hoerbiger.com/fileadmin/files/internet/AN/PDFs/ Veroeffentlichungen/ A5K079D.pdf.

22-57. Fritzson, D., 1990. "Friction of Elastomer Composites: Influence of Surface Temperature, Sliding Speed, and Pressure." *Wear* 139: 17-32.

Chapter 23
Future Directions

There are many methods for predicting the future. For example, you can read horoscopes, tea leaves, tarot cards, or crystal balls. Collectively, these methods are known as 'nutty methods.' Or you can put well-researched facts into sophisticated computer models, more commonly referred to as 'a complete waste of time.'

—*Scott Adams* (American Cartoonist)

ANY PREDICTION THAT ATTEMPTS TO PRECISELY define the direction of the automotive friction material industry may be readily classified as either a "nutty method" or "a complete waste of time." The industry is at a point when external influences are changing the direction and scope of future research efforts. One pertinent example can be seen with recent legislation that restricts the concentration of copper in automotive friction materials. This will undoubtedly result in an alteration to the current design philosophies employed by many of the world's commercial friction material producers.

Additionally, this legislation effectively re-sets the developmental landscape to a point where all current commercial suppliers must devote energy toward the creation of a new class of copper-free materials. If history repeats itself, the process will consist of change in the form of incremental adjustments leading toward the goal, and future products will ultimately attain complete liberation from copper. And this is only one area of change.

Future studies that explore long-term exposure and associated health concerns with the inhalation of small particle materials has lead to significant reductions in permissible exposure levels.

One noteworthy example is the PM2.5 Standard for air quality [23-1]. This recent standard for improving air quality is based on "thousands of scientific studies" that have linked fine particle exposure to a number of serious medical concerns including premature death, especially in individuals with heart and lung disease. PM2.5 has limited the occupational exposure limit of airborne fine particles (many of which are less than $1/30^{th}$ the diameter of a human hair) to a value of only 35 micrograms per cubic meter over a twenty-four hour exposure period.

Another area of interest will be with raw materials that are commonly used today, which may become materials of concern in the future. This has been seen previously with asbestos, lead, antimony compounds, and, most recently, with copper. Future studies may reveal other materials that may ultimately be deemed hazardous. Whether this change will be based on morphological factors, rates of dissolution in biological fluids, or respiratory clearance rates, future science may restrict current materials and once again promote a frantic attempt to find suitable replacements.

Furthermore, as automotive manufacturers continue to explore lighter and faster vehicles, the need for continued friction material development is assured. Changes in system size and the advancement of alternative power sources may reduce the size and complexity of brake systems extensively. Innovations in hybrid power supplies leading to improvement in regenerative braking, as well as advances in polymer science and composite manufacturing technology, are expected to emerge. These will likely shed new light and provide new insights toward future horizons, which again necessitates a change that will push friction materials of today toward extinction.

References

23-1. Environmental Protection Agency. "PM Standards Revision 2006." Accessed June 30, 2010. http://www.epa.gov/oar/particlepollution/naaqsrev2006.html.

Index

About the Author

Roy L. Cox began his career in friction materials immediately after receiving his bachelor's degree in chemistry in 1994. He has been employed by many of the world's leading producers of friction materials during his 17 years in the friction material industry. He has been an affiliate of the Society of Automotive Engineers for sixteen years. Amazed with the dearth of published information on the subject of friction material design and manufacturing, Roy began a personal journey cataloging public domain information from all aspects of available literature in the areas of tribology, wear processes, friction materials, and manufacturing techniques. The desire to compile the information into a single source led him to formalize his vast collection of personal notes into this work.

Roy enjoys spending time with his wife and two children. His hobbies include reading, gaming, and the study of ancient languages, notably Koine Greek. In addition, Roy enjoys exploring the edible delights of the culinary arts and now conducts a great majority of his personal experimentation in the kitchen!